Demystifying Switched-Capacitor Circuits

Demystifying Switched-Capacitor Circuits

Mingliang Liu

ELSEVIER

AMSTERDAM • BOSTON • HEIDELBERG • LONDON
NEW YORK • OXFORD • PARIS • SAN DIEGO
SAN FRANCISCO • SINGAPORE • SYDNEY • TOKYO
Newnes is an imprint of Elsevier

Newnes

Newnes is an imprint of Elsevier
30 Corporate Drive, Suite 400, Burlington, MA 01803, USA
Linacre House, Jordan Hill, Oxford OX2 8DP, UK

Recognizing the importance of preserving what has been written, Elsevier prints its books on acid-free paper whenever possible.

Library of Congress Cataloging-in-Publication Data

Application Submitted

British Library Cataloguing-in-Publication Data
A catalogue record for this book is available from the British Library.

ISBN 13: 978-0-7506-7907-7
ISBN 10: 0-7506-7907-7

For information on all Newnes publications
visit our Web site at www.books.elsevier.com

Transferred to digital printing in 2009.

To my parents

About the author

Mingliang Liu was born in Nanchang, China. He earned a Bachelor of Engineering degree in Electronics Engineering from Beijing Institute of Technology (Beijing, China), and a Master of Science degree in Electrical and Computer Engineering from Oregon State University (Corvallis, OR).

He has held several industrial positions as design engineer, applications engineer, and product development manager. He is a member of IEEE and Sigma Xi.

Contents

Preface

Everything should be made as simple as possible, but no simpler.
Albert Einstein (1879–1955)

As the most common approach for realizing accurate and linear analog signal processing (ASP) in metal-oxide semiconductor (MOS) integrated technologies, switched-capacitor (SC) circuit techniques have dominated the design of high-quality monolithic filters since the 1980s. The incomparable technological adaptability shown by SC circuits has furthermore made them the competent candidate appropriate for a rich variety of applications such as instrumentations, digital audio, wireless communications, power management, and sensors. This near ubiquity is perhaps the aspect of SC circuits that intrigues circuit designers and engineering students the most.

This book presents a unified text that deals with the basic concepts as well as advanced design methodologies of SC circuits. To achieve this goal, the book provides a systematic treatment of each selected subject with the help of technically proven circuit examples. Thus, numerous practical design examples have been included; however, this book is not a plain survey. A conscious effort has been made to choose a well-connected set of topics that are worthy of detailed treatments. While some quantitative analyses are necessarily presented to reveal the underlying ideas, an effort is made to avoid entangling the reader in tedious mathematical equations.

This book is intended for industrial practice as well as classroom adoption. In both cases, it is expected that the reader has been exposed to basic theories of discrete-time signals and systems such as Laplace-transform, z-transform, and the concept of s-to-z mapping. Additionally, the reader should be familiar with basic MOS transistor modeling such as the small-signal analysis and single-transistor amplifiers like common-source, common-gate, and source follower.

In preparing the syllabus for a senior-year undergraduate or first-year graduate analog integrated circuit course, this book can be listed as a supplement to a more exhaustive textbook. For entry-level design engineers or circuit hobbyists, this book can be used as a tutorial. The materials provided in this book can also be tailored for internal training, short courses, or product seminars. For more experienced engineers, this book may serve as a designer's handbook, offering fruitful technical discussions and extensive bibliographies for carrying out further investigations.

The book has eight chapters and is outlined as follows.

Chapter 1 emphasizes the basic physical behavior of MOS transistors, metal-oxide semiconductor field-effect transistor (MOSFET) switches, and MOSFET capacitors. It is intended as a review of basic principles rather than an in-depth treatment of advanced device physics topics.

Chapter 2 discusses the fundamental aspects of two-stage operational amplifiers (op-amps). The basic analysis of op-amp compensation is provided, and cascode-type op-amp topologies such as telescopic and folded-cascode op-amps are investigated.

Chapter 3 examines the fundamental building blocks of SC circuits. The advantages of SC resistor simulations over physical diffused resistor implementations are discussed. The chapter also describes the parasitic capacitances' effects on SC circuits and presents parasitic-insensitive SC configurations. The design issues of the SC integrator, sampled-and-hold, interpolator, and decimator are detailed. Finally, the discussion explores the signal-flow graph (SFG) analysis of SC circuits and introduces the Mason's rule.

Chapter 4 describes the fundamental aspects of active SC filters (SCFs). The cascaded design of high-order SCFs is demonstrated through a step-by-step design example of a sixth-order elliptical low-pass SCF. Also, high-frequency complementary metal-oxide semiconductor (CMOS) SCFs are introduced.

Chapter 5 focuses on CMOS data converters by presenting a number of important performance parameters for specifying integrated analog-to-digital converters (ADCs) and digital-to-analog converters (DACs). The effects of capacitor mismatch on SC data converters are discussed, and mismatch error cancellation techniques are introduced. Various ADCs—including flash, two-step, pipelined, cyclic, successive-approximation, and delta-sigma ($\Delta\Sigma$) ADCs—are investigated.

Chapter 6 exploits a relatively new topic—the design of SC direct-current-to-direct-current (DC-DC) converters. A few practical step-up converter topologies such as the Dickson charge-pump and cross-coupled are presented. An overview of the existing SC step-down DC-DC converters is provided. Finally, the chapter investigates the principles of multiple-gain SC step-down-step-up DC-DC converters.

Chapter 7 investigates two major challenges that are of immediate relevance to modern switched-capacitor circuits. One is to design high-performance SC circuits in the presence of a low power supply voltage ($V_{dd} < 1.5$ V), and the other is to reduce the effect of the imperfections (or nonidealities) of operational amplifiers on SC circuits.

Chapter 8 presents the top-down design of a delta-sigma ($\Delta\Sigma$) modulator suitable for multistandard radio-frequency (RF) signal receptions. The transistor-level circuit implementation of the modulator is detailed, and the postfabrication measurement results are provided.

As mentioned earlier, an extensive bibliography accompanies each chapter, allowing the reader to trace the original treatment of each topic.

Acknowledgments

I would first like to thank Professor Gabor C. Temes for guiding me to the adventurous world of analog integrated circuits and systems, for supporting an environment in which I was allowed to pursue independent avenues of thought, and for sharing his knowledge and experience during my graduate study at Oregon State University.

I would like to express my gratitude and gratefulness to the Elsevier staff, especially Charles B. Glaser and Leslie Weekes for their continual support, as well as the illustration editor for the excellent artwork and literary editor for the hard work. Also, I would like to acknowledge Harry Helms for encouraging me to start writing this book in the first place.

I am greatly indebted to my dearest wife, Dr. Meng-Yin Chen. I would like to thank her for bearing with me through the book-writing process and for her tremendous support.

Mingliang Liu

CHAPTER 1

Basic MOS Device Physics

1.1 Introduction

This chapter focuses on the fundamental aspects of metal-oxide semiconductor (MOS) device behavior that are of immediate relevance to practical integrated circuit (IC) design. It is intended as a review of basic principles rather than an in-depth treatment of advanced topics.

Chapter Outline

This chapter is organized as follows. Section 1.2 describes the fundamental aspects of MOS transistors. The basic properties of MOS switches are discussed in Section 1.3. The behavior of the MOS device as a capacitor is discussed in Section 1.4.

1.2 MOS Transistors

Basic Operation

Perhaps the most widely adopted process technologies in today's IC industry are those that use metal-oxide semiconductor (MOS) transistors. A MOS transistor can also be referred to as a metal-oxide semiconductor field-effect transistor (MOSFET). Other acronyms are MOST (for MOS transistor) and IGFET (for insulated-gate field-effect transistor). The term *metal* in the acronym MOS indicates that the transistor's *gate* is made of metallic materials such as metalsilicon. Nowadays, heavily doped polycrystalline silicon (also known as polysilicon) is usually chosen over metalsilicon because polysilicon can be aligned and scaled with higher geometric precision, resulting in smaller and faster transistors.

There are two complementary types of MOS transistors: N-channel MOS (NMOS) and P-channel MOS (PMOS). NMOS transistors use electrons to deliver charge in the presence of a positive gate voltage, while PMOS transistors use holes (which are equivalent to positive carriers) to conduct current in the presence of a negative gate voltage. Be it of NMOS or of PMOS type, each MOS transistor is a *unipolar* device, meaning there can be only one type of carrier (electrons for NMOS and holes for PMOS) traveling in the channel.

If a *bipolar* MOS device is desired, we can incorporate both NMOS and PMOS transistors onto the same monolithic chip, resulting in what is called a complementary MOS (CMOS) circuit. In practice, there are three different types of CMOS processes: local-oxidation-of-silicon (LOCOS) process, shallow-trench-isolation (STI) process, and silicon-on-insulator (SOI) process. The latter two are known for their immunities to the *latch-up* problem [1]. In this book, we discuss the CMOS transistors that are realized in the LOCOS process only.

A conceptual cross-section diagram of a typical NMOS transistor implemented in a LOCOS-type CMOS process is shown in Figure 1.1.

In the NMOS transistor, the two heavily doped *N+* regions are the source and the drain, respectively. They are diffused into a slightly doped semiconductor body called the *P-substrate.* The distance between the source and the drain is called the *channel length*, which is denoted as *L* in the diagram. It may also be referred to as

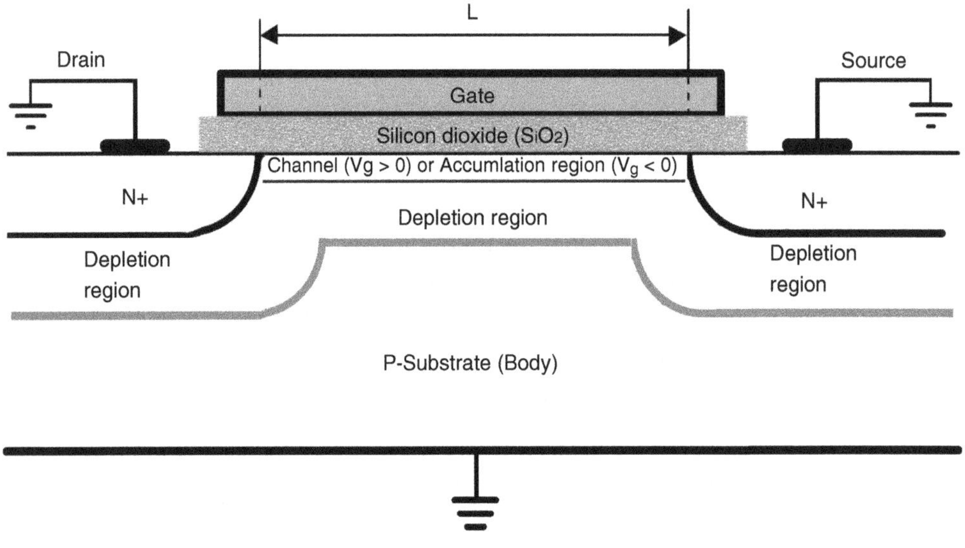

Figure 1.1 Cross-section diagram of an NMOS transistor.

the *effective gate length* and is typically shorter than the physical gate length. In comparison with the NMOS transistor, the PMOS transistor is normally fabricated in an N-well *pocket*. The N-well is not a substrate, but rather an isolated region of higher surface concentration with relatively more free carriers as compared to the P-substrate.

A layer of silicon dioxide (SiO_2) is grown beneath the gate to physically isolate it from the remaining regions in the transistor. In an ideal situation, no charge is leaked from the gate into the channel. However, in reality, when a varying signal (e.g., a clock signal) is applied to the gate, a transient charge is coupled into the channel through the small-signal capacitance that resides between the gate and the channel. Additionally, the gate-source capacitance (C_{gs}) and the gate-drain capacitance (C_{gd}) result in more charge leakages. This phenomenon is called *clock feedthrough*, and its effect becomes more pronounced when the transistor is placed at the input of an open-loop amplifier. In such a case, the input offset caused by the charge leakage may saturate the amplifier.

Once the transistor is turned off, the residual charge stored in the channel is dispersed through the drain and the source to elsewhere in the circuit. This phenomenon is known as *charge injection*, which introduces signal-dependent errors to the circuit. The mechanism of charge injection is treated later in this chapter.

As shown in Figure 1.1, the source, drain, and P-substrate of the NMOS transistor are all connected to ground. When the gate voltage (V_g) is below zero (i.e., $V_{gs} < 0$), positive carriers (P+) accumulate in the region under the gate and the dioxide layer, which is called the *accumulation region*. When V_{gs} is of a sufficiently large positive value, negative carriers such as electrons take over this region and form a channel connecting the drain and source regions. In other words, the accumulation P-region doped with (P+) carriers is converted into an N-region consisting of negative carriers. Hence, the channel is *inverted*, and the transistor is said to be working in the *inversion region*.

Then the question arises: What is the minimum positive value of V_{gs} for which an inverted channel can be formed between the drain and the source? The appropriate response to this question leads to the *threshold voltage*, which is commonly denoted as V_{thn} (or V_{thp} for PMOS transistors). That is, when $V_{gs} \geq V_{thn}$, the device enters into the inversion region. The result of subtracting V_{thn} from V_{gs} is usually called the *effective drain-source voltage* and denoted as V_{eff}. When $0 < V_{gs} < V_{thn}$, or equivalently, $V_{eff} < 0$, both holes and electrons have low density levels, and the transistor is said to be operating in the *depletion region*.

When $V_{gs} > V_{thn}$, the connection between the drain and the source is formed. However, to allow current to flow from the drain to the source, the drain-source voltage (V_{ds}) must be larger than zero. It can be shown that the drain-source current (I_d) gradually increases with V_{ds}, and for a small V_{ds} ($0.1\,\text{V} < V_{ds} \ll V_{eff}$), the relationship between I_d and V_{ds} is given by

$$I_d = \mu_n C_{ox} \frac{W}{L} (V_{gs} - V_{thn}) \cdot V_{ds} \tag{1.1}$$

where μ_n is the electron mobility near the silicon surface (the *skin effect* [1] is negligible since we assume a low frequency configuration here), C_{ox} is the gate capacitance per unit area, W is the gate width, and L is the effective channel length. The transistor is now working in the *weak inversion region.* When $V_{ds} < 0.1\,\text{V}$, the transistor is said to be working in the *subthreshold region.*

It can be shown that for a moderate V_{ds} ($0.5\,V_{eff} < V_{ds} < V_{eff}$), the relationship between I_d and V_{ds} is approximately given by

$$I_d = \mu_n C_{ox} \frac{W}{L} \left[(V_{gs} - V_{thn}) \cdot V_{ds} - \frac{V_{ds}^2}{2} \right] \tag{1.2}$$

The transistor is now working in the *triode region.* Finally, once V_{ds} reaches V_{eff}, the *pinch-off* condition [2] is satisfied, meaning beyond this point I_d remains constant (to a first-order approximation) with respect to V_{ds}. At the pinch-off point where $V_{ds} = V_{eff}$, the resulting $I_d \sim V_{ds}$ relationship is given by

$$I_d = \mu_n C_{ox} \frac{W}{L} \frac{(V_{gs} - V_{thn})^2}{2} = \mu_n C_{ox} \frac{W}{L} \frac{V_{ds}^2}{2} \tag{1.3}$$

This is known as the *square-law I-V characteristic.* The transistor is said to be working in the *active* (or *saturation*) *region.*

The transconductance (g_m), which is commonly used in the small-signal model for a MOS transistor working in the active region, is defined as

$$g_m = \frac{\partial I_d}{\partial V_{gs}} \tag{1.4}$$

In the active region, the transconductance can be obtained based on Equation (1.3)

$$g_m = \frac{\partial I_d}{\partial V_{gs}} = \mu_n C_{ox} \frac{W}{L} (V_{gs} - V_{thn}) = \mu_n C_{ox} \frac{W}{L} V_{eff} \tag{1.5}$$

The transconductance can also be expressed as

$$g_m = \sqrt{2\mu_n C_{ox} \frac{W}{L} I_d} = \frac{2I_d}{V_{eff}} \tag{1.6}$$

Interestingly, Equation (1.6) indicates that g_m can be determined by the ratio of the drain current (I_d) to the effective gate-source voltage (V_{eff}). What's more, it is possible to make the transconductance be independent of the value of (W/L) as long as that ratio is kept as a constant. Hence, to a first-order approximation, the scaling of the transistor geometry does not affect g_m. This is a desirable feature because the voltage gain and accuracy can be maintained while the device is being downscaled (however, this argument does not hold in submicron processes where the short-channel effects become prominent).

Also, from Equation (1.5) we realize that for a given transistor in a given process, the value of g_m is controlled by the gate-source voltage (V_{gs}). Qualitatively speaking, this property is appropriate for analog MOS amplifiers where the linearity performance is reflected by how cohesively the intrinsic gain (i.e., $g_m R_{out}$, where R_{out} is the output impedance) tracks the change in V_{gs} of the MOS transistor.

In the triode region, the transconductance is obtained based on Equation (1.2),

$$g_m = \frac{\partial I_d}{\partial V_{gs}} \cong \mu_n C_{ox} \frac{W}{L} V_{ds} \tag{1.7}$$

where $V_{ds} < V_{eff}$. Since V_{ds} is not governed by V_{gs}, the value of g_m does not accurately reflect the change in V_{gs}; consequently, the linearity of the analog amplifier is degraded as compared to the situation where the transistor is operating in the active region. As a result, in many analog applications where a good linearity is required, all the MOS transistors in the signal path should operate in the active rather than the triode region. By contrast, in digital circuits that make use of MOS transistors to realize digital logic gates, linearity is usually not a concern; thus, the transistors may operate in either triode or active regions, depending on the desired logic function.

Scaling of MOS Transistors

Propelled by the nonstopping advancement of lithography and implantation technologies, the minimum feature size of a MOS transistor has been continually reduced since the 1980s, enabling the unprecedented success of digital CMOS circuits and doubling the system-on-a-chip (SOC) computing capability every 18 to 24 months, which is known as *Moore's law.*

In addition to a higher level of system integration and lower cost, the continual downscaling of MOS transistors brings about a significant increase in the *cutoff frequency* (f_t) of the MOS transistor, opening an avenue to achieve high-speed/high-frequency integrated systems using a pure CMOS technology. Specifically, the cutoff frequency f_t is normally defined to be the frequency at which the transistor's current gain is unity. It can be shown that f_t of a typical NMOS transistor is given by the following expression [3]:

$$f_t = \frac{g_m}{2\pi \cdot (C_{gd} + C_{gs})} \cong \frac{3\mu_n \cdot (V_{gs} - V_{thn})}{4\pi \cdot L^2} \tag{1.8}$$

Thus, f_t is proportional to $1/L^2$.

However, as the value of L continues declining, various short-channel effects become prominent. A common result of these effects is that the transistor enters into the saturation region before V_{ds} becomes sufficiently large (i.e., before the pinch-off condition is met) [4]. For instance, the electric field between the source and drain regions is strengthened due to the shrunk device geometries, which in turn constrains the velocity of the carriers traveling in the channel. Consequently, the saturation drain current (I_d) is smaller than that given by Equation (1.3), and the I-V relationship no longer follows the square law. This particular phenomenon is called *velocity saturation*, which is one of the dominant short-channel effects. It can be shown that in a short-channel MOS device, velocity saturation tends to minimize the dependence of I_d on L [3], and since V_{gs} is not controlled by L, the transconductance (g_m) eventually ceases to depend on L as well. In such a case, as it is known that both of the small-signal capacitances in Equation (1.8), C_{gs} and C_{gd}, are typically proportional to the gate area (i.e., WL), we can thus see that in a short-channel MOS device, f_t is inversely proportional to L rather than to L^2.

The loss of critical device characteristics (e.g., the square-law I-V relationship) due to aggressive device downscaling hampers the technology portability of CMOS circuits, which means the circuits designed in an older CMOS technology with longer channel length may not function properly in a present or future technology with shorter channel length, making a *process-independent* design seem incrementally impracticable. Reduced transistor geometries also further complicate the statistical approximation of many device parameters, posing a major challenge in the research area of MOS device modeling [5].

There are many other short-channel MOS device effects, including *oxide breakdown*, *drain-induced barrier lowering*, *hot carrier effect*, and *punch-through* [3][5].

In addition, second-order phenomena that are relevant to both long-channel and short-channel MOS devices, such as *channel-length modulation* and *subthreshold conduction*, should not be ignored in practice. Due to the space constraint, the detailed analyses of the aforementioned effects are not included in this book, and the reader is encouraged to explore them in the references. A brief discussion on an important second-order effect called the *body effect* can be found in Chapter 6, where the SC Dickson DC-DC converter is treated.

1.3 MOSFET Switches

Switch On-Resistance

In a switched-capacitor (SC) circuit implemented in a CMOS technology, the switch may be realized one of three different ways: by using an NMOS transistor, a PMOS transistor, or a CMOS transmission gate that contains two complementary transistors (i.e., one NMOS and one PMOS).

To turn an NMOS switch on, the clock signal on its gate goes up to the supply voltage (V_{dd}), and the resultant ($V_{gs} - V_{thn}$) becomes larger than V_{ds}. Thus, the transistor is working in the triode region, and the relationship between I_d and V_{ds} can be expressed by Equation (1.2). Ignoring the body effect, we can derive the transistor's equivalent drain-source resistance (i.e., the switch on-resistance) by dividing V_{ds} by I_d, and we have

$$R_{on} = \frac{V_{ds}}{I_d} = \frac{1}{\mu_n C_{ox} \frac{W}{L}\left(V_{gs} - V_{thn} - \frac{V_{ds}}{2}\right)} \cong \frac{1}{\mu_n C_{ox} \frac{W}{L}(V_{gs} - V_{thn})} \tag{1.9}$$

Here, the approximation is based on the assumption that ($V_{gs} - V_{thn}$) >> V_{ds}, and the switch can be modeled as a resistor. By revisiting Equation (1.1), we find that as the clock signal applied to the NMOS transistor's gate approaches V_{dd}, the reasoning through the preceding equation becomes altogether sound.

It is of practical interest to get a rough feel of the numerical value of the switch-on resistance (R_{on}), which often turns out to be relatively large in many processes. For example, if it is assumed that $V_{dd} = 3.3\,\text{V}$, $V_{in} = 1\,\text{V}$, $\mu_n C_{ox} = 30\,\mu\text{A/V}^2$, $(W/L) = 1$ and $V_{thn} = 0.7\,\text{V}$, then based on the preceding equation we obtain a R_{on} of about $21\,\text{k}\Omega$. If the body effect is accounted for increasing the threshold voltage, then a larger switch on-resistance will occur. Additionally, the value of R_{on} increases with the input signal level. For instance, if we let $V_{in} = 2.5\,\text{V}$ in the preceding case, then the value of R_{on} becomes as high as about 333 kΩ. Such a high resistance results in an impractically long settling time needed to charge the capacitor and is unacceptable in most SC applications. For a typical PMOS transistor, the value of $\mu_p C_{ox}$ is approximately one-third that of $\mu_n C_{ox}$ for its NMOS counterpart; therefore the PMOS transistor is less attractive when it comes to realizing a switch of low on-resistance.

The CMOS transmission gate is a better candidate than both NMOS and PMOS transistors in terms of reducing the on-resistance. This makes intuitive sense since the turned-on CMOS transmission gate can be considered as a parallel configuration composed of two resistors, R_{onn} (for the NMOS transistor's on-resistance) and R_{onp} (for the PMOS transistor's on-resistance), apparently on the assumption that both NMOS and PMOS transistors are turned on at the moment. What's more, as long as the supply voltage (V_{dd}) is larger than the sum of the absolute values of the PMOS and NMOS threshold voltages, at least one of the transistors will be on when the clock signal goes to V_{dd}, regardless of the input signal level (V_{in}). However, as we will see in Chapter 7, when this condition is not satisfied, which is often the case in low-supply-voltage applications, additional components or circuit blocks are needed to ensure that the switch be turned on when the clock signal goes to V_{dd}.

kT/C Noise

As mentioned earlier, during the "on" phase, the switch can be modeled as a resistor whose resistance value is given by Equation (1.9). The equivalent thermal noise of this resistor (R_{on}) has a one-sided, white-noise-like power spectral density of [6]

$$\frac{\overline{V_n^2}}{\Delta f} = 4kTR_{on} \tag{1.10}$$

where k is the *Boltzmann's constant* ($1.38 \times 10^{-23}\,\mathrm{JK^{-1}}$), and T is the sampling clock period. As the switch is primarily used in a sampling network where a sampling capacitor (C_s) is charged by the input signal through the switch during the "on" phase, the thermal noise is processed by a first-order low-pass filter composed of R_{on} and C_s, whose transfer function is given by

$$H(j\omega) = \frac{1}{1 + j\omega R_{on} C_s} \tag{1.11}$$

Thus, the total noise power at the filter's output is computed by integrating the filtered (or bandlimited) noise power spectral density from dc to infinity, and we have

$$\overline{V_{out}^2} = \frac{1}{2\pi}\int_0^{\infty}|H(j\omega)|^2\overline{V_{in}^2}d\omega = \frac{1}{2\pi}\int_0^{\infty}\frac{4kTR_{on}}{1+(\omega R_{on}C_s)^2}d\omega$$
$$= \frac{2kTR_{on}}{\pi}\cdot\left(\frac{1}{R_{on}C_s}\right)\cdot\arctan(\omega R_{on}C_s)\Big|_{\omega=0}^{\omega=\infty} \tag{1.12}$$
$$= \frac{kT}{C_s}$$

hence the name *kT/C noise* (sometimes also called the *sampling noise*). Interestingly, the total noise power at the output is independent of the actual on-resistance value (R_{on}). This is because when the thermal noise power density is increased with the on-resistance (R_{on}), the *noise bandwidth* of the filter, which is given by

$$f_n = \frac{\pi}{2}f_{-3dB} = \frac{\pi}{2}\frac{1}{2\pi R_{on}C_s} = \frac{1}{4R_{on}C_s} \tag{1.13}$$

is decreased with the same magnitude. Specifically, the total noise power can be easily calculated by multiplying the spectral density given by Equation (1.10) by the noise bandwidth noted earlier, and we have

$$\overline{V_{out}^2} = \overline{V_{in}^2}f_n = 4kTR_{on}\cdot\frac{1}{4R_{on}C_s} = \frac{kT}{C_s} \tag{1.14}$$

which is the same as the result given by Equation (1.12). Furthermore, for a given value of R_{on}, the total output noise power is decreased with the clock period *T.* That is, if the signal is oversampled by a factor of *M*, then the resultant *kT/C* noise power is also reduced by a factor of *M*. As we will see in Chapter 5, in a similar manner, oversampling (i.e., sampling at a rate much higher than the *Nyquist rate*) is useful for suppressing the white-noise-like quantization noise of an analog-to-digital data converter (ADC).

Charge Injection

When a MOS switch is on, it operates in the triode region and has a very small voltage drop across the drain and source. When the switch is turned off, a finite amount of residual charge in the inverted channel underneath the gate is dispersed into the drain, source, and substrate. It can be shown that the charge injected into the substrate becomes more pronounced when either the turnoff slope of the clock waveform is close to infinite or the transistor channel length is very long [7], both of which seldom happen in practical SC circuits. Thus, we ignore the effect of the substrate charge leakage in our discussion here.

If the switch is used for sampling an input signal (e.g., in a sample-and-hold), the charge injected into the input junction (when the switch is turned off) has no effect on the held voltage, and hence it can also be ignored. However, the charge that flows in the opposite direction is not negligible for it introduces an error to the held voltage. How the total channel charge is divided between the source and drain of the switch is a device-modeling question, which involves multiple parameters such as the clock turnoff slope, the ratio of the input capacitance to the output capacitance (C_{in}/C_{out}), and the drain-source voltage (V_{ds}) [8]. The basic idea is that when the clock waveform has a steep turnoff slope, the channel is cut off before V_{ds} gets a chance to weigh in; consequently, the channel charge is equally divided between the drain and source terminals. On the other hand, a slow falling clock signal gives rise to a capacitive voltage divider where V_{ds} distributes the channel charge based on the capacitor ratio (C_{in}/C_{out}).

Although it is seldom possible to have the clock signal fall off fast enough so that exactly one-half of the channel charge is injected into each terminal, in crude pencil-and-paper calculations it is usually assumed that the charge could be equally divided. Based on this assumption, various techniques have been reported to compensate the charge injection that flows away from the input. The use of a CMOS transmission gate is one option, but it usually requires the input signal to stay at or near $V_{dd}/2$. Figure 1.2 shows an alternative charge injection compensation configuration where a dummy switch is added next to the main switch [9]. As shown in Figure 1.2, the size of the dummy switch is set as one-half that of the main switch, and the dummy's drain and source are shorted. According to the preceding discussion, when the main switch is turned off, half of the channel charge flows through the dummy switch toward the hold capacitor (C_s). Note that the signal applied to the dummy's gate is complementary to *clk*, and also, it should be slightly delayed as compared to *clk* so that when the dummy is turned off, the main switch is already on, and as a result, the injected charge by the dummy will be *overpowered* by the input signal (V_{in}),

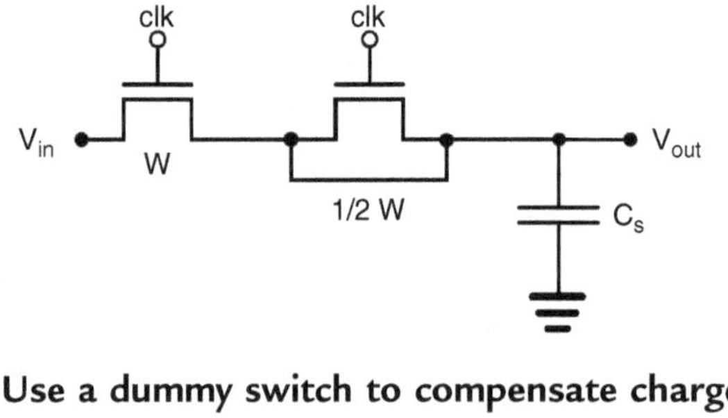

Figure 1.2 Use a dummy switch to compensate charge injection.

thereby having no effect on the output voltage (V_{out}). As *clk* falls, a channel is formed under the dummy's gate despite the shorted drain and source. Ideally, the charge injected by the full-size main switch would be canceled out by the intrinsic channel charge (with a polarity opposite to that of the former) induced by the half-size dummy. However, in reality, the size ratio of "1/2" rarely works out to be the optimum required for achieving the perfect charge injection cancellation, due to circuit imperfections such as the finite clock slope and body effect [2][9]. In fact, the optimization of the size ratio is a rather complicated modeling problem that involves tedious calculations based on various device parameters, and computer simulation is usually utilized in practice to find the optimal solution. We will discuss several other techniques of minimizing charge injection in Chapter 3, where sample-and-hold (S&H) circuits are treated.

1.4 MOSFET Capacitors

This section discusses the behavior of the MOS transistor as a capacitor. Similar to what we've seen earlier, the operation of a MOSFET capacitor can be divided into three different regions: *accumulation*, *depletion*, and *inversion*.

Assuming an NMOS transistor is used, we can say that it operates in the accumulation region when the gate voltage is negative. In this region, the negative gate voltage attracts holes from the substrate to the oxide-gate interface. Thus, the gate and the substrate electrode (i.e., the oxide) form a capacitor, which has a thickness of t_{ox} and holds a voltage difference between its two plates (i.e., V_{gb}). It can be shown that its capacitance is given by [10]

$$C_{equ} = C_{ox}WL \tag{1.15}$$

where C_{ox} is the oxide capacitance per unit area, W is the transistor gate width, and L is the effective gate length.

When the transistor is operating in the depletion region, its gate voltage is relatively small (i.e., near zero). Although the gate voltage is positive, not many electrons are attracted to the oxide-gate interface. As a result, the "battery" cannot hold much electronic charge; hence the equivalent capacitance is relatively small.

When a large positive voltage (typically larger than twice the Fermi potential or approximately equal to the threshold voltage, V_{thn}) is applied to the gate, the negative carriers (i.e., electrons) dominate the channel underneath the gate-oxide interface, and a larger equivalent capacitance is recovered. Figure 1.3 shows the MOSFET capacitance plotted against the gate-source voltage. Note that V_{fb} denotes the *flat-*

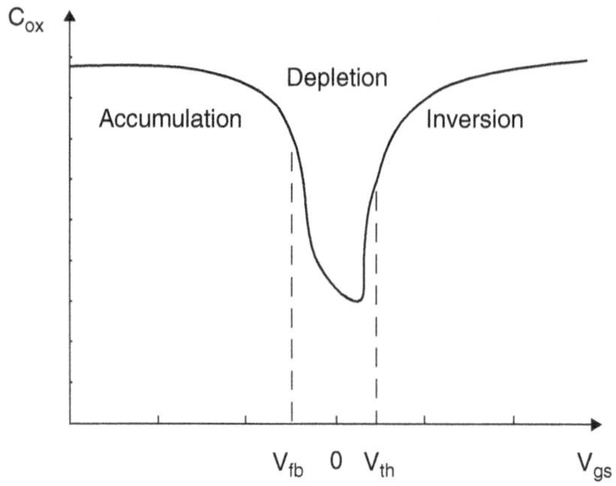

Figure 1.3 Capacitance versus voltage characteristic of a MOSFET.

band voltage [10], which separates the accumulation and depletion regions, and V_{th} is the threshold voltage, which (as mentioned earlier) divides between the depletion and inversion regions.

We will see in Chapter 3 that switched-capacitor (SC) circuits normally require accurate capacitor ratios. As concerns the realization of capacitors in SC circuits, we should notice that standard MOSFETs cannot meet the accuracy requirements, for the capacitances will vary considerably with the corresponding gate-source voltages. In practice, easy-to-scale polysilicon layers are almost always chosen to realize integrated capacitors. However, physical CMOS processing imperfections such as *overetching, parasitic coupling,* and *excessive lateral diffusion* [1][2][6] post an obstacle to precisely realizing the desired capacitor ratio, and the resultant error is often called the *capacitor mismatch error.* The effects of capacitor mismatching on the circuit's accuracy can be effectively reduced on the mask development level (also known as the circuit layout level), where the circuit's geometry and wiring configurations are determined in accordance with certain design rules. Details on layout design rules useful for compensating capacitor mismatch errors introduced by overetching and other CMOS device nonidealities can be found in [11]. For SC applications that require high accuracy (e.g., SC data converters), in addition to compensating capacitor mismatch errors on the layout level, extra circuit elements are often incorporated into the core to further alleviate the errors' effects (i.e., compensating mismatch errors on the circuit design level). The operations of these extra circuit elements embody what is called the *mismatch error compensation* (or *cancellation*), the concept of which will be treated in Chapter 5.

References

[1] J.-Y. Chen, *MOS devices and technologies for VLSI*, Prentice Hall, Englewood Cliffs, NJ, 1990.

[2] Y. Tsividis, *Operation and modeling of the MOS transistor* (2nd Ed.), McGraw-Hill, New York, 1999.

[3] T. H. Lee, *The design of CMOS radio-frequency integrated circuits*, Cambridge University Press, Cambridge, UK, 1998.

[4] R. H. Dennard et al., "Design of ion-implanted MOSFET's with very small physical dimensions," *IEEE Journal of Solid-State Circuits*, Vol. 9, No. 5, pp. 256–268, October 1974.

[5] C. Hu, "IC reliability simulation," *IEEE Journal of Solid-State Circuits*, Vol. 27, No. 3, pp. 241–246, March 1992.

[6] R. Gregorian and G. C. Temes, *Analog MOS integrated circuits for signal processing*, John Wiley & Sons, New York, 1986.

[7] G. Wegmann, E. A. Vittoz, and F. Rahali, "Charge injection in analog MOS switches," *IEEE Journal of Solid-State Circuits*, Vol. SC-22, No. 6, pp. 1091–1097, December 1987.

[8] B. Sheu, J. Shieh, and M. Patil, "Modeling charge injection in MOS analog switches," *IEEE Trans. on Circuits and Systems*, Vol. CAS-34, No. 2, pp. 214–216, February 1987.

[9] J. McCreary and P. R. Gray, "All MOS charge redistribution analog-to-digital conversion techniques—Part 1," *IEEE Journal of Solid-State Circuits*, Vol. SC-10, No. 6, pp. 371–379, December 1975.

[10] S. M. Sze, *Semiconductor devices, physics and technology*, John Wiley & Sons, New York, 1985.

[11] A. Hastings, *The art of analog layout* (2nd Ed.), Prentice Hall, Englewood Cliffs, NJ, 2005.

CHAPTER 2

Operational Amplifiers

2.1 Introduction

Operational amplifiers (op-amps) are perhaps the most important element in analog circuits and deserve the most attention from the circuit designer. The design of op-amps has been one of the most thoroughly covered topics in the field of integrated circuits (ICs), and a plentiful number of textbooks and handbooks are available that provide in-depth discussions and fruitful circuit design examples. Thus, the following context shall be positioned as an introductory overview, rather than a comprehensive study, of the modern complementary metal-oxide semiconductor (CMOS) op-amp topologies.

Chapter Outline

This chapter is organized as follows. Section 2.2 describes the fundamental aspects of the two-stage op-amp. Cascode-type op-amp topologies such as the telescopic and the folded-cascode op-amps are discussed in Section 2.3.

2.2 Two-Stage Op-Amps

A two-stage op-amp is typically composed of two cascaded stages: the first is a fast open-loop stage, while the second is a high-gain and yet slower closed-loop stage. The total op-amp voltage gain is the product of the two stage gains. A typical *compensated* two-stage CMOS op-amp is shown in Figure 2.1 (the output buffer for resistive loading is not shown). The ultimate goal of compensating an op-amp is to ensure that the op-amp is not only stable (which is often reflected by *phase margin* and *gain margin*), but also capable of promptly settling its output signal to the desired value in response to a fast changing (or busy) input signal (which is reflected by *unity-gain bandwidth* and *slew rate*).

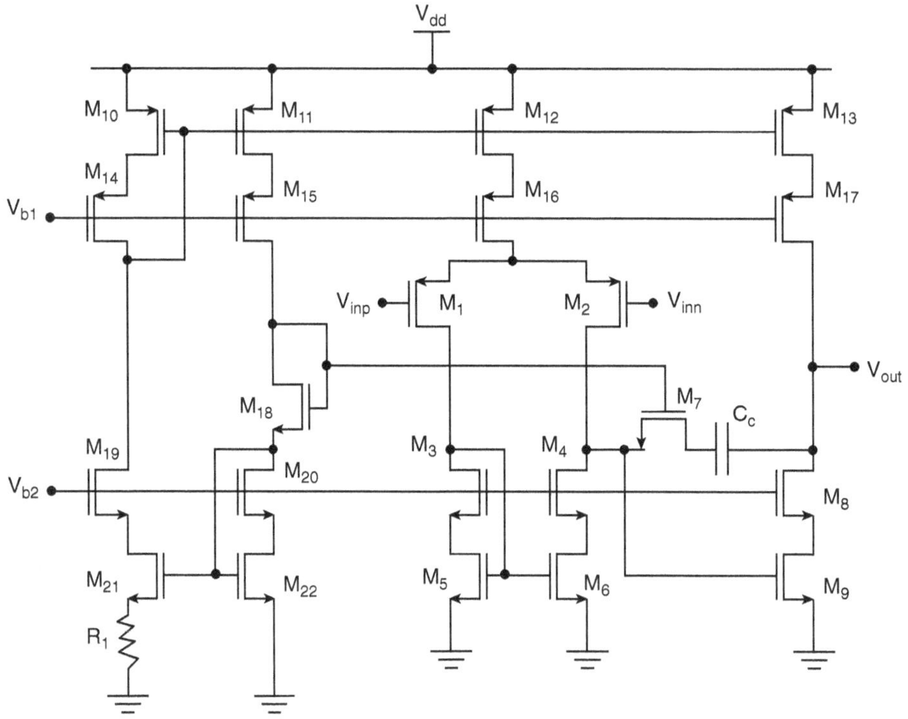

Figure 2.1 Compensated two-stage CMOS op-amp.

Two different compensations are applied to the op-amp in Figure 2.1: one is called *Miller compensation* (or *pole-splitting compensation*), and the other is called *lead compensation* (or *right-half-plane zero cancellation*). The former is realized by the compensation capacitor (or Miller capacitor) denoted C_c, while the latter is realized by the N-channel MOS (NMOS) transistor denoted M_7, which operates in the triode region and has an on-resistance of

$$R_c = \frac{1}{\mu_n C_{ox}\left(\frac{W}{L}\right)_7 (V_{gs7} - V_{thn})} \tag{2.1}$$

The mechanism of compensating the two-stage op-amp may be better understood by investigating the high-frequency small-signal op-amp model [1]. The basic idea is as follows. If the NMOS transistor M_7 is shorted (i.e., $R_c = 0$), then it can be shown that the two-stage op-amp has two poles and one zero, which are approximately expressed as

$$\omega_{p1} \cong \frac{1}{g_{m9} R_1 R_2 C_c}, \quad \omega_{p2} \cong \frac{g_{m9}}{C_1 + C_2 + C_1 C_2 / C_c}, \quad \omega_z \cong -\frac{g_{m9}}{C_c} \tag{2.2}$$

where ω_{p1} is the low-frequency pole or the dominant pole, ω_{p2} is the high-frequency pole or the nondominant pole, ω_z is a zero located in the right half of the s-plane ($\omega_z > 0$), R_1 is the output resistance of the first stage, R_2 is the output resistance of the second stage, C_1 is the total capacitance at the first stage's output, and C_2 is the total capacitance at the second stage's output.

As g_{m9} is increased, ω_{p1} and ω_{p2} are moved away from each other. This phenomenon is called *pole splitting*. Pole splitting can also be done by increasing C_c, which is less effective as compared to increasing g_{m9}. From basic feedback systems theories we should know that the separation of these two poles (ω_{p1} and ω_{p2}) tends to cause the value of phase margin to increase. Since phase margin is one of the most commonly used quantitative criteria of stability (although not the perfectly reliable measure) in practice, we can say that in general pole splitting helps improve the op-amp's stability.

On the other hand, from Equation (2.2), if the output load capacitance, C_2, is increased, then ω_{p2} will be moved closer to ω_{p1}. As a consequence, the two-stage op-amp topology is normally not chosen to directly drive large capacitive loads (e.g., larger than 10 pF).

The right-half-plane (RHP) zero, ω_z, is not desirable because it introduces a phase lag to the system, resulting in a smaller phase margin. Hence, the system may become unstable. The simplest way to reduce the effect of the RHP zero is to make use of M_7 or, equivalently, R_c, to introduce a phase lead so that the phase lag is compensated; hence, the name *lead compensation*. After inserting a nonzero R_c into the circuit, we can find that the zero of the op-amp's transfer function is now given by

$$\omega_z = -\frac{1}{C_c\left(\dfrac{1}{g_{m9}} - R_c\right)} \tag{2.3}$$

To eliminate the RHP zero, we may equal R_c to $(1/g_{m9})$. Next, by utilizing the result of Equation (2.1) and one of the well-known transconductance formulae, we obtain

$$\frac{\left(\dfrac{W}{L}\right)_7}{\left(\dfrac{W}{L}\right)_9} = \frac{(V_{gs9} - V_{thn})}{(V_{gs7} - V_{thn})} = \frac{V_{eff9}}{V_{eff7}} \tag{2.4}$$

Note that M_9 must always operate in the active region. In Appendix 2.1, the processes of pole splitting and RHP zero cancellation for a standard two-stage op-amp are investigated using MATLAB programs.

Note that the voltage across the resistor (R_1) in the biasing network should be designed to equal the effective drain-source voltage of M_9 (V_{eff9}). Also, V_{eff9} should be equal to V_{eff5} and V_{eff6}. This is partially due to the consideration of minimizing the inherent op-amp offset voltage, which is defined as the output voltage in the presence of a zero differential input voltage [2].

Furthermore, it is known that the linear settling performance of an op-amp is reflected by its *unity-gain bandwidth* (UGBW). It can be shown that the unity-gain bandwidth of the compensated two-stage op-amp presented in Figure 2.1 is approximately given by (g_{m1}/C_c). This approximation is made based on the assumption that the output impedance of the first gain stage is dominated by C_c (i.e., the Miller capacitor); consequently, the overall gain response can be simplified to

$$A_{overall}(s) = A_1 A_2 \cong g_{m1}\left(\frac{1}{sA_2C_c}\right)A_2 = \frac{g_{m1}}{sC_c} \tag{2.5}$$

Next, the unity-gain bandwidth can be obtained by equating the absolute magnitude of the preceding to unity, and the result is expressed as $\omega_{UGBW} = g_{m1}/C_c$.

Another important factor that contributes to the op-amp's overall settling performance is the *slew rate* (SR). In practice, the slew rate is often referred to as the *nonlinear* settling rate, and is defined as the maximum rate at which the op-amp's output voltage changes in the presence of a large input signal. It can be expressed as (assuming C_c is much larger than both C_1 and C_2)

$$SR = \left.\frac{dV_{out}}{dt}\right|_{\max} = \left.\frac{d(Q/C_c)}{dt}\right|_{\max} = \frac{I_{C_{c\_\max}}}{C_c} \tag{2.6}$$

Here, $I_{Cc\_\max}$ is the maximum current that can flow through the Miller capacitor (C_c). When a large differential signal is applied to the input devices, M_1 and M_2, either one of them may be turned off depending on the polarity of the input signal. Hence, the maximum current available to charge or discharge the Miller capacitor is equal to the current flowing in the branch that contains transistors M_{12} and M_{16}, and the resultant slew rate is approximately given by (I_{d12}/C_c) or, equivalently, by (I_{d16}/C_c).

From the preceding development, we can see that there is a practical tradeoff between the stability criteria and the settling characteristics (i.e., speed). Specifically, for a given current budget, increasing C_c tends to make the op-amp more stable, but in the meantime it reduces the unity-gain bandwidth as well as the slew rate.

2.3 Telescopic and Folded-Cascode Op-Amps

Telescopic and folded-cascode op-amps belong to the cascode op-amp topologies, where multiple transistors are stacked up to increase the total output impedance, which in turn boosts the op-amp's intrinsic dc gain value. A telescopic CMOS op-amp and a folded-cascode CMOS op-amp are shown in Figures 2.2 (a) and (b), respectively. Both are fully differential configurations while the *common-mode feedback* (CMFB) circuits are not shown for simplicity.

Let us start with the telescopic op-amp shown in Figure 2.2(a) [3]. Specifically, note that the unity-gain bandwidth (i.e., the dominant pole frequency) is determined by the transconductance of the input transistor (g_{m1} or g_{m2}) and the load capacitance (not shown), while the nondominant pole frequency is inversely proportional to the time-constant product of the impedance and parasitic capacitance at the source of the third-layer P-channel MOS (PMOS) cascode transistor, M_3 or M_4. The impedance

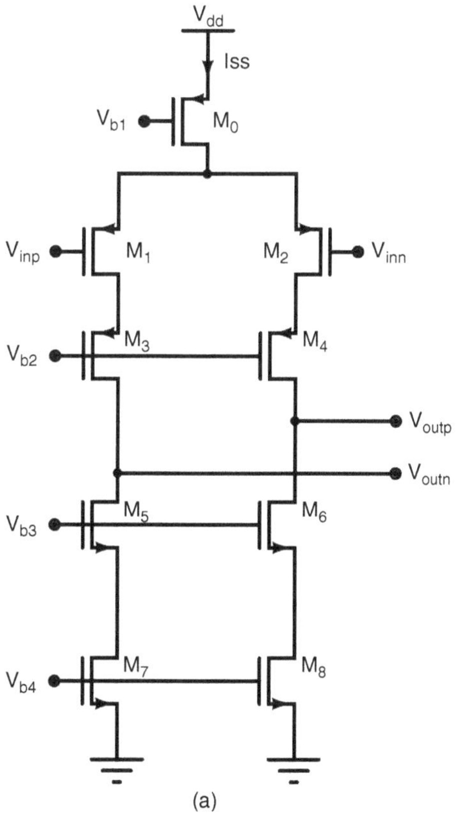

Figure 2.2 (a) Telescopic CMOS op-amp. (b) Folded-cascode CMOS op-amp.

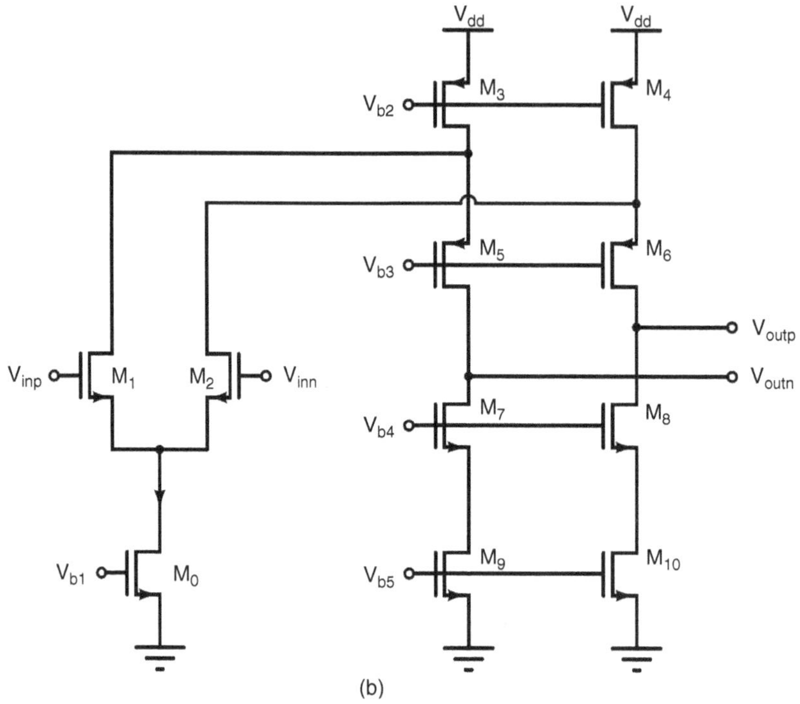

Figure 2.2 ***Continued***

value is approximately given by one over the cascode transistor's transconductance, and the parasitic capacitance is composed of its C_{gs}, the corresponding input transistor's C_{gd} and C_{db}. In other words, both the dominant and nondominant poles depend on the same type of transistors (PMOS in this case).

In practice, given the same current budget and sizing limit, NMOS transistors are usually a better choice than PMOS transistors to realize the telescopic op-amp's input devices because the NMOS transconductance will turn out to be larger than the PMOS transconductance, and a large input transconductance is essential to realizing a high voltage gain and a large unity-gain bandwidth. However, the telescopic op-amp suffers from the reduced output voltage swing, which is a common limitation on all cascode-type op-amp topologies (including the folded-cascode op-amp). Moreover, in the op-amp shown in Figure 2.2(a), it is not possible to bias the input and output at the same dc voltage level due to the inherent voltage drop from the gate of M_1 (or M_2) through M_3 (or M_4) to V_{outn} (or V_{outp}), which is approximately given by $|V_{thp}| + V_{eff1} + V_{eff3}$ (or $|V_{thp}| + V_{eff2} + V_{eff4}$).

The folded-cascode op-amp circumvents this issue of inherent voltage drop by taking the input devices out of the cascode branches, such that their effective drain-source voltages do not contribute to the overall voltage drop of the op-amp. As shown in Figure 2.2(b), a pair of NMOS transistors, M_1 and M_2, are used as the input devices, and another NMOS transistor, M_0, is used as the current sink for M_1 and M_2. Also, in this op-amp, there are four rather than five cascode layers between V_{dd} and ground, which results in a higher output swing than that of the telescopic op-amp presented in Figure 2.2(a).

Although the input pair is taken out of the cascode branches, the resulting folded-cascode op-amp is still a single-stage op-amp since no additional gain stage is added. In the folded-cascode op-amp shown in Figure 2.2(b), the unity-gain bandwidth is determined by the transconductance of the NMOS input device (g_{m1} or g_{m2}) and the load capacitance, while the nondominant pole frequency is determined by the impedance and parasitic capacitance at the source of the PMOS cascode transistor, M_5 or M_6. This pole allocation is nearly the same as that in the preceding telescopic case except that the dominant pole and the nondominant pole of the op-amp now depend on different types of transistors. Alternatively, a PMOS input pair can be used in the op-amp. As a result, the dominant pole and the nondominant pole of the op-amp depend on the PMOS input pair and the NMOS cascode transistors, respectively.

The NMOS input pair should be chosen over the PMOS input pair if large unity-gain bandwidth and small silicon area have the highest priority. On the other hand, if large phase margin and good flicker noise suppression are the most important merits, the PMOS input pair should be adopted. Also, the PMOS input-pair configuration has an added benefit of allowing the op-amp's input common-mode level to be set to a negative power supply [2]. Nevertheless, the decision of which input pair to choose for the folded-cascode op-amp is not absolute and highly depends on the application's requirements.

An example of folded-cascode op-amp implementation is shown in Figure 2.3 (the CMFB circuit is not shown). This is a fully differential op-amp, and it uses six cascode layers to achieve high dc gain.

Note that this op-amp makes use of cascode current mirrors [4]. The purpose of using cascode current mirrors is to provide the appropriate bias voltages so that a rail-to-rail output swing is obtained. The bias current I_{bias} is set to $50\,\mu$A. A standard MOSIS 5-V 0.6-μm CMOS process is used, and the transistor-sizing for the op-amp is listed in Table 2.1.

With a total load capacitance of 4 pF (i.e., $C_1 = C_2 = 2\,$pF), SPICE simulations show that at the standard room temperature (27 degrees centigrade), the op-amp can achieve a maximum dc gain of about 79 dB and a UGBW of about 217 MHz. The

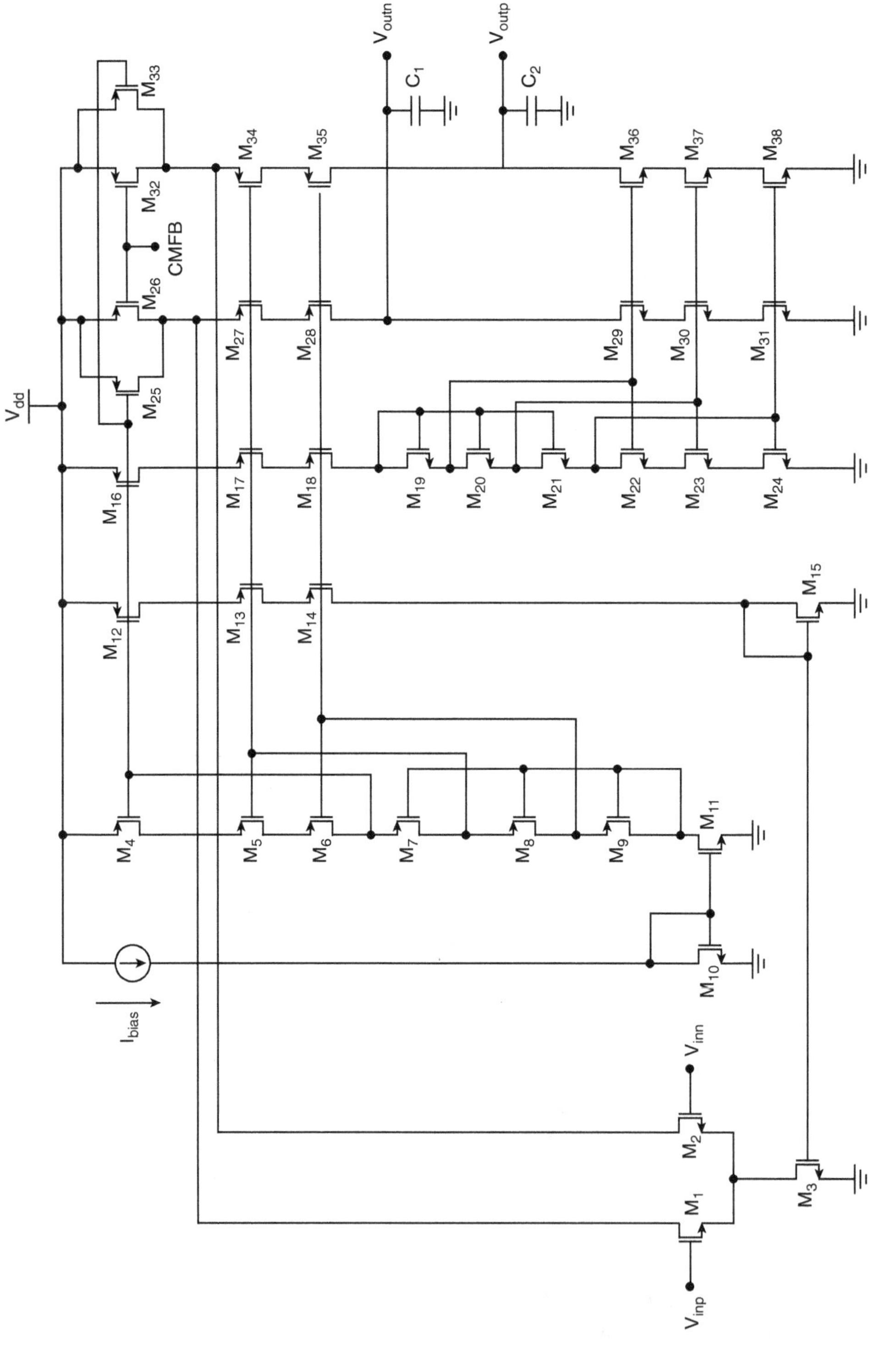

Figure 2.3 Practical folded-cascode CMOS op-amp design.

Table 2.1 Transistor-sizing for the folded-cascode op-amp.

Devices	Type	Size (μm)
$M_{1,2}$	NMOS	400/1
M_3	NMOS	200/1
$M_{4,5,6}$	PMOS	20/1
M_7	PMOS	20/5
M_8	PMOS	20/3
M_9	PMOS	20/1
$M_{10,11}$	NMOS	20/1
$M_{12,13,14}$	PMOS	20/1
M_{15}	NMOS	10/1
$M_{16,17,18}$	PMOS	20/1
M_{19}	NMOS	8/1
M_{20}	NMOS	8/3
M_{21}	NMOS	8/5
M_{22}	NMOS	12/1
M_{23}	NMOS	15/1
M_{24}	NMOS	18/1
M_{25}	PMOS	300/1
M_{26}	PMOS	100/1
$M_{27,28}$	PMOS	120/0.6
$M_{29,30,31}$	NMOS	100/1
M_{32}	PMOS	100/1
M_{33}	PMOS	300/1
$M_{34,35}$	PMOS	120/0.6
$M_{36,37,38}$	NMOS	100/1

maximum loop phase margin (measured along with the CMFB circuit) is about 57 degrees, and the range of output voltage swing is from 0.37 V to 4.98 V. This is not intended as an optimal design, and the interested reader is encouraged to improve the performance merits.

In applications such as high-speed and high-accuracy data converters, the gain requirement for the op-amp often exceeds what can be readily achieved by any conventional single-stage topologies. Moreover, in some cases the unity-gain bandwidth of the op-amp must be very large, which thus often hampers the use of multistage or cascaded architectures due to the inherent speed penalty caused by interstage compensations. One of the most widely adopted gain-enhancement techniques is using a fast regulated amplifier (or auxiliary amplifier) to improve the output impedance of the main single-stage amplifier, thereby increasing the overall gain without significantly affecting the original op-amp's speed [3][5].

Appendix 2.1

Use the MATLAB program to analyze the pole-splitting phenomenon and RHP zero cancellation of a standard differential-input, single-ended-output, two-stage op-amp. For the pole-splitting analysis, use $R_c = 0$ and sweep C_c from 0 to 30 pF. For RHP zero cancellation, use $C_c = 10$ pF and sweep R_c from 0 to 300 Ω. It is assumed that the input transconductance capacitor $g_m = 0.01$, the first-stage output impedance $R_1 = 100\,\text{k}\Omega$, the second-stage output impedance $R_2 = 20\,\text{k}\Omega$, $C_1 = 1$ pF, and $C_2 = 3$ pF. The source codes are as follows:

```
% MATLAB analysis of pole splitting, using Rc = 0 and sweep Cc.%
clear all;
close all;
gm1=0.01;
gm2=0.01;
R1=100000;
R2=20000;
C1=1e-12;
C2=3e-12;

for Cc=0:1e-13:30e-12
  a1=R1*R2*(C2*C1+Cc*C2+Cc*C1);
  a2=(C2+Cc)*R2+(C1+Cc)*R1+gm2*R1*R2*Cc;
  A=[a1, a2, 1];
  plot(real(roots(A)), imag(roots(A)), 'X');
  title('POLE SPLITTING: THE MOTION OF POLE (SWEEPING Cc FROM 0 TO
30pF)');
  xlabel('REAL');
  ylabel('IMAGINARY');
  hold on;
% MATLAB analysis of RHP zero cancellation, using Cc = 10pF and
sweep Rc. %

close all;
```

```
Cc=10e-12;

for Rc=0:10:300

  a1=R1*R2*Rc*C1*C2*Cc;
  a2=R1*R2*(C1*C2+Cc*C1+Cc*C2)+Rc*Cc*(R1*C1+R2*C2);
  a3=R2*(C2+Cc)+R1*(C1+Cc)+Rc*Cc+gm2*R1*R2*Cc;
  A=[a1, a2, a3, 1];
  figure(1)
  subplot(2,1,1),
  plot(real(roots(A)), imag(roots(A)), 'x');
  title('RHP ZERO CANCELLATION: THE MOTION OF POLE (SWEEPING Rc FROM
0 TO 300)');
  xlabel('REAL');
  ylabel('IMAGINARY');
  hold on;
  b1=-gm1*Cc*R1*R2+gm1*gm2*R1*R2*Cc*Rc;
  b0=gm1*gm2*R1*R2;
  B=[b1, b0, 1];
  subplot(2,1,2),
  plot(real(roots(B)), imag(roots(B)), 'o');
  title('RHP ZERO CANCELLATION: THE MOTION OF ZERO (SWEEPING Rc FROM
0 TO 300)');
    xlabel('REAL');
  ylabel('IMAGINARY');
  hold on;
end
```

References

[1] P. R. Gray and R. G. Meyer, *Analysis and design of analog integrated circuits*, John Wiley & Sons, New York, 1993.

[2] D. A. Johns and K. Martin, *Analog integrated circuits design*, John Wiley & Sons, New York, 1997.

[3] G. Nicollini, P. Confalonieri, and D. Senderowicz, "A fully differential sample-and-hold circuit for high-speed applications," *IEEE Journal of Solid-State Circuits*, Vol. 24, pp. 1461–1465, October 1989.

[4] N. Sooch and AT&T Bell Lab., "MOS cascode current mirror," U.S. Patent 4550284, 1985.

[5] K. Bult and G. Geelen, "A fast-settling CMOS Op Amp for SC circuits with 90-dB DC gain," *IEEE Journal of Solid-State Circuits*, Vol. 25, pp. 1379–1383, December 1990.

CHAPTER 3

Switched-Capacitor Building Blocks

3.1 Introduction

In this chapter we study basic switched-capacitor building blocks, which are the requisites for realizing analog sampled-data functions.

Switched-capacitor (SC) building blocks are often classified into two categories: *passive* and *active*. A passive SC building block (or element) is defined as a network composed of switches and capacitors only, whereas its active counterpart is not only built from switches and capacitors, but also active devices such as op-amps. Strictly speaking, a MOS switch is an active device, for it is composed of one or more transistors, which must be driven by the system clock or its derivatives. However, as we will see shortly, the primary use of SC-only elements is the simulation (or approximation) of the physical resistor; and in analog integrated circuits (ICs), the op-amp is usually regarded as the dividing line between passive and active integrated implementations. Therefore in this book we consider SC-only elements as *passive* and the others using op-amps as *active*.

As we have mentioned, the major application of passive SC elements is the physical resistor simulation in a monolithic configuration. Sometimes passive SC elements can also be used to construct passive filters and voltage converters. In comparison, as we will see in the following sections and chapters, active SC elements are widely adopted in the design of integrators, active filters, data converters, and so on.

Chapter Outline

This chapter is organized as follows. Section 3.2 presents different types of passive SC resistor simulations. The advantages of SC simulations over the physical resistor in terms of silicon area and accuracy are described. It is emphasized that an SC circuit's frequency response is determined by capacitance ratios rather than by individual capacitances [1]. The relationships of capacitance ratios and critical circuit parameters are investigated. Section 3.3 introduces single-ended as well as fully differential SC integrators. The effects of parasitic capacitances on the circuit's transfer function are discussed, and a few parasitic-insensitive solutions are provided. Section 3.4 presents the design principles of CMOS sample-and-hold (S&H) circuits. The key performance parameters of S&H circuits are investigated, and several design examples are presented. Section 3.5 presents an overview of SC interpolators and decimators. Finally, a brief tutorial on the *signal-flow-graph* (SFG) technique and the Mason's rule is included in Section 3.6.

3.2 Switched-Capacitor Resistor Simulation

SC Resistor Simulations

A physical resistor can be simulated by different passive SC elements. We start with the *periodically reverse-switched capacitor* [2], which is better known as the *bilinear SC resistor simulation* [3] (we shall see shortly why the term *bilinear* is used here). The SC configuration is illustrated in Figure 3.1(a). Controlled by two non-overlapping pulses Φ_1 and Φ_2, the switches change positions only momentarily at discrete instants t_n, $n = 1, 2, 3, 4, 5, 6 \ldots$ as shown in Figure 3.1(b).

Before proceeding, we should note that to simulate a physical resistor accurately (i.e., with less than 1% relative error) using an SC resistor simulation, both V_1 and V_2 must vary slowly in relation to the sampling clock. In fact, there is a rule of thumb suggesting that $f_{max} \leq f_{clk}/100$ (f_{max} is the highest possible signal frequency and f_{clk} is the sampling clock frequency). To simplify the analysis, for this moment we assume that all switches are ideal with zero *on-resistances*, and as a result, the time taken for charging and discharging C can be considered negligible. In other words, C is charged or discharged to a specified voltage *instantaneously* at each sampling instant. Additionally, note that in this section we ignore *parasitic capacitances* or *stray capacitances*, and we will study their effects on circuit performance in Section 3.3.

The following is a time-domain analysis of the SC network shown in Figure 3.1(a). Initially, all four switches are open, and the capacitor C is not charged. At time t_1, Φ_1 is on, and switches 1 and 3 are closed. By the end of Φ_1 or time t_2, the voltage across C will become $V_1(t_2) - V_2(t_2)$. Since V_1 and V_2 are assumed to remain constant during the clock phase period (i.e., $f_1, f_2 \ll f_{clk}$), the index t_2 is considered

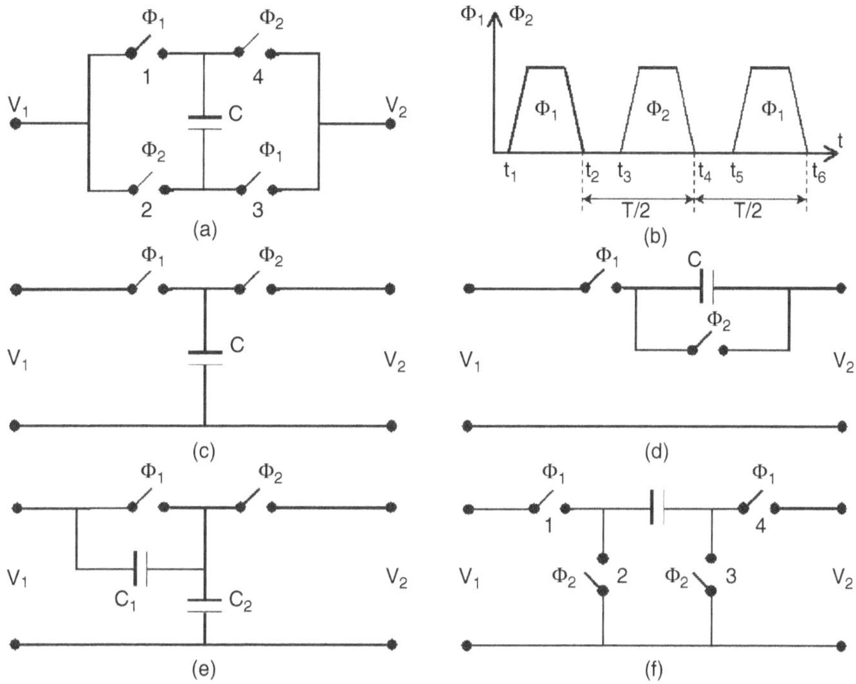

Figure 3.1 (a) Bilinear resistor simulation. (b) Nonoverlapping clock pulses. (c) Parallel resistor simulation. (d) Series resistor simulation. (e) Series-parallel resistor simulation. (f) Parasitic-insensitive resistor simulation.

insignificant, and the capacitor voltage can be approximated to a dc voltage ($V_1 - V_2$). The electronic charge (in coulombs) transferred from node V_1 to node V_2 is thus given by

$$\Delta q_1 = C(V_1 - V_2) - 0 = C(V_1 - V_2) \tag{3.1}$$

At time t_3, Φ_1 is already off and Φ_2 is just turned on, and switches 1 and 3 are open whereas 2 and 4 are closed. By the end of Φ_2 or time t_4, the capacitor will be charged to ($V_2 - V_1$), and the charge transferred from V_2 to V_1 is thus given by

$$\Delta q_2 = C(V_2 - V_1) - C(V_1 - V_2) = 2C(V_2 - V_1) \tag{3.2}$$

At time t_5, again, Φ_1 is turned on instantly and Φ_2 is already off, and switches 1 and 3 are closed whereas 2 and 4 are open. Similarly, by the end of the subsequent Φ_1 or time t_6, the capacitor C will be charged to ($V_1 - V_2$), and the charge transferred from V_1 to V_2 is given by

$$\Delta q_3 = C(V_1 - V_2) - C(V_2 - V_1) = 2C(V_1 - V_2) \tag{3.3}$$

From now on, the sequence of charge transfer between time t_2 and t_6 is repeated every T, and the system reaches a steady state. Under a steady-state condition, the average current that flows through the system is determined by the amount of charge transferred between, for example, time t_2 and t_6, which is equal to T. Therefore, if the direction of the electric current flow is assumedly from V_1 to V_2, then the average current can be expressed as

$$I_{avg} = \frac{\Delta q_3 - \Delta q_2}{(t_6 - t_2)} = \frac{4C(V_1 - V_2)}{T} \tag{3.4}$$

Apply Ohm's law, and the equivalent resistance is thus given by

$$R_{equ} = \frac{(V_1 - V_2)}{I_{avg}} = \frac{T}{4C} = \frac{1}{4Cf_{clk}} \tag{3.5}$$

Interestingly, the preceding development implies that the polarity of the voltage across the capacitor changes *twice* from t_2 till t_6 (i.e., within a clock period T). In other words, the effective cycle of polarity change is $T_1 (T_1 = T/2)$. As a result, to *keep up with* these changes, all switches must operate (i.e., open/close) at a rate of $2f_{clk}$. Hence, the actual sampling frequency required for operating the circuit shown in Figure 3.1(a) is given by $2f_{clk}$.

Next we investigate why this configuration is often referred to as the *bilinear SC resistor simulation*. Let us first take a look at the frequency-domain description of the charge distribution by a resistor. Consider the charge transferred by a continuous-time resistor R_0, which is given by

$$Q_0(t) = \int_{-\infty}^{t} i_0(\tau)d\tau = \int_{-\infty}^{t} \frac{V_0(\tau)}{R_0} d\tau = \frac{1}{R_0} \cdot \int_{-\infty}^{t} V_0(\tau)d\tau \tag{3.6}$$

Here V_0 is the voltage difference between two nodes, V_1 and V_2. We then obtain the s-domain expression of Q_0 by applying the Laplace-transform to the preceding equation,

$$Q_0(s) = \frac{1}{R_0} \cdot \frac{V_0(s)}{s} \tag{3.7}$$

It has been proven that an s-domain equation relating $Q_0(s)$, $V_0(s)$, and R_0, such as Equation (3.7), can be mapped directly to its corresponding z-domain equation that relates $Q_0(z)$, $V_0(z)$, and C by simply replacing s in the s-domain equation with its z-domain equivalence, $F(z)$ [3]. Next, we need to identify the right s-to-z transformation to map Equation (3.7) to its z-domain counterpart.

In the following frequency-domain analysis, we will adopt most assumptions made for the preceding time-domain derivation of R_{equ}, for instance, all switches operate periodically at a rate much faster than that of the variations in V_1 and V_2. However, the particular assumption of V_0 (i.e., the voltage difference between V_1 and V_2) being a dc signal defeats the purpose of frequency-domain analysis and causes loss of generality. Therefore, we no longer assume V_0 as a dc voltage from this point on.

Now, let us revisit the charge behavior during the time interval from t_2 till t_4, whose pattern repeats every half period or T_1. Specifically, at time t_2, C is charged to $CV_0(t_2)$, which is retained on C till time t_3, when C is charged to $CV_0(t_3)$. Likewise, at the next instant $t_4 = t_2 + T_1$, C is instantaneously discharged and recharged to $CV_0(t_4)$. According to the charge conservation principle, a difference equation can be established to relate the corresponding charges sampled at t_2, t_3, and t_4, which is given by

$$\begin{aligned} Q_0(t_4) - Q_0(t_4 - T_1) &= C[V_0(t_4) + V_0(t_3)] - C[V_0(t_3) - V_0(t_4 - T_1)] \\ &= C[V_0(t_4) + V_0(t_4 - T_1)] \end{aligned} \tag{3.8}$$

The preceding difference equation shows that the charge transfer depends on samples taken at discrete-time instants (e.g., t_4) in a periodical fashion (on a T_1 cycle); hence we can apply z-transform to both sides of the equation, and we have

$$Q_0(z) = C \cdot \frac{1 + z^{-1}}{1 - z^{-1}} \cdot V_0(z) \tag{3.9}$$

The right-hand side of Equation (3.9) may remind the reader of the *bilinear mapping* concept introduced in basic discrete-time signal theories. The well-known formula of the bilinear mapping scheme is provided here for convenience:

$$s \rightarrow \frac{2}{T_1} \cdot \frac{1 - z^{-1}}{1 + z^{-1}} \tag{3.10}$$

The right-hand part of the preceding formula is the transformed result of s in z-domain, $F(z)$. Again, note that the effective clock period is indeed T_1 (which is equal to half T). By comparing Equations (3.7), (3.9), and (3.10), we can conclude that Equation (3.7) is *bilinear-mapped* to its z-domain counterpart, Equation (3.9) (assuming the frequency warping effects have been alleviated by *prewarping* operations), only if the following condition is satisfied:

$$R_0 = \frac{T_1}{2C} = \frac{T}{4C} = \frac{1}{4Cf_{clk}} \tag{3.11}$$

A comparison between Equations (3.5) and (3.11) indicates that $R_0 = R_{equ}$. The equality thus verifies the aforementioned time-domain derivation of R_{equ}. To conclude, the forgoing treatment shows that Equation (3.7) can be mapped directly to Equation (3.9) using the bilinear s-to-z transformation, which in turn explains the term *bilinear* used in the naming of the SC resistor simulation shown in Figure 3.1(a).

However, to the derivation of a bilinear SC circuit from its continuous-time RC prototype, simply replacing each branch resistor with its corresponding SC resistor simulation is not adequate, and a sample-and-hold (S&H) clocked at $2f_{clk}$ must be added before the SC circuit's input. As reported by Temes et al. [4], the main purpose of this S&H is to eliminate a charge leakage. This leakage is neither due to *charge injection* (see Chapter 1) nor caused by *voltage-dependent nonlinear capacitance* [5] because so far ideal switches and capacitor have been assumed. In fact, this leakage can be identified as an *inter-phase-input-dependent charge leakage*, meaning the charge behavior of the circuit during the odd ($\Phi_1 = 1$) or even ($\Phi_2 = 1$) phase depends on the input during the even or odd phase. As a result, such a bilinear circuit cannot be decomposed into odd-phase and even-phase subcircuits.

It can be proved that the leakage reported in [4] will occur if either of the following conditions is satisfied: (1) the input signal has a continuous-time waveform; or (2) the neighboring Φ_1s and Φ_2s in Figure 3.1(b) are located very close to each other, such that the two nodes (V_1 and V_2) appear to be coupled with each other by the capacitor C *all the time*, except for the moments t_n, $n = 1, 2, 3, 4, 5, 6 \ldots$

Intuitively speaking, the added S&H alleviates the nonlinearity effects (or distortions) due to the leakage by periodically decoupling the output from the input. Furthermore, if it is assumed that the input is a true continuous-time signal, then the frequency response of the S&H, which is basically a *sinc response*, helps attenuate the higher-frequency signal replicas (at $2f_{clk}$ and beyond). The interested reader is referred to [3][4] for mathematical analysis of the added S&H's influence on the SC circuit's frequency response.

Now that we have practiced analyzing the first SC resistor simulation in both time and frequency domains, it is straightforward to deal with the remaining circuits in Figure 3.1 since we can simply follow the same steps. A second SC resistor simulation is shown in Figure 3.1(c) [6]. The two switches are turned on and off repeatedly under the control of pulses Φ_1 and Φ_2, which are shown in Figure 3.1(b). As a result, the amount of charge that flows from V_1 to V_2 is equal to $C(V_1 - V_2)$ during each clock period T (assuming there is no parasitic capacitance). Thus, the average current that flows into V_2 is given by

$$I_{avg} = \frac{C(V_1 - V_2)}{T} \tag{3.12}$$

The equivalent continuous-time resistor is given by

$$R_{equ} = \frac{(V_1 - V_2)}{I_{avg}} = \frac{T}{C} = \frac{1}{Cf_{clk}} \tag{3.13}$$

This SC circuit is often called the *parallel resistor simulation*, since V_1 and V_2 are never directly coupled together (i.e., they are connected in parallel).

Figures 3.1(d) and (e) show the *series resistor simulation* and the *series-parallel resistor simulation*, respectively. In comparison with the parallel simulation, in both settings, a serial capacitor is used to couple V_1 to V_2. These two circuits can be analyzed in a similar fashion as before. For the sake of brevity, the analysis is left to the reader as an exercise. The equivalent resistor result from the *series simulation* is given by

$$R_{equ} = \frac{T}{C} = \frac{1}{Cf_{clk}} \tag{3.14}$$

And the equivalent resistor from the *series-parallel simulation* shown in Figure 3.1(e) is

$$R_{equ} = \frac{T}{C_1 + C_2} = \frac{1}{(C_1 + C_2) f_{clk}} \tag{3.15}$$

Interesting enough, once we replace T with T_1 and make $C_1 = C_2$ in Equation (3.15), the resulting equivalent resistance turns out to be exactly the same as that given by Equation (3.5). This indicates a conditional equivalence between the bilinear and series-parallel resistor simulations. The element in Figure 3.1(e) can in fact be used to build a bilinear SC circuit, for example, an integrator (for $C_1 = C_2$), whose input signal should also be sampled and held [7].

Lastly, the SC circuit illustrated in Figure 3.1(f) [8][9] has been one of the most widely adopted solutions to the simulation of a continuous-time resistor so far. Circuit designers prefer it over the remaining configurations in Figure 3.1 mainly because of its insensitivity to parasitic/stray capacitances in the circuit, from which the others suffer. In an integrated capacitor, the troublesome parasitic capacitances usually reside between certain nodes on the lower polysilicon layers, which are neither grounded nor connected to a low impedance point such as an op-amp's output in most cases. Among them, the critical ones are the top-plate parasitic (C_{tp}) and the bottom-plate parasitic (C_{bp}), whose absolute values tend to vary, and are not controllable even through careful fabrications. Depending on circuit layout and process technology, C_{tp} varies from 0.1% to 1%, while C_{bp} varies from 5% to as much as 20% of the desired capacitance C [10]. In Section 3.3, we will learn in detail how this particular SC element (or its modification, discussed later) can be used to effectively eliminate the effects of parasitic capacitances.

By following the same procedure employed throughout this section, we can derive the equivalent resistance simulated by the circuit in Figure 3.1(f), which is given by $R_{equ} = 1/Cf_{clk}$. A modification of this SC circuit can be realized by simply interchanging the clock phases of switches 3 and 4 in Figure 3.1(f). It can be shown that such phase reversals will introduce a sign change to the transfer function as well as to the equivalent resistance value. In fact, the resulting resistance is given by $R_{equ} = -1/Cf_{clk}$ (i.e., a negative resistor). The derivation of these resistances is left to the reader as an exercise.

The Advantages of SC Resistor Simulations

The most significant advantage of SC resistor simulations is the considerable saving of silicon chip area. To understand this, consider the circuit shown in Figure 3.1(c), if $C = 0.5\,\text{pF}$ and $f_{clk} = 200\,\text{kHz}$, then an R_{equ} as large as $10\,\text{M}\Omega$ is realized by simply using two clock-controlled switches and one relatively small capacitor (0.5 pF), which consume a silicon area of less than $0.01\,\text{mm}^2$ in a standard CMOS process. In comparison, to realize a 10-MΩ diffused resistor in the same technology, a silicon area of up to $1.0\,\text{mm}^2$ will be required, which is 1.0/0.01 = 100 times larger!

Another important advantage is that the frequency response of the SC circuit is controllable by adjusting capacitance ratios. This feature enables us to improve the overall accuracy of the circuit at much less physical design cost and effort. First, let us look into an important parameter called *time constant*, which is often denoted as τ. In a first-order RC network, the time constant is expressed as $\tau = R_0C_0$ (R_0 is the diffused resistor, and C_0 is the *unswitched* capacitor). In the modern IC fabrication process, capacitors and resistors are made in rather different steps. So, there is no referable relation between the capacitor error and that of the resistor [11]. In general, these two types of errors are considered to be uncorrelated from each other. Thus, the accuracy of τ can be given by

$$\frac{d\tau}{\tau} = \frac{d(R_0C_0)}{(R_0C_0)} = \frac{dR_0}{R_0} + \frac{dC_0}{C_0} \tag{3.16}$$

The preceding equation indicates that the accuracy of τ is determined by the inherent precisions of R_0 and C_0, which tend to vary considerably with temperature and technology tolerance [1]. In practice, the absolute errors of integrated R_0 and C_0 due to temperature variations and fabrication inaccuracies are normally on the order of 10%. This implies that in a worst-case scenario, the variance of τ obtained from Equation (3.16) may amount to as large as 20%, which is apparently unacceptable to most signal processing applications.

If R_0 is replaced by, say, the parallel simulation shown in Figure 3.1(c), then the resulting time constant τ_1 is given by (assuming that ideal switches with zero on-resistances are used, and that T is a two-phase clock generated by a precise crystal oscillator)

$$\tau_1 = \frac{1}{Cf_{clk}} \cdot C_0 = T \cdot \frac{C_0}{C} \cong R_0 \cdot C_0 \tag{3.17}$$

Equation (3.17) indicates that τ_1 is now determined by the capacitance ratio in the circuit (i.e., C_0/C) and T. Hence, the accuracy of the time constant is given by

$$\frac{d\tau_1}{\tau_1} = \frac{d\left(\frac{T}{C}C_0\right)}{\left(\frac{T}{C}C_0\right)} = \frac{dT}{T} - \frac{dC}{C} + \frac{dC_0}{C_0} \cong \frac{dC_0}{C_0} - \frac{dC}{C} = \frac{d\left(\frac{C_0}{C}\right)}{\left(\frac{C_0}{C}\right)} \tag{3.18}$$

Note that in the preceding equation, the variance of T is ignored because the output frequency value of such a crystal-based clock is usually very precise (with an absolute error of less than 0.001%). The rightmost side of Equation (3.18) reveals that in the absence of the physical resistor, the accuracy of τ_1 is dominated by the *tracking* or *matching accuracy* between C_0 and C. In other words, the time constant τ_1 depends on the capacitance ratio rather than the individual capacitances (i.e., C_0 and C).

Furthermore, C_0 and C are often located close to each other in the chip layout. With careful layout, the mismatch error between two capacitors fabricated within the same area of a chip can be kept as low as 0.1% in a standard CMOS process [5] and even lower than 0.05% in the cutting-edge CMOS technologies. Therefore, the accuracy of τ_1 given by Equation (3.18) is a significant improvement (e.g., 200 times more accurate) to that given by Equation (3.16).

Capacitance Ratios versus Circuit Parameters

In the previous development, we learned the relationship between the time constant τ_1 and the capacitance ratio (C_0/C). Furthermore, it is of practical interest to explore the relationships of the following critical parameters: *damping factor* (σ_p), *quality factor* (Q), *pole* or *center frequency* (ω_0), and *maximum magnitude* or *voltage gain* (G), with respect to capacitance ratios.

Let us begin our discussion with the SC circuit's transfer function. Recall from signals and systems theories that a high-order transfer function can be realized by factoring its numerator and denominator into a group of first- or second-order (*biquadratic* or *biquad*) subfunctions. Once these low-order building blocks are determined, it is easy to reassemble the overall transfer function by cascading them. The calculation of the capacitance ratio in a first-order SC network is straightforward since only one op-amp is used. Therefore, without loss of generality, we will focus our discussion on second-order transfer functions.

In general, a second-order continuous-time system can be described by an s-domain transfer function in the form of

$$H(s)=\frac{a_0+a_1s^{-1}+a_2s^{-2}}{b_0+b_1s^{-1}+b_2s^{-2}}=\frac{a_0/b_0+(a_1/b_0)s^{-1}+(a_2/b_0)s^{-2}}{1+(b_1/b_0)s^{-1}+(b_2/b_0)s^{-2}} \tag{3.19}$$

For synthesis purposes, the foregoing transfer function is often rewritten to

$$H_A(s)=\frac{a_0/b_0+(a_1/b_0)s^{-1}+(a_2/b_0)s^{-2}}{1+(\omega_0/Q)s^{-1}+\omega_0^2s^{-2}}=\frac{a_0/b_0+(a_1/b_0)s^{-1}+(a_2/b_0)s^{-2}}{(1-p_1s^{-1})(1-p_2s^{-1})} \tag{3.20}$$

Assuming p_1 and p_2 are two conjugate-complex s-plane poles of $H_A(s)$, we have

$$p_{1,2}=-\sigma_p\pm j\omega_p,\quad 0\le\sigma_p\le\infty,\quad 0\le\omega_p\le\infty \tag{3.21}$$

Based on Equations (3.20) and (3.21), we can write σ_p and ω_p as functions of Q and ω_0,

$$\sigma_p=\frac{\omega_0}{2Q}\quad\text{and}\quad\omega_p=\sigma_p\sqrt{4Q^2-1}=\omega_0\sqrt{1-\frac{1}{4Q^2}} \tag{3.22}$$

where σ_p is the *damping factor* and ω_p is the *overshoot* or *peak frequency* of the magnitude response. The expression of σ_p given by Equation (3.22) may remind the reader of the *half-bandwidth* of a band-pass (BP) system, which is expressed as $\sqrt{2}\sigma_p=\omega_{-3dB}=1/\tau$. What's more, this reveals a linear relation between σ_p and τ. According to Equation (3.18), if we utilize SC resistor simulations to realize all the required resistances in a continuous-time system, then the accuracy of the resultant time constant (τ) will depend heavily on the capacitance ratios. Thus, based on the aforementioned relation between τ and σ_p, we can say that the realized σ_p is also determined by the capacitance ratios. Furthermore, from Equation (3.22) we can find that the ratio of ω_0 to Q depends on the capacitance ratios as well.

According to Equation (3.22), if Q is increased then (ω_p/σ_p) is increased, and $\omega_p \to \omega_0$. As a consequence, the magnitude of $H_A(j\omega)$ will have a sharper peaking or overshoot around ω_0, meaning a higher quality factor tends to result in a more aggressing peaking in the magnitude response. Two benchmarks are often used in practice to evaluate the peaking characteristics:

1. $Q = 0.5$, $\omega_0 = \sigma_p$, and $\omega_p = 0$—that is, the pole frequency is lower than ω_{-3dB} and no peaking occurs in the pass-band.
2. $Q = 1/\sqrt{2}$, $\omega_0 = \sqrt{2}\sigma_p$, and $\omega_p = \sigma_p = \omega_0/\sqrt{2}$—that is, the pole frequency is equal to ω_{-3dB} with a mere 4.32% overshoot at ω_0, which imposes no danger of instability on the overall response.

Next, we should derive the z-domain biquad transfer function using one of the four standard s-to-z transformations (i.e., forward-Eular, backward-Eular, bilinear, and lossless-discrete-integrating). From discrete-time signal theories, we know that if a high sampling rate ($f_{clk}/f_0 >> 2\pi$ or $\omega_0 T << 1$) is assumed, then all four transformations can be roughly expressed as

$$z = e^{sT} \cong 1 + sT \tag{3.23}$$

Inserting the previous approximation of s into Equation (3.19), we obtain the equivalent z-domain transfer function in the form of

$$H(z) = \frac{c_0 + c_1 z^{-1} + c_2 z^{-2}}{1 + d_1 z^{-1} + d_2 z^{-2}} = \frac{c_0 + c_1 z^{-1} + c_2 z^{-2}}{(1 - P_1 z^{-1})(1 - P_2 z^{-1})} \tag{3.24}$$

Also, it is known that the z-plane poles are given by

$$P_{1,2} = e^{p_{1,2}T} = e^{-\omega_0 T/2Q}(\cos\omega_p T \pm j\sin\omega_p T) \tag{3.25}$$

Inserting the z-plane poles (P_1 and P_2) into the denominator of the transfer function given by Equation (3.24), after some algebraic treatments we obtain the expressions of d_1 and d_2 in terms of ω_0, T, and Q:

$$d_1 = -2e^{-\omega_0 T/2Q}\cos\left(\omega_0 T\sqrt{1 - \frac{1}{4Q^2}}\right) \quad \text{and} \quad d_2 = e^{-\omega_0 T/2Q} \tag{3.26}$$

For most low-frequency applications (particularly audio signal processing), a high sampling rate ($f_{clk}/f_0 >> 2\,\pi$ or $\omega_0 T << 1$) is usually required. In these cases, we can apply approximation to Equation (3.26) and obtain the following (assuming Q is not smaller than 0.5):

$$d_1 \cong -2 + \frac{\omega_0 T}{Q} + (\omega_0 T)^2 \quad \text{and} \quad d_2 \cong 1 - \frac{\omega_0 T}{Q} \tag{3.27}$$

And we can write the expression of Q in terms of d_1 and d_2:

$$Q \cong \frac{\sqrt{1+d_1+d_2}}{(1-d_2)} \tag{3.28}$$

Equation (3.28) is very useful for estimating the quality factor of an SC biquad based on its z-domain transfer function. The higher the value of Q, the more accurate the preceding estimation will be in relation to the actual value. Furthermore, the foregoing equation can considerably simplify the calculation of the sensitivity of Q to capacitance ratios [12][13].

Before proceeding, let us revisit Equations (3.19), (3.23), and (3.24). After some algebraic efforts, we can rewrite $H(z)$ of Equation (3.24) to

$$H(z) = \frac{c_0 + c_1 z^{-1} + c_2 z^{-2}}{1 + d_1 z^{-1} + d_2 z^{-2}} \cong \frac{a_0/b_0 - \left(\frac{2a_0 - a_1 T}{b_0}\right) z^{-1} + \left(\frac{a_0 - a_1 T + a_2 T^2}{b_0}\right) z^{-2}}{1 - \left(2 - \frac{b_1 T}{b_0}\right) z^{-1} + \left(1 - \frac{b_1 T}{b_0} + \frac{b_2 T^2}{b_0}\right) z^{-2}} \tag{3.29}$$

Utilizing the results given by Equation (3.27), we derive the following approximations:

$$\frac{b_1}{b_0} \cong \frac{\omega_0}{Q} + \omega_0^2 T \quad \text{and} \quad \frac{b_2}{b_0} \cong \omega_0^2 \tag{3.30}$$

Comparing them with the denominator's coefficients in Equation (3.20), we realize that an error in the form of $\omega_0^2 T s^{-1}$, which is primarily due to the approximation given by Equation (3.23), has been introduced. In many cases, this error is not negligible due to its effects on both magnitude and phase responses of the transfer function [12]. Thus, the approximations of d_1, d_2, and Q derived earlier, although useful for lending insights, are inadequate for the exact design and analysis, where high precision is required. For exact representations of Q and ω_0, we must use more precise mapping formulae, one of which is the bilinear s-to-z transformation given by $z = (1 + s)/(1 - s)$.

However, deriving the precise expressions of Q and ω_0 in terms of capacitance ratios is slightly involved. Fortunately, a few established results with respect to this topic can be found in the literature [14][15]. Here we adopt the formulae from [15], which are listed as follows (assuming ω_0 is the *prewarped* frequency):

$$H_d(z) = \frac{N(z)}{1-(2-\alpha-\beta)z^{-1}+(1-\beta)z^{-2}} \tag{3.31}$$

$$\omega_0 = \frac{1}{T}\sqrt{\frac{\alpha}{1-\alpha/4-\beta/2}} \quad \text{and} \quad Q = \frac{\sqrt{\alpha(1-\alpha/4-\beta/2)}}{\beta} \tag{3.32}$$

where $H_d(z)$ is the exact bilinear-transformed result of $H(s)$ given by Equation (3.19). The system coefficients α and β are given by

$$\alpha = \frac{4}{X(z)} \quad \text{and} \quad \beta = \frac{\alpha}{\omega_0 TQ}, \quad \text{with } X(z) = 1 + \frac{2}{\omega_0 TQ} + \frac{4}{(\omega_0 T)^2} \tag{3.33}$$

Ki and Temes [15] demonstrated that both α and β can be found in terms of capacitance ratios rather than individual capacitances. Thus, based on the preceding development, we conclude that for a given T, both Q and ω_0 can be controlled by simply adjusting capacitance ratios in the circuit. In addition, from Equation (3.33) we find out that α and β are indeed correlated to each other, and $\alpha \to (\omega_0 T)^2$ and $\beta \to (\omega_0 T/Q)$ given a high sampling rate ($\omega_0 T \ll 1$).

Interestingly enough, Equations (3.32) and (3.33) indicate that Q and ω_0 cannot be adjusted *independently*, making the SC filter discussed earlier unsuitable for applications such as adaptive filter and vocal tracking. As a solution, Allstot et al. [16] proposed a parametric programmable SC filter implementation. The basic mechanism of the filter is described as follows: the independent programmability of Q and ω_0 is obtained by scaling the equivalent resistance R of the filter such that $R = 1/Q$, which in essence removes the inherent correlation between Q and ω_0. In addition, two different types of programmable capacitor arrays are used to implement the parametric control (a binary-weighted type for programming Q and G; a logarithmic type for programming ω_0), thereby further separating Q and ω_0.

However, the usage of programmable capacitor arrays inevitably increases the total silicon area and power consumption. Also, if a high quality factor is required ($Q \gg 1$), the switched capacitor required to implement the $1/Q$ impedance becomes quite small, making it difficult to keep the capacitance spread as close to minimum as possible (note that *capacitance spread* is the ratio of the largest to the smallest capacitance, which affects the total silicon area and accuracy of capacitor matching). Although several novel modifications have been reported to circumvent these drawbacks [17][18], the results are not comparable to that of fully digital adaptive filters in terms of cost and performance. As a result, we are not going to study parametric programmable SC filters in detail; the interested reader is referred to [16][17][18][19]. Sometimes a programmable capacitor array may also be used as the input stage for binary-weighted data converters.

Lastly, it can be proved that the maximum magnitude/voltage gain (G) of an SC biquad also depends on capacitance ratios (assuming T is a constant). For the sake of brevity, the proof is left to the interested reader as an exercise (*Hint*: Consider the dimensions of system coefficients such as α and β [20]). As a rule of thumb in practice, for band-pass, low-Q ($Q < 1/\sqrt{2}$) low-pass, and low-Q high-pass biquads, the maximum gain is *normalized* to unity; while for high-Q ($Q > 3$) low-pass and high-Q high-pass biquads, the maximum gain is given by $G \approx Q$ [13][15][21].

To conclude, Section 3.2 explored various SC resistor simulations and their advantages over the conventional diffused resistor implementation. The relationships of several circuit performance parameters with respect to capacitance ratios were explained. The discussion emphasized that capacitance ratios play a critical role in the frequency and magnitude characterizations of an SC circuit. The subject of the next section is one of the most important active elements for building filters and data converters—the SC integrator.

3.3 Switched-Capacitor Integrators

Parasitic-Sensitive SC Integrators

One of the most important building blocks in a continuous-time active-RC filter is the inverting analog integrator, which is illustrated in Figure 3.2(a). It is assumed that all the op-amps used in this section are ideal—that is, dc-offset-free, with infinite gain and bandwidth, powered by sufficient supply voltages and so on. The time-domain expression of the circuit is given by

$$V_{out}(t) = -\frac{1}{R_0 C_0}\int_{-\infty}^{t} v_{in}(\tau)d\tau \tag{3.34}$$

Utilizing the Laplace-transform, we obtain the s-domain transfer function of the active-RC integrator as follows:

$$H(s) = \frac{V_{out}(s)}{V_{in}(s)} = -\frac{1}{sR_0C_0} \quad \text{or} \quad H(j\omega) = -\frac{1}{j\omega R_0 C_0} = -\frac{\omega_{-3dB}}{j\omega} \tag{3.35}$$

where ω_{-3dB} is the –3 dB frequency, and it can be shown that the magnitude response and phase shift of the active-RC integrator are given by (ω_{-3dB}/ω) and $\pi/2$, respectively.

The simplest way to realize an SC integrator is to replace the resistor R_0 in the active-RC integrator with one of the SC simulations introduced in the previous section. Ideally, the resulting SC integrator should retain the magnitude of (ω_{-3dB}/ω) as well as the phase shift of $\pi/2$. However, we will see later that an SC integrator only approximates its continuous-time counterpart.

Before delving into the analysis of the SC integrator, as a convention, we define the sampling instants by the end of clock phase Φ_1 to be $(n-1)T$, nT, $(n+1)T$, and so on (T is the clock period), whereas those by the end of clock phase Φ_2 are deemed to be $(n-3/2)T$, $(n-1/2)T$, $(n+1/2)T$, and so on. This denotation of the nonoverlapping clock facilitates the writing of difference equations. Nevertheless, in most cases it is not important that the falling edge of Φ_1 be precisely half-T apart from that of Φ_2, as long as the duration of each clock phase is long enough for the signal to settle properly. Lastly, we denote a switch that is on when $\Phi_1 = 1$ ($\Phi_2 = 1$) to be a Φ_1 (Φ_2) switch for convenience.

The circuit in Figure 3.2(b) is called the *parasitic-sensitive bilinear SC integrator* for it is the result of replacing the resistor R_0 with the bilinear simulation in Figure 3.1(a), and as we will see later, its performance is sensitive to parasitic capacitances in the circuit. The input is a sampled-and-held signal clocked at $2f_{clk}$ (sampled when Φ_1 is on). Note that a Φ_2 switch, which is turned on when $\Phi_2 = 1$, is added before the output, and thus the output jumps only when $\Phi_2 \rightarrow 1$. This extra Φ_2 switch also indicates that the subsequent circuitry should sample the integrator's output when $\Phi_2 = 1$. Alternatively this switch can be turned on when $\Phi_1 = 1$; however, this change simply introduces a delay element.

The next step is to derive the z-domain transfer function of the bilinear integrator. To simplify the analysis, let us ignore all parasitic capacitances for the moment. We shall understand the effects of parasitic capacitances on the transfer function shortly. Recalling Equation (3.9) and assuming V_0 is equal to V_{in}, we rewrite Equation (3.9) to

$$Q(z) = C\frac{1+z^{-1}}{1-z^{-1}}V_{in}(z) = C\frac{z+1}{z-1}V_{in}(z) \tag{3.36}$$

where $Q(z)$ is the charge transferred from the input to the top plate of the feedback capacitor (C_0) every half clock cycle (i.e., $T/2$). Apparently, this charge will build a voltage across C_0, whose polarity is always opposite to that of V_{out}. Hence, we obtain the following equations:

$$V_{out}(z) = -\frac{Q(z)}{C_0} = -\frac{C}{C_0}\frac{z+1}{z-1}V_{in}(z) \Rightarrow H_{BL}(z) = \frac{V_{out}(z)}{V_{in}(z)} = -\frac{C}{C_0}\frac{1+z^{-1}}{1-z^{-1}} \quad (3.37)$$

One thing to remember here is that in both Equations (3.36) and (3.37), $z = e^{j\omega T/2}$. This is due to the fact that the effective clock period of a bilinear SC circuit is indeed $T/2$ (i.e., all the switches operate at $2f_{clk}$). As a consequence, a time delay of $T/2$ is equivalent to a full-cycle delay to the bilinear SC integrator. This is a unique property of the bilinear SC integrator since other integrators (e.g., Euler) are normally clocked at f_{clk}.

The inquisitive reader may ask about how to derive the z-domain transfer function of an SC integrator based on its charge transfer behavior in the time domain. In the following context, the *parasitic-sensitive parallel SC integrator* in Figure 3.2(c) is used as an example to demonstrate how this is done. Similar to the bilinear SC integrator, an extra Φ_1 switch is added before the output of the parallel SC integrator (i.e., the output changes only when $\Phi_1 = 1$). Hence, the subsequent circuitry should sample the integrator's output on Φ_1.

To simplify the analysis, all parasitic capacitances (including C_{p2}) are ignored for the moment. Once Φ_1 is off, the charge on the feedback capacitor C_0 will remain the same till Φ_1 is on again. In other words, the charge on C_0 at the instant nT (i.e., by the end of Φ_1) is equal to that at the instant $(nT - T/2)$ (i.e., by the end of the previous Φ_2), hence we can write the following difference equation:

$$C_0 V_{out}(nT) = C_0 V_{out}(nT - T/2) \quad (3.38)$$

Similarly, at time $(nT - T)$, C is charged to $CV_{in}(nT - T)$. Next, Φ_1 is off and Φ_2 is on; this charge $CV_{in}(nT - T)$ is thus sent to C_0 and combined with the existing charge across C_0, which is given by $C_0V_{out}(nT - T)$. The outcome is the charge $C_0V_{out}(nT - T/2)$. Note that the polarity of $CV_{in}(nT - T)$ is opposite to that of $C_0V_{out}(nT - T)$. In view of the charge behavior of the integrator from $(n - 1)T$ till $(n - 1/2)T$, we obtain the following charge equation:

$$C_0 V_{out}(nT - T) - CV_{in}(nT - T) = C_0 V_{out}(nT - T/2) \quad (3.39)$$

Combine Equations (3.38) and (3.39), and apply the z-transform to both sides of the resulting difference equation. We get

Figure 3.2 (a) Active-RC integrator (inverting). (b) Parasitic-sensitive bilinear SC integrator. (c) Parasitic-sensitive parallel SC integrator. (d) Parasitic-sensitive series SC integrator. Assume ideal op-amps are used. Only the troublesome parasitic capacitances are shown.

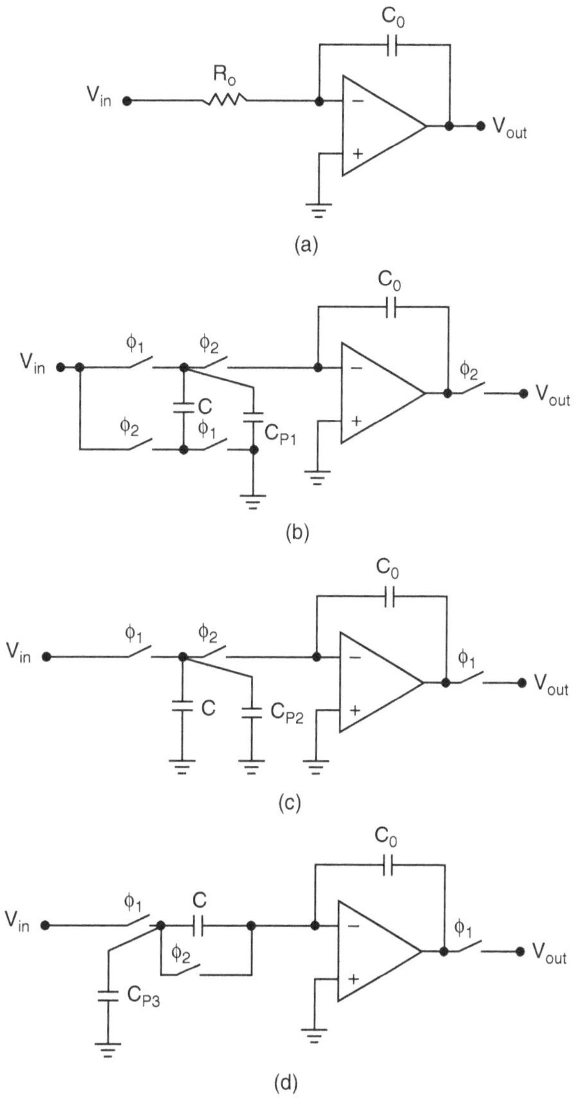

$$C_0 V_{out}(z) = C_0 V_{out}(z) z^{-1} - C V_{in}(z) z^{-1} \tag{3.40}$$

We can then write the transfer function:

$$H_{FE}(z) = \frac{V_{out}(z)}{V_{in}(z)} = -\frac{C}{C_0}\frac{1}{z-1} = -\frac{C}{C_0}\frac{z^{-1}}{1-z^{-1}} \tag{3.41}$$

Utilizing well-known principles such as $C = T/R_0$, $z = e^{j\omega T}$, we can find that for frequencies much lower than the clock sampling frequency (i.e., $\omega T \ll 1$),

Equation (3.41) approximates the transfer function of an ideal continuous-time active-RC integrator, whose transfer function is given by Equation (3.35).

EXAMPLE 3.1

Determine the corresponding phase shifts of the ideal (parasitic-free) bilinear and parallel SC integrators based on Equations (3.37) and (3.41), respectively. Explain why the latter is called the *forward-Euler integrator* (FEI).

Solution: Replacing z by $e^{j\omega T}$ in both equations, we notice that

$$H_{BL}(e^{j\omega T}) = -\frac{C}{C_0}\frac{1+e^{-j\omega T}}{1-e^{-j\omega T}} = -\frac{C}{C_0}\frac{e^{j\omega T/2}+e^{-j\omega T/2}}{e^{j\omega T/2}-e^{-j\omega T/2}} \tag{3.42}$$

and

$$H_{FE}(e^{j\omega T}) = -\frac{C}{C_0}\frac{e^{-j\omega T}}{1-e^{-j\omega T}} = -\frac{C}{C_0}\frac{e^{-j\omega T/2}}{e^{j\omega T/2}-e^{-j\omega T/2}} \tag{3.43}$$

Applying the well-known Euler's formula to these results, we have

$$H_{BL}(e^{j\omega T}) = -\frac{2\omega_{-3dB}}{\omega}\frac{\dfrac{\varpi T}{2}}{\tan\dfrac{\omega T}{2}}\cdot\left(\frac{1}{j}\right) \tag{3.44}$$

and

$$H_{FE}(e^{j\omega T}) = -\frac{\omega_{-3dB}}{\omega}\frac{\dfrac{\varpi T}{2}}{\sin\dfrac{\omega T}{2}}\cdot\left(\frac{e^{-j\omega T/2}}{j}\right) \tag{3.45}$$

where $\omega_{-3dB} = C/TC_0$ (–3 dB frequency). From elementary signals and systems courses, we know that the phase shift of each transfer function is determined by the term in parentheses:

$$Arg H_{BL}(e^{j\omega T}) \equiv Arg\left(\frac{1}{j}\right) = \frac{\pi}{2} \tag{3.46}$$

and

$$Arg H_{FE}(e^{j\omega T}) \equiv Arg\left(\frac{e^{-j\omega T/2}}{j}\right) = \frac{\pi}{2} - \frac{\omega T}{2} \tag{3.47}$$

Comparing the preceding four equations with Equation (3.35), which describes an ideal inverting analog integrator, we find that the bilinear SC integrator has introduced a gain nonlinearity error but no phase error, whereas the parallel SC integrator has introduced both a gain nonlinearity error and a phase lag. These errors are functions of ωT and can be ignored when $\omega T \ll 1$. Given the same pole frequency ω_0 and clock sampling period T, the bilinear integrator's magnitude gain is approximately twice that of the parallel SC integrator (both are sampled at a low frequency). However, given a fixed T, as the frequency of interest ω increases, gain and phase errors become increasingly problematic. Note that the denotation T in Equation (3.44) represents the effective clock period of a bilinear SC integrator.

Substituting R_0 in Equation (3.35) with the expression of its parallel SC simulation, we can rewrite $H(s)$ in terms of C_0, C, and T. Associate the revised $H(s)$ with $H(z)$ given by Equation (3.41), and we have

$$s \rightarrow \frac{z-1}{T} \tag{3.48}$$

This is known as the forward-Euler mapping formula and explains why this SC integrator is called the *forward-Euler integrator*.

What if the output jumps when $\Phi_2 \rightarrow 1$ in Figure 3.2(c)? Without writing tedious charge equations, we can see that the output is advanced by half T in time, which in the frequency domain essentially means adding a leading element of $(z^{1/2})$. The new transfer function is thus given by

$$H_{LDI\_I}(z) = H_{FE}(z) \cdot z^{1/2} = -\frac{C}{C_0}\frac{z^{-1/2}}{1-z^{-1}} = -\frac{C}{C_0}\frac{z^{1/2}}{z-1} \tag{3.49}$$

An SC integrator with such half-delay property is called an *inverting Type I* or *forward-Euler lossless discrete integrator* (LDI), which was introduced by Bruton [22]. It can be found by using the approach of Example 3.1 that the phase shift of this LDI is equal to $\pi/2$. In practice, LDI is especially useful for realizing a single-cycle (i.e., z^{-1}) loop delay in a fully differential SC circuit.

Thus far we have ignored the parasitic capacitances in Figures 3.2(b) and (c). Now that we have obtained their ideal (i.e., parasitic-free) transfer functions, we shall take the effects of parasitic capacitances into consideration. Let us start with Figure 3.2(b), where only one parasitic capacitance C_{p1} is shown. Note that C_{p1} consists of the parasitic capacitance of the top plate of C as well as the nonlinear capacitances associated with the two MOSFET switches. Apparently, C_{p1} is not the only parasitic capacitance residing in the circuit; however, it is more troublesome than the others. For example, the parasitic capacitance between the integrator's input V_{in} and ground is *neutralized*—that is, the electric charge on it plays no role in the signal transmission path and has no effect on the transfer function. This is because ideally V_{in} is from a voltage-controlled voltage source (VCVS) and its nodal impedance is very low [20][21].

Similar arguments apply to the parasitic capacitance on the bottom plate of C, which is connected to either V_{in} or ground. In addition, the parasitic capacitance at the virtual ground of the op-amp is negligible if an ideal op-amp is assumed.

To investigate the effect of C_{p1} on the integrator's transfer function, we employ the principle of *superposition*. The circuit in Figure 3.2(b) can be decomposed into two subcircuits: an ideal (i.e., parasitic-free) bilinear SC integrator with a switched capacitor of C and a parallel SC integrator with a switched capacitor of C_{p1}. However, it should be noticed that in the latter, all switches now change positions practically twice every T, the input is sampled and held at a clock rate of $2f_{clk}$, and the output jumps only when $\Phi_2 \rightarrow 1$! (Recall what we just learned from Equation (3.49) and the standard rule of $z = e^{j\omega T/2}$ in bilinear SC circuits.)

Taking the preceding into full account, we can substitute C in Equation (3.41) with C_{p1} and multiply the equation by a leading factor (i.e., z). Combining the result with Equation (3.37), we obtain the transfer function of the parasitic-sensitive bilinear SC integrator, which is given by

$$H_{BL}(z) = \left(-\frac{C_{p1}}{C_0}\frac{z^{-1}}{1-z^{-1}}\right)\cdot z + \left(-\frac{C}{C_0}\frac{1+z^{-1}}{1-z^{-1}}\right) = -\left(\frac{C+C_{P1}}{C_0}\right)\cdot\frac{1+\dfrac{C}{C+C_{p1}}z^{-1}}{1-z^{-1}} \tag{3.50}$$

Comparing this with Equation (3.37), we realize that C_{p1} introduces both magnitude and phase errors. In a standard CMOS, C_{p1} is normally on the order of 0.05 pF, or up to 1% of the desired capacitance C (when C > 10 pF), and it varies with temperature. Thus, in this situation, to achieve a 1% accuracy of C, we must specify the capacitance C to be at least 5 pF, which leads to a C_0 as large as 50 pF (for $\omega T = 0.1$)! Furthermore, if a larger time constant is required (e.g., $\omega T = 0.01$), then the size of C_0 will become inapplicably large for monolithic realizations, and as the result, extra capacitance-spread reduction techniques will be required.

In many cases, the exact mathematical analyses of SC circuits are quite tedious, and computer-based simulation tools are employed. SWITCAP [23] is one of many well-established *above-transistor-level* SC circuit simulation programs. At this point, a good exercise for the interested reader may be to simulate using SWITCAP the parasitic-sensitive bilinear SC integrator (for $C_0 = 5\,\text{pF}$, $C = 0.5\,\text{pF}$, $C_{p1} = 0.05\,\text{pF}$, $f_{in} = 48\,\text{kHz}$, and $f_{clk} = 1.2\,\text{MHz}$), plot the frequency responses (with and without C_{p1}, respectively) between 1 and 600 kHz, and utilize the results to verify Equation (3.50). For the sake of brevity, the illustrations are waived here; however, the SWITCAP source codes are provided in Appendix 3.1 as a reference.

In accounting for all the parasitic capacitances of the circuit shown in Figure 3.2(c), we adopt the arguments used in the bilinear integrator case, and we find that all parasitic capacitances, except for C_{p2}, do not affect the operation of the circuit. Since C_{p2} is in parallel with C, their capacitances can be added up. Substituting C in Equation (3.41) with (C_{p1} + C), we obtain the transfer function of the parasitic-sensitive parallel SC integrator, which is given by

$$H_{FEp}(z) = -\frac{C_{p1} + C}{C_0}\frac{z^{-1}}{1 - z^{-1}} \tag{3.51}$$

The circuit in Figure 3.2(d) is called the *parasitic-sensitive series SC integrator.* Ignoring parasitic capacitances for the moment and assuming the output jumps only when $\Phi_1 \rightarrow 1$, we find the transfer function of this integrator to be

$$H_{BE}(z) = -\frac{C}{C_0}\frac{1}{1 - z^{-1}} \tag{3.52}$$

Adopting the approach of Example 3.1, we can show that the preceding integrator facilitates the following s-to-z transformation:

$$s \rightarrow \frac{1 - z^{-1}}{T} \tag{3.53}$$

Thus, it is often referred to as the *backward-Euler integrator* (BEI). And the phase shift of the integrator can be found to be

$$ArgH_{BE}(e^{j\omega T}) \equiv Arg\left(\frac{e^{j\omega T/2}}{j}\right) = \frac{\pi}{2} + \frac{\omega T}{2} \tag{3.54}$$

Furthermore, if the output jumps when $\Phi_2 \rightarrow 1$, from Equation (3.52) we can derive an *inverting Type II* or *backward-Euler lossless discrete integrator* (LDI), as reported by Bruton [22]. And the transfer function is given by

$$H_{LDI\_II}(z) = H_{BE}(z) \cdot z^{-1/2} = -\frac{C}{C_0}\frac{z^{-1/2}}{1 - z^{-1}} = -\frac{C}{C_0}\frac{z^{1/2}}{z - 1} \tag{3.55}$$

It can be seen that this is identical to Equation (3.49). Thus, this LDI also has a phase shift of $\pi/2$.

To investigate the behavior of the series SC integrator with respect to C_{p3} (i.e., the only parasitic capacitance that matters), we again employ the principle of superposition. The circuit shown in Figure 3.2(d) is essentially composed of two subcircuits: an ideal series SC integrator with a switched capacitor of C and an ideal parallel SC integrator with a switched capacitor of C_{p3}. Note that the outputs of both circuits jump only when $\Phi_1 \rightarrow 1$. From Equations (3.41) and (3.52) we can obtain the transfer function of the parasitic-sensitive series SC integrator, which is given by

$$H_{BEp}(z) = \left(-\frac{C_{p3}}{C_0}\frac{z^{-1}}{1-z^{-1}}\right) + \left(-\frac{C}{C_0}\frac{1}{1-z^{-1}}\right) = -\left(\frac{C}{C_0}\right)\cdot\frac{1+\frac{C_{p3}}{C}z^{-1}}{1-z^{-1}} \tag{3.56}$$

Thus, C_{p3} causes the zero to deviate from the desired location. The interested reader is encouraged to simulate the parasitic-sensitive integrator in Figure 3.2(d) using tools such as SWITCAP and verify the transfer function derived here.

Parasitic-Insensitive SC Integrators

Thus far all the SC integrators that we have seen are parasitic sensitive. That is, their transfer functions are affected by certain parasitic capacitances, which is highly undesirable because these parasitic capacitances are not well controlled and can cause errors so serious that the consequent penalty may fail the overall IC design (e.g., the large capacitance C_0).

To overcome this deficiency, researchers have come up with some ingenious solutions, such as the parasitic-insensitive parallel SC integrator reported by Jacobs et al. [8] in Figure 3.3(a), the parasitic-insensitive series SC integrator reported by Martin [9] in Figure 3.3(b), and the parasitic-insensitive bilinear integrator reported by Knob [24] in Figure 3.3(c). Note that in these schematics, the outputs of the integrators jump only when $\Phi_1 \rightarrow 1$, and the input of Figure 3.3(c) shall be a sampled-and-held signal (sampled when Φ_1 is on). All op-amps used here are ideal.

Looking into the first integrator in Figure 3.3(a), we focus on the parasitic capacitance C_{p1}, since the other parasitic capacitances in the circuit can be immediately discarded by repeating the same arguments as before. When Φ_1 is on ($\Phi_1 = 1$), C_{p1} is connected with C in parallel, and both capacitors are charged to the instantaneous input voltage. When Φ_2 is on ($\Phi_2 = 1$), different from C, C_{p1} will pass no electric charge through the feedback capacitor C_0 to the output, because both of its plates are

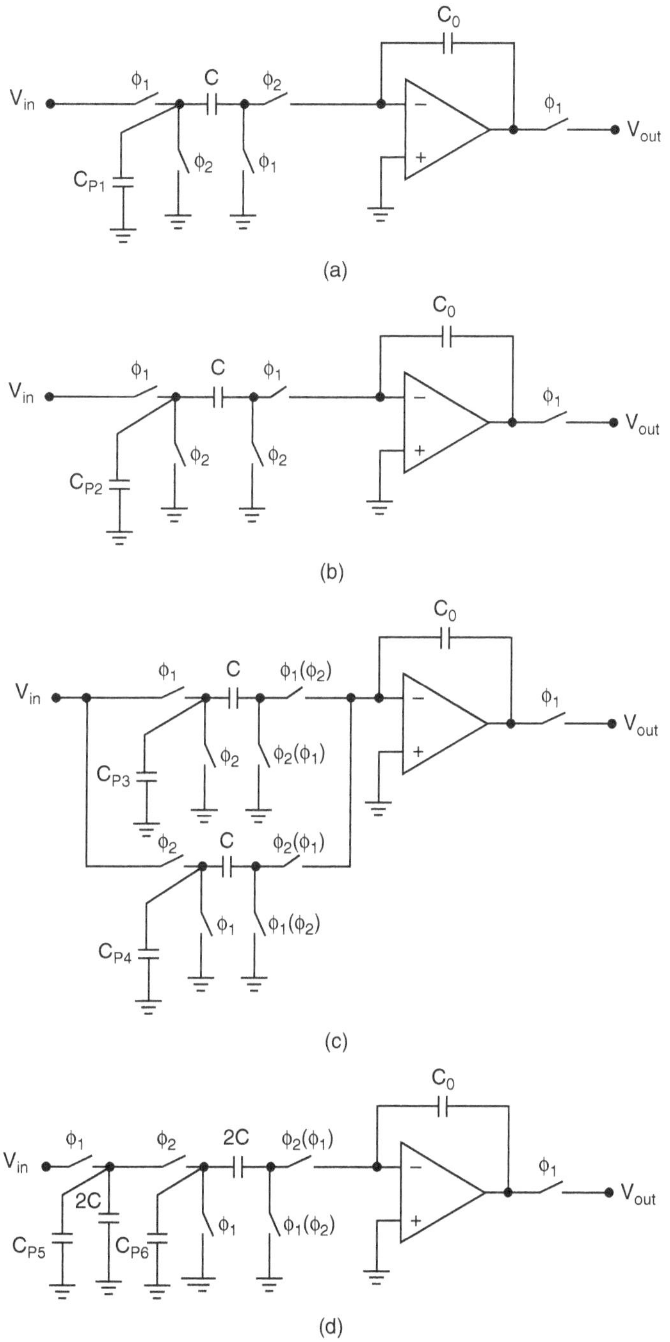

Figure 3.3 (a) Parasitic-insensitive parallel SC integrator. (b) Parasitic-insensitive series SC integrator. (c) Parasitic-insensitive bilinear SC integrator. (d) Parasitic-compensating SC integrator. Only the troublesome parasitic capacitances are shown.

connected to ground (i.e., it is neutralized). Thus, C_{p1} does not affect the transfer function, and the integrator is parasitic insensitive.

It is instructive to compare this integrator with the *parasitic-sensitive parallel SC integrator* illustrated in Figure 3.2(c). Interestingly enough, it can be found by inspection that besides a few changes in nodal and switch configurations, the key difference between these two integrators lies in the polarity of the net charge transferred from V_{in} to V_{out}. This difference is due to the following: during Φ_1 on, in Figure 3.2(c) the input V_{in} is charged onto the top plate of C, while in Figure 3.3(a) the input is charged onto the bottom plate of C. Since the polarity of the transfer function is determined by that of the net charge, we can conclude that the transfer function of this integrator, as shown in Figure 3.3(a), has a sign opposite to that of the transfer function given by Equation (3.41)—that is, the circuit realizes a noninverting and parasitic-insensitive FEI. The transfer function is thus given by

$$H_1(z) = -H_{FE}(z) = \frac{C}{C_0}\frac{z^{-1}}{1-z^{-1}} \tag{3.57}$$

With this sign change, the phase shift of this integrator is

$$ArgH_1(e^{j\omega T}) = ArgH_{FE}(e^{j\omega T}) - \pi = -\frac{\pi}{2} - \frac{\omega T}{2} \tag{3.58}$$

Note that in Figure 3.3(a), V_{in} and V_{out} are never coupled in series. Thus, the circuit is essentially a parallel SC integrator that is parasitic insensitive, hence the name parasitic-insensitive parallel SC integrator. In addition, if the output jumps only when $\Phi_2 \to 1$, then a noninverting and parasitic-insensitive Type I LDI results, which has a phase shift of $(-\pi/2)$.

Likewise, the transfer function of the *parasitic-insensitive series SC integrator* shown in Figure 3.3(b) can be derived through a comparison between itself and the *parasitic-sensitive series SC integrator* in Figure 3.2(d) (*Hint:* Consider the "Φ_2" switches in both circuits). As we can find out, ideally these two integrators have identical transfer functions; in other words, they both are inverting BEIs, with one being parasitic sensitive and the other not. The derivation of its transfer function is left to the interested reader as an exercise. The result is provided in the following:

$$H_2(z) = H_{BE}(z) = \frac{-C}{C_0}\frac{1}{1-z^{-1}} \tag{3.59}$$

And the phase shift of this integrator is

$$ArgH_2(e^{j\omega T}) = ArgH_{BE}(e^{j\omega T}) = \frac{\pi}{2} + \frac{\omega T}{2} \tag{3.60}$$

If the output of the integrator jumps only when $\Phi_2 \rightarrow 1$, then the circuit realizes an inverting and parasitic-insensitive Type II LDI, whose phase shift is equal to $\pi/2$. However, it should be noticed that the noninverting and parasitic-insensitive BEI (and Type II LDI) are not realizable by any single-ended SC circuits.

Comparing Equation (3.60) with Equation (3.58), we realize that the overall phase shift can be canceled out (i.e., zero phase shift) by cascading these two parasitic-insensitive integrators. This is useful for building a frequency-tuning circuit with zero loop-phase-shift such as an LDI resonator or a ring oscillator, which is widely used in band-pass systems.

Figure 3.3(c) illustrates a parasitic-insensitive bilinear SC integrator in its inverting version. A noninverting version can be realized if the four switches near the virtual ground are controlled by the clock phases in parentheses. By inspection we realize that these two circuits basically utilize the previous parasitic-insensitive techniques to deal with the effects of parasitic capacitances. For example, the inverting bilinear integrator is the result of substituting the bilinear SC input or sampling stage used in Figure 3.2(b) with two complementary versions of the series SC resistor simulation seen in Figure 3.3(b), one of which is working in the even ($\Phi_2 = 1$) phase and the other in the odd ($\Phi_1 = 1$) phase. This configuration is equivalent to superposing two subcircuits: a parasitic-insensitive series integrator shown in Figure 3.3(b) and its half-T delayed version. Therefore, utilizing Equation (3.59), we can write the transfer function of the *parasitic-insensitive bilinear inverting integrator*, which is given by

$$H_3(z) = H_2(z) \cdot (1 + z^{-1}) = -\frac{C}{C_0} \frac{1 + z^{-1}}{1 - z^{-1}} \tag{3.61}$$

Note that in the foregoing equation, $z = e^{j\omega T/2}$ as before. The transfer function of the *parasitic-insensitive bilinear noninverting integrator* can be found in a similar manner by superposing the parasitic-insensitive parallel integrator shown in Figure 3.3(a) and its half-T delayed version. The derivation is left to the reader as an exercise. As a further step, it will be an insight-gaining experience for the interested reader to prove all the transfer functions derived here by investigating the charge equations in detail.

Finally, it can be shown that the parallel-type (also called *toggle*) SC input stage, for example, as in Figure 3.2(c) or 3.3(a), has an inherent sample-and-hold (S&H).

This feature is desirable in practice because it helps circuit designers save a dedicated S&H at the input.

Thanks to the inherent S&H property, the aforementioned parallel SC integrators are less sensitive to the transient responses of practical op-amps, in comparison with their series SC counterparts shown in Figures 3.2(d) and 3.3(b). To understand this, consider a situation in which V_{in} is a continuous-time signal, and no dedicated S&H or buffer is provided at the input. We can see that in both series SC integrators (parasitic-sensitive and insensitive), V_{in} is charged directly to the inverting input of the op-amp during Φ_1. Although thus far it has been assumed that the op-amps used here are ideal (i.e., with perfectly fast settling), in practice they are not. As a result, the unbuffered V_{in} may not be settled by the op-amp to a sufficient extent within a single phase (Φ_1 or Φ_2), thereby distorting the output signal. The output distortion due to insufficient settling becomes more pronounced for higher-frequency input signals. In a word, in the absence of a dedicated S&H at the input, the parallel integrator is superior to its series counterpart for it can buffer the practical op-amp inherently.

On the other hand, among all the integrators that we have seen so far, only the one in Figure 3.3(a) is of parallel type *and* parasitic insensitive. Moreover, it can only realize transfer functions of the noninverting *and* delaying integrators, such as that given by Equation (3.57), which are often not adequate in practice (for single-ended SC designs in particular since they do not have the flexibility of cross-coupling op-amps' input or output wires for sign changes, whereas fully differential designs do).

To provide more design options, Fleischer et al. [25] presented an ingenious *parasitic-compensating* structure, as shown in Figure 3.3(d), where two different toggle switched capacitors are connected in tandem to form a new input stage. Apparently, the inherent S&H quality is maintained as before. In addition, it is proved that the integrator can be made parasitic insensitive by carefully matching the layout of the two switched capacitors, such that their top-plate parasitic capacitances cancel each other out—that is, $C_{p5} = C_{p6}$ [25]. Furthermore, this parasitic-insensitive structure can be used to realize inverting *and* delaying integrators such as the following:

$$H_M(z) = -\frac{C}{C_0}\frac{z^{-1}}{1-z^{-1}} \quad \text{and} \quad H_{M\_LDI}(z) = -\frac{C}{C_0}\frac{z^{-1/2}}{1-z^{-1}} \qquad (3.62)$$

where $H_M(z)$ has a full delay in the forward signal path, and $H_{M\_LDI}(z)$ is an inverting LDI with a half delay.

Alternatively, we may create a noninverting integrator by simply swapping the two switches near the virtual ground in Figure 3.3(d) (i.e., using the clock phases in parentheses). However, comparing Equations (3.62) and (3.57), we realize that this parasitic-compensated integrator basically clones the transfer function of the circuit presented in Figure 3.3(a), yet at the cost of more switches and larger capacitors. Thus, it is seldom used in practice.

Making use of the inherent S&H property and the parasitic-compensating concept developed earlier, researchers have created a few interesting *parasitic-compensated bilinear SC integrators.* One applicable example was reported by Eriksson and Akhlaghi [26]. In addition to saving a dedicated S&H circuit at the input, the circuit of [26] realizes a *noninverting* bilinear SC integrator. From Equation (3.61), we can derive the integrator's transfer function, which is given by

$$H_{BLM}(z) = -H_3(z) = \frac{C}{C_0}\frac{1+z^{-1}}{1-z^{-1}} \tag{3.63}$$

where $z = e^{j\omega T/2}$ as before. The interested reader is referred to [26] for more circuit details.

In summary, Equations (3.57), (3.59), and (3.61–3.63) represent all the possible parasitic-insensitive single-ended SC integrators. These transfer functions can be realized by the corresponding parasitic-insensitive single-ended structures, which are illustrated respectively in Figure 3.3(a–d), except that for realizing Equation (3.63) [26]. In addition, we found that the noninverting and parasitic-insensitive BEI and Type II LDI cannot be realized by any of the single-ended SC circuits derived earlier. However, as we shall see shortly, they can be easily realized by fully differential SC circuits.

Fully Differential Integrators

Fully differential integrators are more popular than the single-ended ones in modern SC circuits because they not only improve the circuit's common-mode noise (e.g., power supply noise, voltage offsets) performance, but also facilitate the sign changes for transfer functions through simple wire-crossing operations.

A standard fully differential Euler SC integrator is shown in Figure 3.4(a). As we can see, the circuit is composed of the integrator in Figure 3.3(a) and its horizontally flipped-over duplicate. In general, a fully differential SC integrator is built to be parasitic insensitive by default since it is usually a combination of two (or more) parasitic-insensitive single-ended subcircuits.

It can be proved that if both the input and output are sampled when $\Phi_1 = 1$ (as illustrated in the schematic), the resulting transfer function will be given by Equation (3.57). On the other hand, if the output is sampled when $\Phi_2 = 1$, the delay element (z^{-1}) in the numerator of Equation (3.57) will be removed. In other words, depending on the sampling moments of the input and output, the circuit in Figure 3.4(a) may realize either a forward- (FEI) or backward-Euler integrator (BEI).

Furthermore, by simply cross-wiring the two output nodes (i.e., swapping V_{outp} and V_{outn}), we can achieve sign inversions for the transfer functions derived earlier. This feature indicates that the noninverting BEI and Type II LDI can now be realized based on the differential SC circuit shown in Figure 3.4(a). The interested reader is encouraged to prove this statement.

Note that in Figure 3.4(a), both nodes A and B are shown to be connected to ground. In practice, however, they are seldom connected to ground or with each other. Rather, node A is literally divided into two nodes, one for the upper part and the other for the lower part of the circuit. The former is connected to a positive reference voltage V_{refp}, and the latter is connected to a negative reference voltage V_{refn}. In effect, the voltage difference between the input $V_{inn(p)}$ and reference $V_{refp(n)}$ will be sampled by the capacitor C. Alternately, the positive and negative references can be connected respectively to nodes C and D, rather than to upper and lower A nodes. There are *give-and-take* performance tradeoffs between these sampling configurations in terms of sampling noise and input-dependent signal leakage to the feedback path [27]. Thus, they have been seen equally often in practice.

In addition, the differential reference (i.e., $V_{ref} = V_{refp} - V_{refn}$) is often used to specify the dynamic range requirement for the circuit. In practice, V_{ref} is often set to (or slightly less than) half the power supply voltage for enabling all switches to operate properly.

As for node B, it is usually connected to a voltage source V_{CM1}, which can be adjusted conveniently through an on-chip regulator or a biasing circuit. The ultimate goal of adjusting V_{CM1} is to ensure the following common-mode voltage tracking condition is met [28][29]:

$$V_{CM1} - C_{CM2} = V_{CM3} - V_{CM4} \tag{3.64}$$

Here, V_{CM2} is the input-common mode voltage of the op-amp, and V_{CM3} and V_{CM4} are the common-mode voltages of V_{in} and V_{ref}, respectively.

Similar rules of thumb apply to the fully differential integrator shown in Figure 3.4(b). However, compared to the preceding, this circuit is more desirable for its

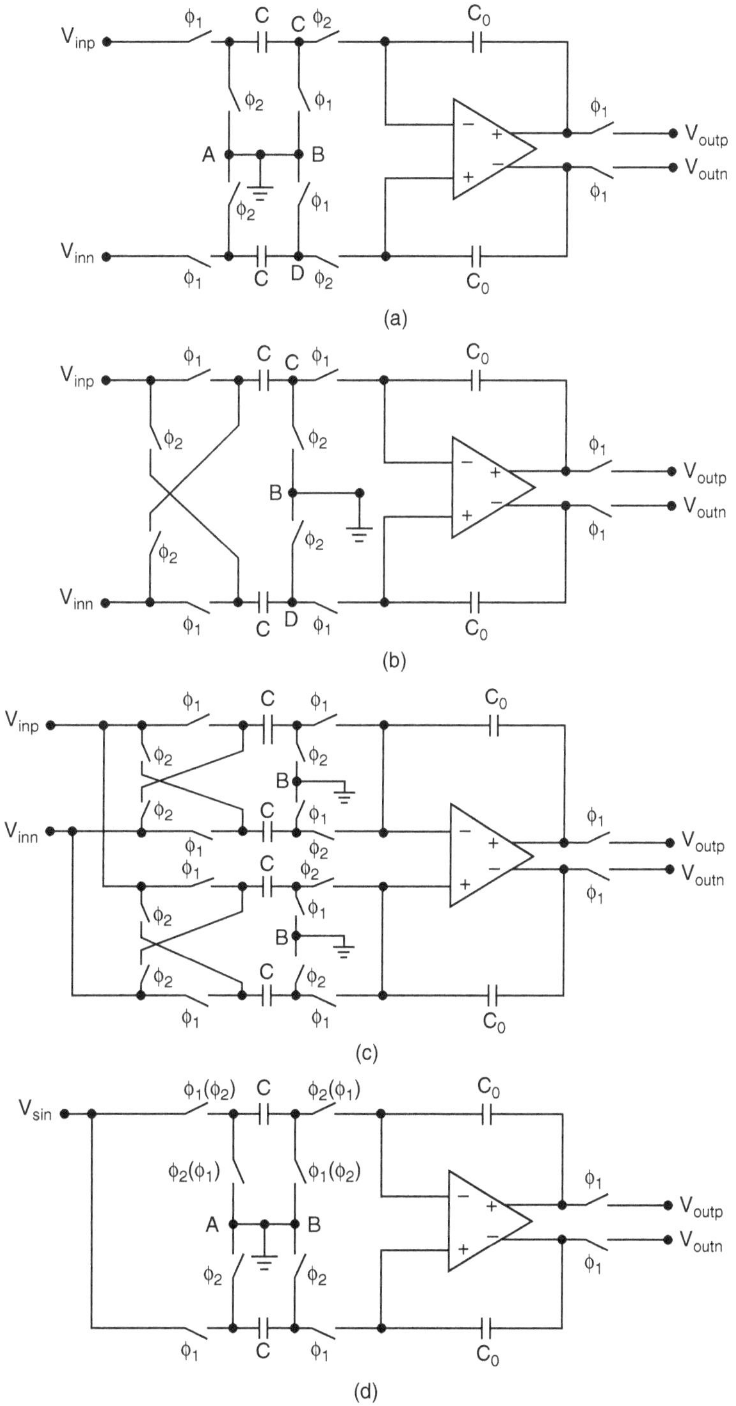

Figure 3.4 (a) Fully differential Euler SC integrator. (b) Fully differential bilinear/forward-Euler SC integrator. (c) Double-sampled fully differential bilinear/Euler SC integrator. (d) Fully differential SC integrator with single-ended-to-differential convertibility. All op-amps shown are ideal.

capability of realizing both parasitic-insensitive bilinear and forward-Euler integrators [30]. Specifically, if both the input and output are sampled when $\Phi_1 = 1$ (as illustrated in the schematic), the circuit realizes a bilinear SC integrator, whose transfer function may be identical to that given by Equation (3.61) or (3.63), depending on whether or not output cross-wiring is used. On the other hand, if the input and output are sampled when $\Phi_2 = 1$ and $\Phi_1 = 1$, respectively, then the circuit realizes the following forward-Euler integrating function (inverting or noninverting):

$$H_{BLFE}(z) = \pm\frac{2C}{C_0}\frac{z^{-1}}{1-z^{-1}} \tag{3.65}$$

Note that in both cases (bilinear and forward-Euler), $z = e^{j\omega T/2}$ since the effective clock period is $T/2$ as before (i.e., all switches operate at a clock frequency of $2f_{clk}$). In practice, it is desirable to reduce the clock frequency for the sake of low power. One interesting solution is to duplicate the SC input stage in Figure 3.4(b) and connect it to the original one such that the resulting circuit is *double-sampled*, meaning that the circuit's input is equivalently sampled at $2f_{clk}$ although the actual clock frequency is only f_{clk} [30]. A double-sampled fully differential SC integrator is shown in Figure 3.4(c). In the following example, we derive its transfer function using the superposition concept.

EXAMPLE 3.2

Assuming both the input and output are sampled when $\Phi_1 = 1$ and the clock frequency is f_{clk}, show that the circuit in Figure 3.4(c) realizes a double-sampled bilinear integrator.

Solution: Based on the superposition concept, the integrator in Figure 3.4(c) can be decomposed into four fully differential subintegrators, whose transfer functions are equivalent to those of a BEI, a Type II LDI, a Type I LDI, and a FEI, respectively (all four transfer functions are inverting). Although the two LDIs have identical transfer functions, it is instructive to draw a distinction between them (i.e., Type II and I), since their corresponding realizations (subintegrators) are different. Utilizing the results that we have derived thus far, we can obtain the subintegrators' transfer functions, which are listed as follows:

$$H_{BEI} = -\frac{C}{C_0}\frac{1}{1-z^{-1}}, \quad H_{LDI\_II} = -\frac{C}{C_0}\frac{z^{-1/2}}{1-z^{-1}},$$

$$H_{LDI\_I} = -\frac{C}{C_0}\frac{z^{-1/2}}{1-z^{-1}}, \quad H_{FEI} = -\frac{C}{C_0}\frac{z^{-1}}{1-z^{-1}}$$

Combining all four transfer functions, we have

$$H(z) = H_{BEI} + H_{LDI\_II} + H_{LDI\_I} + H_{FEI} = -\frac{C}{C_0}\frac{\left(1+z^{-1/2}\right)^2}{1-z^{-1}} = -\frac{C}{C_0}\frac{1+z^{-1/2}}{1-z^{-1/2}} \quad (3.66)$$

Note that all the switches operate at f_{clk}—that is, the effective clock period is T, in the preceding equation $z = e^{j\omega T}$, and as a result, $z_1 = z^{1/2} = e^{j\omega T/2}$. Substituting $z^{1/2}$ with its new denotation z_1, we can rewrite the transfer function to

$$H(z) = -\frac{C}{C_0}\frac{1+z_1^{-1}}{1-z_1^{-1}}, \quad \text{where } z_1 = e^{j\omega T/2} \quad (3.67)$$

The preceding transfer function has the same format as that given by Equation (3.61), thus the circuit in Figure 3.4(c) also realizes a bilinear SC integrator. However, in comparison with all the other bilinear integrators that we have seen so far, this integrator manages to realize the same bilinear transfer function while being sampled at only one-half the clock frequency, thereby alleviating the speed and power requirements.

Double-sampled SC integrators are widely used in sampled-data signal processing circuits such as delta-sigma (or $\Delta\Sigma$) data converters (see Chapter 5), where the resulting *oversampling ratio* (OSR) can be reduced by half. A good design example of a delta-sigma A/D converter based on double-sampled SC integrators was reported by Senderowicz et al. [31].

Sometimes the input may be a single-ended signal. In this case, to utilize one of the fully differential integrators discussed earlier, we need to precede the integrator with a dedicated single-ended-to-differential converter. Alternatively, we can use the circuit shown in Figure 3.4(d), which is essentially an SC integrator with an inherent single-ended-to-differential converter. As shown in the schematic, assuming both the input and output are sampled when $\Phi_1 = 1$, the upper half portion of the circuit realizes a noninverting FEI (V_{outp}/V_{sin}), while the lower half portion realizes an inverting BEI (V_{outn}/V_{sin}) and accomplishes the sign inversion. Since the output V_{out} is differential and $V_{out}/V_{sin} = (V_{outp} - V_{outn})/V_{sin}$, it can be shown that the circuit realizes a noninverting and parasitic-insensitive bilinear SC integrator. The clock phases in parentheses are optional and can be used to control the corresponding switches so that the realized transfer function will have a full delay (i.e., z^{-1}). Note that this design of single-ended-to-differential SC integrator is not unique, and the reader is referred to the literature for alternatives.

To conclude, we have combed through various types of SC integrators, including parasitic-sensitive/insensitive single-ended integrators, and fully differential integrators. However, note that there are many other feasible SC integrator realizations than what have been presented here, such as damping (lossy) SC integrators [30] and very-large-time-constant (VLTC) SC integrators [32][33]. Different realizations lay emphasis on resolving different design issues. The reader is referred to the literature for more design examples.

Finally, recall that all the op-amps used earlier are ideal—that is, the errors due to dc offset voltage, finite op-amp gain and bandwidth, low-voltage power supply, and so on, have *not* been taken into account yet! In Chapter 7, the design techniques able to overcome these deficiencies will be described in detail.

3.4 CMOS Sample-and-Hold Circuits

In this section we study basic CMOS sample-and-hold (S&H) circuits. The S&H is an essential active building block of many sampled-data circuits and systems, the majority of which are A/D converters (ADC).

Very often, an S&H serves as a front-end circuit that captures or *samples* the value of the analog input signal at the sampling instant and *holds* it for a certain time interval, within which the ADC accomplishes a cycle of digitization operation, hence the name *sample-and-hold.* What's more, an S&H capable of *tracking* the input signal—that is, a *track-and-hold* (T&H)—is often required in practice. In a T&H, the output *tracks* the input during the sampling or tracking mode and *holds* the sampled value at the end of the sampling/tracking mode through the subsequent holding mode until the next sampling instant, hence the name *track-and-hold.*

In comparison with the *sample-and-hold,* the T&H dedicates a prescribed time interval (e.g., half the clock period) to tracking the input signal. Except for very high frequency applications (e.g., wireless or wireline communications), most sampling circuits operate as in the T&H scheme [34]. Nevertheless, in this book we will keep using the conventional denotation of *sample-and-hold* (S&H).

A simple S&H is shown in Figure 3.5. V_{in} is the input signal, M_1 is the sampling switch that samples the input signal, C_h is the holding capacitor that holds the sampled voltage, *clk* is the clock signal, and V_{out} is the output signal. The sampling switch M_1 can be implemented as a MOS (either NMOS or PMOS) transistor, a CMOS (a combination of NMOS and PMOS) transmission gate, a GaAs transistor or a diode bridge, and so on. In this book, only low-cost MOS transistors and CMOS transmission gates are considered for implementing the switches.

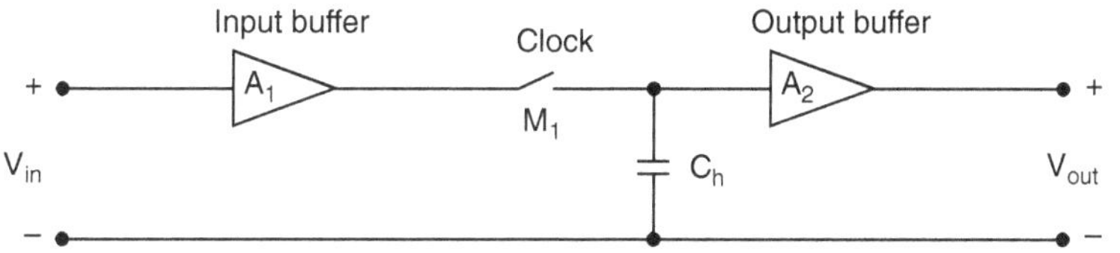

Figure 3.5 Open-loop S&H with input and output voltage buffers.

In the schematic, A_1 is the input voltage buffer that prevents the input from being affected by the switching transient due to the switched-capacitor (i.e., M_1 and C_h) operations, and A_2 is the output voltage buffer that protects the voltage held by C_h from being corrupted by the transient in the subsequent circuit during the holding mode.

The operation of the S&H circuit is described as follows: During the sampling mode, the sampling switch M_1 is turned on, and the input voltage V_{in} is sampled onto C_h. During the holding mode, the sampling switch is turned off such that the connection between the hold capacitor's top plate and the input is broken. Ideally the capacitor voltage V_{out} will remain at the last value of V_{in} before M_1 is turned off until it is turned on again. The time intervals of the sampling and holding modes are controlled by the system clock, *clk*.

In practice, component imperfections including nonzero on-resistance of the sampling switch M_1, charge injection, dc offset, and speed limitations of the input/output buffers would introduce various types of inaccuracies to the S&H circuit. To evaluate these inaccuracies, it is necessary to define the critical performance parameters of the S&H circuit, which is the subject of the following subsection.

Performance Parameters

In this subsection, several critical performance parameters of an S&H will be defined. Among these parameters, whether or not one has a higher priority over another greatly depends on the specific application that the S&H circuit is used for [34].

Acquisition time is the time interval from the beginning of sampling mode to the moment when the voltage across C_h settles within a specified error range of the input signal level, such as 0.1% or 0.01%. Acquisition time is a dynamic (also known as ac) parameter representing how fast the input signal can be successfully tracked. In practice, acquisition time is determined by the settling performance of A_1, the on-resistance of M_1, the holding capacitor C_h, and the input impedance of A_2. Since the characteristic of acquisition time is quite similar to that of settling time, it is also referred to as *sampling-mode settling time.*

Holding-mode settling time is the time interval from the beginning of holding mode to the moment when the output voltage V_{out} settles within a specified error range of its final value, such as 0.1% or 0.01%. Holding-mode settling time is also a dynamic parameter, and it is determined by the settling performance of A_2.

Aperture jitter or *aperture uncertainty* is perhaps the most critical dynamic parameter of an S&H circuit, especially for high-speed applications. It is defined as the random variation in the time required for the sampling switch M_1 to effectively turn on or off once a sampling or holding command is asserted. Aperture jitter is mainly due to the jitter in the clock signal, which controls the open/close operations of M_1.

In effect, aperture jitter contributes to the errors in the output signal and hence degrades the overall *signal-to-noise ratio* (SNR). Specifically, consider a sinusoid input $V_{in}(t) = A\cdot\sin(2\pi ft)$ sampled at instants $t = nT + \sigma$, $n = 1, 2, 3, \ldots$, where σ is the instantaneous aperture jitter observed after nT. With t deviating from its ideal spot by σ, it can be shown that the error in the sampled voltage is proportional to the slope of the input waveform (i.e., dV_{in}/dt). Thus, the error voltage due to σ can be expressed as

$$\varepsilon_{aperture\_jitter}(t) = \sigma\cdot\left|\frac{dV_{in}(t)}{dt}\right| = A\cdot 2\pi\cdot f\cdot\sigma\cdot\left|\cos(2\pi\cdot f\cdot t)\right| \tag{3.68}$$

From this equation we derive the error power and calculate the SNR limit set by the aperture jitter using the following formula [35]:

$$SNR_{aperture\_jitter} = -20\log_{10}(2\pi\cdot f\cdot\sigma_t) \tag{3.69}$$

where f is the input signal frequency and σ_t is the RMS value of aperture jitter. The foregoing formula indicates that the error due to the aperture jitter becomes more pronounced for high-speed input signals. The reader is referred to [35] for the mathematical proof of the formula. However, it should be noticed that the preceding formula is based on the assumption of the instantaneous aperture jitter σ being a random signal and independent of the input voltage $V_{in}(nT)$. This assumption implies that the aperture jitter is not held responsible for introducing *harmonic distortion* to the output.

In accounting for the sources of the harmonic distortion that appear at the S&H's output, we find three important items: *charge injection* or *clock feedthrough, input-dependent switch on-resistance*, and *imperfect clock waveform with a finite slope.*

The mechanism of charge injection can be briefly described as follows: during the holding mode, when the sampling switch is turned off, a negative charge due to the memory effect of MOSFET is distributed from the channel to both source and drain junctions [36]. Assuming M_1 in Figure 3.5 to be an NMOS switch, we find that the negative voltage (the *hold step*) injected into one of its junctions can be given by

$$\Delta V = -\frac{C_{ox}WL\cdot(|V_{gs}| - V_{th})}{2C_h} = -\frac{C_{ox}WL\cdot(V_{dd} - V_{in1} - V_{th})}{2C_h} \tag{3.70}$$

where V_{dd}, V_{in1}, and V_{th} are the power supply voltage, the output voltage of the input buffer A_1 at the time M_1 is off, and the threshold voltage of M_1, respectively. The foregoing equation indicates that charge injection introduces a linear gain error and a distortion to the circuit. The gain error is attributed to the linear relationship shown by the preceding equation between the injected voltage ΔV and the input voltage V_{in1}. The distortion is due to the nonlinear link from V_{in1} to ΔV through the threshold voltage V_{th}, which is linearly related to ΔV (as shown in the preceding equation) but nonlinearly related to V_{in1} since in a given CMOS process, V_{th} can also be given by [37]

$$V_{th} = V_{th-0} + \gamma\left\{\sqrt{(V_{in1} - V_{sub}) + 2|\phi_F|} - \sqrt{2|\phi_F|}\right\} \tag{3.71}$$

where V_{th-0} is the inherent threshold voltage when $V_{sb} = 0$, γ is the body effect coefficient, and Φ_F is the Fermi potential of the substrate. They are all constants for a given CMOS technology. V_{sub} is the substrate voltage, which is usually tied to ground or the negative supply voltage V_{ss}; thus it is also a constant. Being the only two variables remaining in Equation (3.71), V_{th} and V_{in1} form a nonlinear relationship between themselves; consequently, the hold step ΔV has a nonlinear relationship with the input signal (V_{in1}) based on Equation (3.70), thereby introducing a distortion error.

Moreover, note that a gain error typically causes the degradation in the output voltage amplitude and hence reduces the SNR of the circuit. On the other hand, the nonlinear relationship between the hold step and the input signal introduces distortion to the output signal, thereby reducing the *signal-to-distortion-ratio* (SDR).

In contrast to the charge injection, the switch on-resistance takes effect during the sampling mode. It can be expressed as (see Chapter 1)

$$R_{on} = \frac{1}{\mu_n C_{ox}(W/L)(V_{dd} - V_{in1} - V_{th})} \tag{3.72}$$

From the equation we realize that R_{on} is nonlinearly dependent on V_{in1}. As a result, given a fixed holding capacitor C_h, the S&H's time constant varies with the instantaneous input voltage, thereby causing distortions.

Distortion may stem from an imperfect clock signal whose waveform has a finite slope. As we know, the sampling switch M_1 in Figure 3.5 is made of an NMOS transistor; thus, when the value of $(V_{clk} - V_{in1})$ reaches the transistor's threshold voltage V_{th}, the switch is turned on. If the transition of V_{clk} from "high" to "low" (and vice versa) is not steep—that is, the waveform of V_{clk} changes with a finite slope, then neither the value of $(V_{clk} - V_{in1})$ nor the switch open/close operation changes abruptly, and thereby the actual sampling or holding instant deviates from the ideal. As intuition suggests, the deviation is related to the input signal frequency f (i.e., how fast V_{in1} changes) and the slope of the clock waveform (i.e., how fast V_{clk} changes), thus it contributes to both harmonic distortion and *deterministic* aperture jitter (in comparison with the random aperture jitter derived before).

Moreover, it can be shown that the signal-to-distortion ratio (SDR) is incrementally (and logarithmically, since it is defined in decibels) proportional to the value of $V_{clk}/(V_{in1} \cdot f \cdot t_{ct})$, where t_{ct} is the time it takes for the clock to transit from "high" to "low" (or vice versa).

Taking both the signal-to-distortion ratio (SDR) and signal-to-noise ratio (SNR) performance of the S&H circuit into account, we can define another very useful parameter, *signal-to-noise-plus-distortion ratio* (SNDR). The measured value of SNDR represents the limits imposed by all the preceding phenomena (aperture jitter, charge injection, and finite clock slope) on the circuit's performance.

Dynamic range (DR) is defined as the difference in decibels between the largest and the smallest possible input voltages that the S&H is capable of processing properly. DR can also be interpreted as the ratio of the largest possible over the smallest possible output signals. DR is mainly limited by the power supply voltage, the threshold of the sampling switch, and the input-referred noise power.

Finally, there are many other important parameters, such as pedestal error, droop rate, and holding-mode feedthrough. However, in practice they are seldom measured or quantified individually. Rather, their performance merits are often incorporated into the measurable ones such as SNR or SNDR for the sake of simplicity.

Testing S&H Circuits

The most popular test setup used to characterize S&H circuits is the *back-to-back test* [38], which is very useful for measuring critical performance parameters such as SNR, SNDR, and aperture jitter. In this setup, two identical S&H circuits are fabricated onto the same chip, with one being used to test the other. During the test, the S&H under test operates at the maximum allowable clock frequency f_{max}, while the other S&H is driven by a lower clock frequency of f_{max}/N or thereabout (N is an integer ranging from 1 to 16 in practice). A sinusoidal input signal with a frequency of $f_{max}/N - \Delta f$ is first applied to the S&H under test, and the output is then processed by the other S&H, which is now clocked at $f_{max}/N + \Delta f$.

When $N = 1$, the test is also called the *beat-frequency test*, with Δf being the beat frequency (usually on the order of 100 kHz). When $N > 1$, the test is often referred to as the *envelope test*, since it chooses one out of N possible sinusoidal waveforms (envelopes), all with a frequency of Δf, for spectral analysis, thereby shifting the center of the frequency window from f_{max} to f_{max}/N ($N > 1$).

The basic idea behind the scheme of *back-to-back frequency test* is to utilize the inherent modulation/demodulation capability of the sampling switch to investigate noise and harmonic distortion properties of an S&H circuit operating at high frequencies. This technique is highly desirable in practice because it helps circuit designers avoid the trouble of sending the actual high-frequency signals to the measurement equipment (e.g., spectrum analyzer), otherwise the S&H circuit must be carefully designed such that its output can drive 50-ohm loads or balun.

Consider the case when $N = 1$, the input is demodulated by the first S&H (under test) to a low-frequency signal (i.e., $f = \Delta f$), which is then passed through the second S&H (identical to the one under test but clocked at $f_{max} + \Delta f$) to compensate the *sinc* response [38]. As a result, the first S&H circuit's spectral characteristics such as noise floor and spikes (i.e., harmonic distortions) within the high-frequency window $[f_{max} - \Delta f, f_{max} + \Delta f]$ are literally *copied* to the low-frequency one: $[-\Delta f, +\Delta f]$, which can be handled with a simple differential-to-single-ended converter before being sent to the spectrum analyzer for measurements.

Furthermore, the back-to-back frequency test is also widely used for the characterization of ADCs. However, given the ADC under test has a resolution of around M bits, the auxiliary ADC shall have a higher resolution (e.g., $M + 2$ bits).

Finally, revisiting Equation (3.68), we realize that given a fixed f, the worst-case error induced by aperture jitter σ is observed when the input waveform is around the zero-crossing points (i.e., with the steepest slope). In contrast, little or no error due to aperture jitter is detected when the input waveform is at its peak (i.e., flat or with small slope). Thus, to some extent there is a correlation between the measured error voltage due to jitter and where the measurement is taken.

In addition, the measured result becomes very sensitive to this correlation when both clock and signal frequencies reach a few hundreds of megahertz (MHz).

However, the back-to-back test is not capable of revealing this correlation since its measurement does not distinguish between random and deterministic jitters. Moreover, the maximum precision of the back-to-back test is limited by the jitter performance of the clock generator, which is often worse than 5 ps (in RMS values) by itself. Therefore, it should be noted that the back-to-back frequency test is incapable of accurately measuring the *subpicosecond jitter error* (i.e., less than 3 ps in RMS). New test methods are needed to circumvent these limits and to facilitate accurate subpicosecond jitter measurements for characterizing high-speed and high-resolution circuits such as flash or pipelined ADCs [39][40].

CMOS S&H Circuits

Now that we have reviewed the key performance parameters and testing techniques useful for characterizing S&H circuits, let us investigate the design of CMOS S&H circuits through several practical examples.

The simplest practical S&H circuit is shown in Figure 3.5, which is often referred to as the *open-loop S&H architecture*. The advantages of this open-loop S&H architecture are the high speed and the unconditional stability, since no global feedback is used. However, the accuracy of an open-loop S&H circuit is limited by signal-dependent errors due to the charge injection, switch on-resistance, and finite clock slope. Despite some early proposals to minimize signal-dependent errors (e.g., cascading a dummy transistor with the sampling switch or using a CMOS transmission gate for switching [1][20]), the maximum achievable linearity of a standard open-loop CMOS S&H is limited to about 8 bits (i.e., with an overall SDR of about 48 dB) [41].

A few fast (clocked at 100 MHz or above) open-loop CMOS S&H prototypes with a 10-bit resolution have been reported [42][43], which use either stacked source followers [42] or switched source followers [43] to alleviate signal-dependent charge injection errors and body effects. Moreover, the *bootstrapping* technique has been employed by most low-voltage SC S&H implementations to further reduce charge injection errors and desensitize the switch on-resistance from the input signals [44] (the mechanism of bootstrapped clock generation will be discussed in Chapter 7). In practice, low-leakage and high-speed bipolar diode bridges are often adopted to build the sampling switches in open-loop S&H circuits for very high speed (up to 1.25 GHz) and high-accuracy (9 to 12 bits) sampling applications [45][46].

As an alternative to special source followers and bootstrapped clocks, a seemingly straightforward solution to the low accuracy of the CMOS S&H circuit in

Figure 3.5 is to create a negative feedback path by connecting the output of A_2 to the negative input terminal of A_1. However, this modification does not really remove harmonic distortions since the voltages on both sides of M_1 still depend on the input signal. Moreover, the slew rate requirement for the input buffer A_1 becomes rather stringent, because its output voltage level has to change significantly from one phase to the next.

These problems can be alleviated by using the *closed-loop S&H architecture* illustrated in Figure 3.6 [47]. As shown, the op-amp G_2 creates a virtual ground in the loop, and the holding capacitor C_h is connected between this virtual ground and V_{out}. Once M_1 is turned off and M_2 is connected to ground, the channel charge of M_1 is distributed toward both directions: the charge to the left is grounded via M_2, while the one to the right is absorbed by the virtual ground. Thus, the majority of input-dependent charge injection errors from M_1 are effectively neutralized. Also, the output node of G_1 is connected to the virtual ground via M_1 during the sampling mode ($\Phi_1 \rightarrow 1$), whereas it is grounded via M_2 during the holding mode ($\Phi_2 \rightarrow 1$). As a result, the time required for G_1 to slew from the holding to sampling mode (and vice versa) is reduced. The output voltage level of G_2 is also kept within a small range of its nominal value by the inner loop consisting of C_h and G_2.

Additionally, an extra circuit is connected to the positive input terminal of G_2, where the capacitance of C_{h2} is set equal to that of C_h, and M_3 is identical to M_1. Hence, the extra circuit is indeed a dummy compensating the voltage changes across C_h or M_1 due to signal-dependent leakages, dc offsets, and parasitic capacitances. This concept naturally leads to a fully differential implementation of the circuit in Figure 3.6, which is popular for designing high-resolution (>10-bit) S&H circuits in practice. The reader is referred to the paper by Nayebi and Wooley [47] for more details.

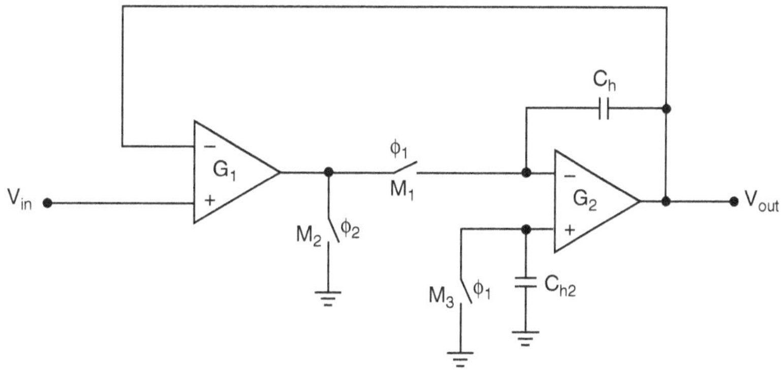

Figure 3.6 Closed-loop S&H with two op-amps.

Although the accuracy of the closed-loop S&H circuit in Figure 3.6 is improved in comparison with that of its open-loop counterpart, the use of a feedback loop encompassing two high-gain op-amp stages entails a basic tradeoff between accuracy and speed. To ensure the stability and accuracy of the circuit in Figure 3.6, an adequate phase margin is required for good op-amp settling characteristics, thereby limiting the overall operating speed.

Variations of the circuit in Figure 3.6 have been proposed to increase the operating speed while maintaining a sufficient accuracy. Most of these circuits include only one op-amp in the loop. A simple switched-capacitor S&H circuit that can easily achieve a 9-bit resolution (clocked at 0.5 MHz) is shown in Figure 3.7(a). Note that

Figure 3.7 (a) Simple SC S&H. (b) SC S&H with a relaxed slew rate requirement. (c) Fast S&H with a Miller holding capacitance.

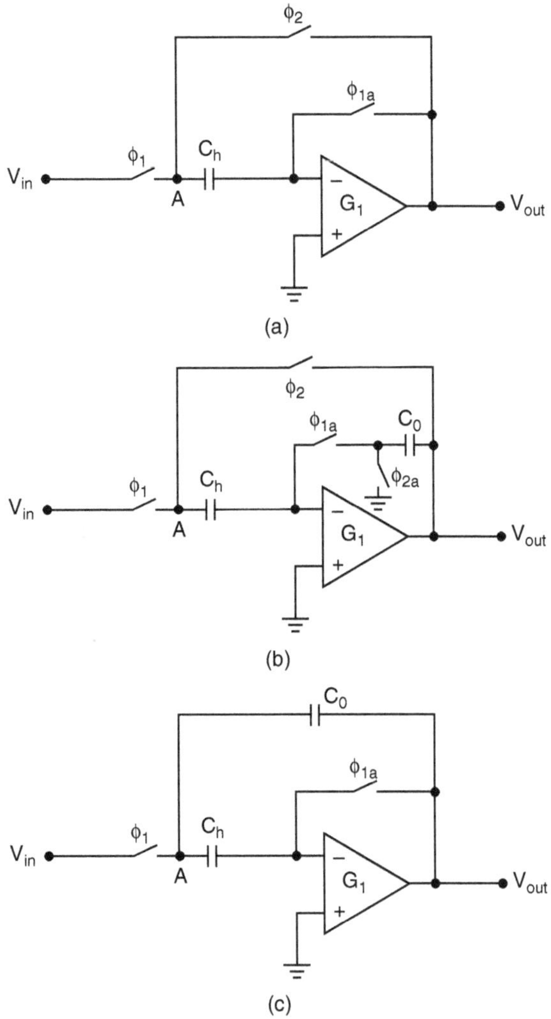

the switch denoted Φ_{1a} (i.e., advanced Φ_1) is turned off slightly earlier than is the switch denoted Φ_1.

The main purpose of this advanced-Φ_1 configuration is to prevent the input-dependent errors injected by the Φ_1 and Φ_{1a} switches from appearing in the output during the holding mode. Specifically, when the Φ_1 switch is turned off (slightly after Φ_{1a} goes off), its negative channel charge does not change the voltage across C_h because the inverting input of G_1 is floating (i.e., open circuit). Both sides of the Φ_{1a} switch are very close to ground when it is just turned off, and hence its channel charge is also neutralized.

The circuit operates as follows. During the sampling mode, if G_1 has a small input-offset voltage V_{off}, then its output level is also equal to V_{off} due to the unity-gain feedback, and the voltage across C_h is equal to $(V_{in} - V_{off})$. During the holding mode, Φ_1 is off and Φ_2 is on, the output is given by $(V_{in} - V_{off} + V_{off}) = V_{in}$ (i.e., V_{off} does not affect the output). Thus, this SC circuit is also referred to as the *input-offset-free* S&H (low-frequency noise signals such as $1/f$ or *flicker noise* can be canceled in a similar manner). In addition, note that a dummy SC network similar to that of Figure 3.6 is often connected to the positive input terminal of G_1, reducing errors due to clock feedthrough and parasitic capacitances.

Thanks to its simplicity and low power consumption, this type of S&H (often in its fully differential configuration) is one of the candidates that are able to meet the needs of pipelined ADCs, where many S&H circuits are required. It can also be used as a half-period delay or unity-gain buffer (UGB) [20]. However, a major drawback of this S&H circuit is that during the sampling mode the output must be reset to V_{off} regardless of V_{in}, and as a result, the output cannot track the input. Moreover, in between modes, the output of G_1 needs to slew from V_{off} to V_{in} (and vice versa), which places a stringent slew rate requirement on G_1 especially when V_{in} is a relatively large signal as compared to V_{off}.

To relax the slew rate requirement for the op-amp, an interesting modification of the preceding circuit is shown in Figure 3.7(b) [48]. Note that the switch denoted Φ_{1a} (i.e., advanced Φ_1) is turned off slightly earlier than are the switches denoted Φ_1, whereas the switch denoted Φ_{2a} (i.e., advanced Φ_2) is turned *on* slightly earlier than are the switches denoted Φ_2. The arguments employed earlier can be repeated here to explain how the switching configurations help alleviate charge injection errors.

The operation of this circuit is described as follows. During the sampling mode, the voltage across C_h is equal to $(V_{in} - V_{off})$, and the voltage across C_0 is equal to $[V_{outp} - V_{off} - (V_{in} - V_{outp})\, C_h/C_0]$, where V_{outp} is the output signal held till the beginning of the current sampling mode. Note that $(V_{in} - V_{outp})\, C_h$ represents the charge transferred from the right side of C_h to the left side of C_0 once $\Phi_1 \rightarrow 1$. As the result, during the holding-to-sampling transition, the op-amp's output changes from V_{outp} to $[V_{outp} - V_{off} - (V_{in} - V_{outp})\, C_h/C_0]$, and the change in the output voltage is thus given by

$$|\Delta V_{h\rightarrow s}| \cong |V_{outp} - V_{off} - (V_{in} - V_{outp})\cdot C_h/C_0 - V_{outp}| = (V_{in} - V_{outp})\cdot C_h/C_0 + V_{off} \quad (3.73)$$

V_{off} is usually less than 15 mV in a standard CMOS technology. During the holding mode, the right side of C_h is floating, and C_0 is connected between the output and ground. Ignoring the dc offset introduced by turning off the Φ_{1a} switch, we can write the change in the output voltage during the sampling-to-holding transition as follows:

$$|\Delta V_{s\rightarrow h}| \cong (V_{in} - V_{off}) - [V_{outp} - V_{off} - (V_{in} - V_{outp})\cdot C_h/C_0] = (V_{in} - V_{outp})(1 + C_h/C_0) \quad (3.74)$$

The approximations derived here indicate that compared to the circuit in Figure 3.7(a), whose maximum intermode output voltage jump is as large as $(V_{in} - V_{off})$, this S&H circuit greatly relaxes the slew rate requirement for the op-amp, especially when $C_h << C_0$ or V_{in} is a slow-varying signal compared to the sampling clock (i.e., V_{in} is oversampled; hence, ΔV_{in} is small). Thus, this circuit is able to track the input and its maximum operating speed is higher than that of the circuit in Figure 3.7(a).

As Wang and Temes [48] reported, with a 20-pF load capacitance, this S&H circuit can function up to about a 2.8-MHz clock rate, whereas the circuit of Figure 3.7(a) can function up to only about 0.6 MHz (in both circuits the op-amp is implemented as a cascode inverter in a 3-μm CMOS technology). This type of S&H circuit (often in its fully differential configuration) is widely used for implementing oversampled audio ADCs, thanks to its low slew rate and relatively high resolution (up to 12 bits at a 2.5-MHz clock).

Based on Equations (3.70) and (3.72), we realize that in a given CMOS technology, the accuracy of the S&H circuit can be improved by using either a wider sampling switch (i.e., a larger W) or a larger effective holding capacitance in the holding mode (i.e., reducing charge injection errors). On the other hand, the operating speed of the circuit is limited by the time-constant product of R_{on} and C_{h_s}, where C_{h_s} is the effective holding capacitance in the sampling mode.

Furthermore, both the accuracy and speed of the S&H would benefit from an increased W (i.e., descreased R_{on}), thus the accuracy-speed tradeoff is essentially governed by the corresponding holding capacitances in the sampling and holding modes. The key concept is, for a *win-win* situation between accuracy and speed, we should design the S&H circuit in such a way that the effective holding capacitance appearing in the holding mode is maximized; while that appearing in the sampling mode is minimized.

A simple CMOS S&H circuit built based on this concept is shown in Figure 3.7(c) [49]. Note that this circuit looks identical to that of Figure 3.7(a) except for the capacitor C_0 between node A and the output. During the sampling mode ($\Phi_1 \rightarrow 1$), the input V_{in} is sampled by an effective holding capacitance $C_{h\_s} = (C_h + C_0)$. During the holding mode, the *effective Miller holding capacitance* that appears at node A is given by

$$C_{h\_Miller} = (1 + G_1)\left(\frac{C_h C_0}{C_h + C_0}\right) \tag{3.75}$$

where G_1 is the op-amp's maximum dc gain, and C_{h_Miller} is normally much larger than $(C_h + C_0)$. As we can see, small *physical* capacitors can be used to build a large holding-mode capacitance, thereby realizing a relatively high accuracy and a high speed. It can be proved that the slew rate requirement for the op-amp here is comparable to that for the aforementioned circuit. As Lim and Wooley [49] reported, fabricated in a 1-μm CMOS technology, this S&H circuit can achieve an 8-bit sampling resolution at up to about 100-MHz clock frequency (assuming the acquisition time takes up half the clock period).

In summary, we studied key performance parameters and testing methods of S&H circuits. A few basic S&H circuits were presented, and the tradeoff between the achievable accuracy and operating speed was emphasized. Most recently, a few S&H circuits have become available that have better accuracy-speed merits than those seen here. Such examples include the double-sampled CMOS S&H with offset and finite-gain compensation [50] and the "flip-around" CMOS S&H with bootstrapped switches [51]. In addition, CMOS S&H circuits using the switched-op-amp technique [52], which is suitable for low-voltage applications, can also be found in the literature.

3.5 Switched-Capacitor Interpolators and Decimators

SC Interpolators

An SC interpolator is an analog sampled-data selective filter that increases the sampling clock frequency from f_{clk} to Nf_{clk} (N is an integer greater than 1) and suppresses the input signal's replicas between f_n and $(Nf_{clk} - f_n)$, where f_n is the Nyquist rate.

The SC interpolator is mainly used as the output stage of an SC filter, allowing the SC filter to operate at a lower clock rate and hence to be less complex [53]. In some classical Nyquist-rate audio D/A converters (DACs) such as that presented in [54], an SC interpolator is placed before the DAC to suppress frequency images centered at f_n, $2f_n$, . . . , $(Nf_{clk} - f_n)$; therefore, it eases the task of the analog antialiasing filter (AAF) following the DAC. However, if a very large N is required (e.g., for an oversampled audio ΔΣ DAC [55]), a fully digital linear interpolator is often chosen over the SC type for the sake of digital programmability and the accuracy versus power ratio.

A standard SC interpolator is equivalent to a low-pass SC filter preceded by an analog upsampler (or vice versa). The input to an SC interpolator is usually a sampled-and-held signal operating at a clock rate of f_{clk}. It may be directly upsampled to Nf_{clk} using an N-step or N-phase capacitor array. An SC interpolator using a four-step capacitor array as the analog upsampler is shown in Figure 3.8(a) [53][56]. Note that $C_1 = C_2 = C_3 = C_4 = C$, $C_0 = 4C$, and $T = 1/f_{clk}$.

Although ideally this circuit would be able to realize a four-fold interpolation, it is seldom used in practice because there is no dc feedback between the inverting input and the output of the op-amp [20][56]. As a result, if there is a small parasitic capacitance C_p between the inverting input of the op-amp and ground, then the channel charge generated by turning off the switches will accumulate on this parasitic capacitor, which will eventually saturate the op-amp.

A modification of the foregoing circuit is shown in Figure 3.8(b). It employs the same clocking scheme as before. Compared to the preceding circuit, two extra switched capacitors C_m and C_n are added ($C_m = C_n = C$). During phase Φ_4, the clock-feedthrough charge on C_p is absorbed by C_m and C_n, which is then discharged to ground during the next phase Φ_1. Also, both C_m and C_n send charges to C_0 when Φ_4 is on; however, these charges do not affect the interpolator's transfer function since they cancel each other out.

A noteworthy disadvantage of this circuit is that a large number of capacitors will be required if the interpolation factor is large. Moreover, the overall capacitance grows with N, as do the total power consumption and the silicon active area.

An SC interpolator consisting of an integrator and a unity-gain S&H feedback stage is shown in Figure 3.9(a) [56]. Note that only three capacitors are used for an interpolation factor of 4 ($C_1 = C_2 = C$ and $C_3 = 4C$), and the integrator is operating at a clock rate four times as fast as that of the S&H stage. And as we can see, the S&H stage is similar to that presented in Figure 3.7(a).

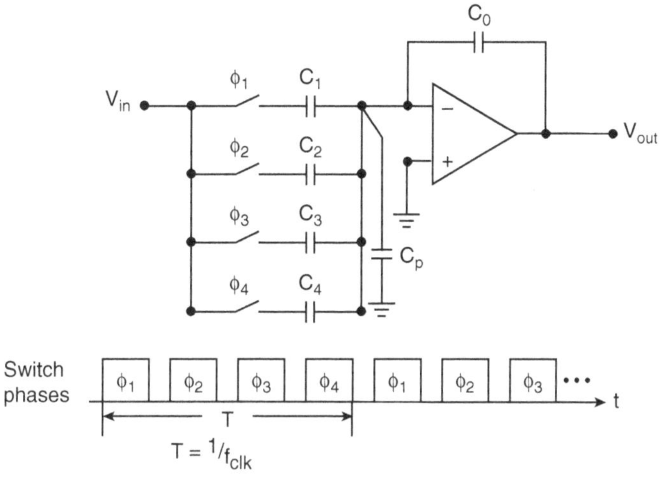

(a)

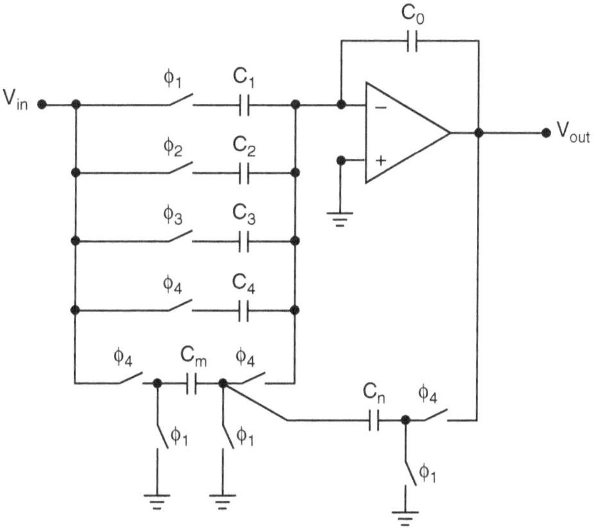

(b)

Figure 3.8 (a) Easy-to-saturate SC interpolator. (b) Improved SC interpolator with a multistep capacitor array ($N = 4$).

The interpolator operates as follows. During phase Φ_1, the input voltage $V_{in}(nT)$ is sampled onto C_1. During phase Φ_2, a charge proportional to the difference between $V_{in}(nT)$ and the sampled-and-held output voltage $V_{out}(nT)$, which ideally should be equal to $V_{in}(nT - T)$, is transferred onto C_3, thereby causing the output to jump by $\Delta V_{out} = [V_{in}(nT) - V_{in}(nT - T)]/4$. This operation repeats every $0.25T$, and the *input-output-difference* charges acquired at different time instants are accumulated on C_3

until the beginning of the next phase Φ_3, when the output voltage would be equal to $V_{in}(nT)$.

By inspection we can find that ideally the transfer function of this configuration is equivalent to that of the four-step capacitor array shown in Figure 3.8(b), where $[V_{in}(nT) - V_{in}(nT - T)]$ is chopped into four equal pieces by multiplexing four input capacitors. However, the overall capacitance is significantly reduced in comparison with that of the previous circuit, especially when N is large ($N > 3$). Furthermore, it can be shown that in this configuration the number of capacitors is always equal to 3, regardless of the value of N. That is, only the values of $(C_3/C_{1,2})$ and switching phases change with the interpolation factor.

However, a common drawback of all the interpolators seen thus far is that the capacitance spread, which is equal to N in all the preceding cases, may become very large. An interesting modification resolving this issue is shown in Figure 3.9(b) [20][54][56] ($N = 4$). Note that four capacitors are used here. Instead of using a unity-gain S&H stage to capture the difference between the input and output, this interpolator places an SC gain stage (voltage gain = C_1/C_2) with reset before the delayed integrator, which has a capacitance ratio of C_3/C_4. As a result, during each fast clock cycle (i.e., Φ_3 and Φ_4) the output voltage also changes by $\Delta V_{out} = [V_{in}(nT) - V_{in}(nT - T)]/N$, as long as the following condition is met:

$$\frac{C_1}{C_2} \cdot \frac{C_3}{C_4} = \frac{1}{N} \tag{3.76}$$

Thus, the capacitance spread can be reduced to as small as the square root of N.

Another advantage of this interpolator is that the fast clock frequency driving Φ_3 and Φ_4 needs to be equal to only $(N - 1)f_{clk}$ instead of Nf_{clk}, due to the extra delay element provided by the gain stage. For instance, a clock rate of only $3f_{clk}$ is required to achieve a four-fold interpolation. This benefit becomes more pronounced when f_{clk} is very high (e.g., in an oversampled audio data converter). Moreover, the speed of the first op-amp stage is determined by the lower sampling rate f_{clk}, thereby relaxing the circuit implementation.

Lastly, recall that the input to an SC interpolator is usually a sampled-and-held signal. As intuition suggests, the frequency response of the SC interpolator could suffer from the *sinc* effect of the S&H circuit, which might cause aliasing problems especially when the value of (f_{clk}/f_n) is not much greater than 2. This consideration becomes realistic in some high-speed applications such as video data converters and wireless communications, where SC interpolators are used to upsample the signals. A

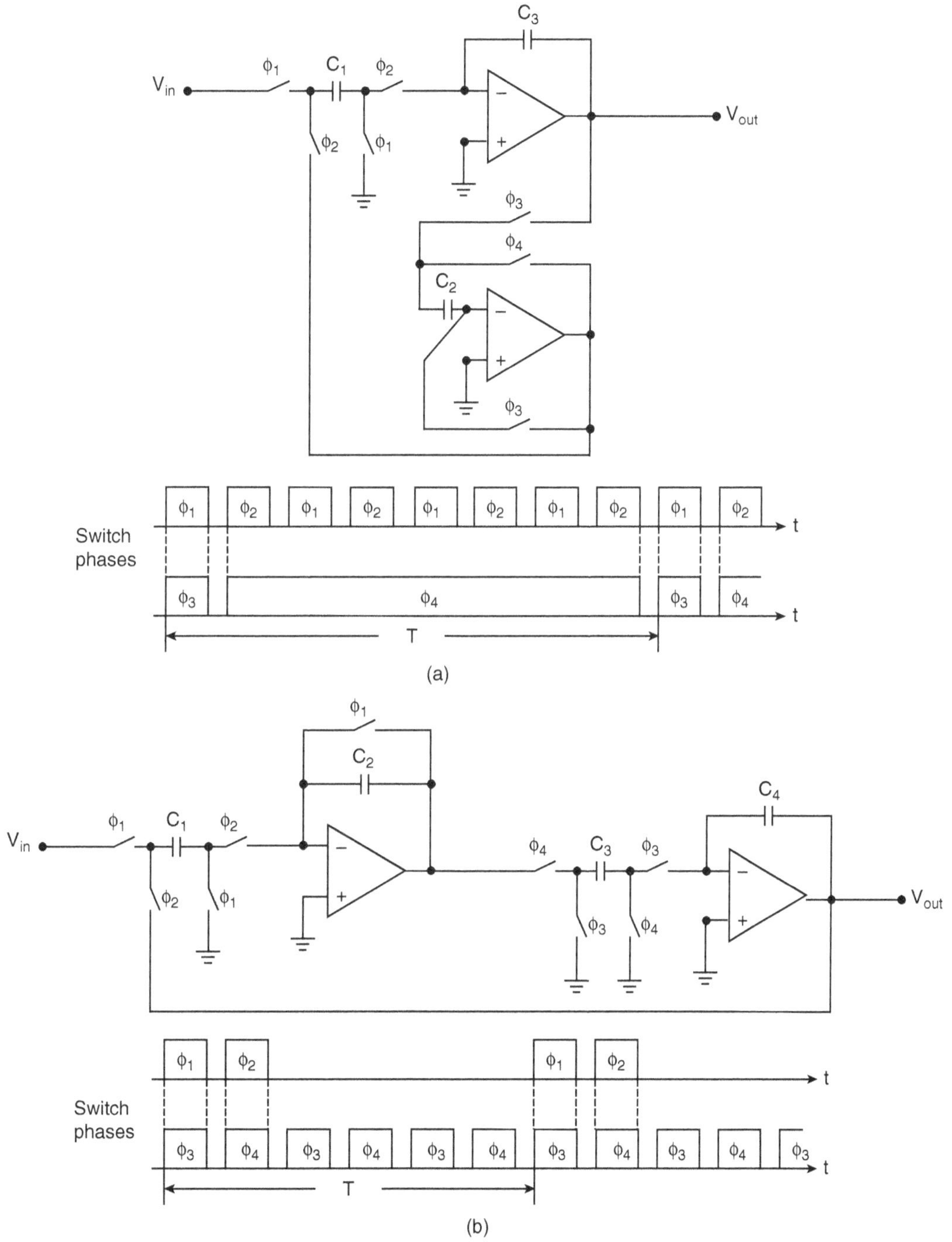

Figure 3.9 (a) SC interpolator with only three capacitors. (b) SC interpolator with a reduced capacitance spread ($N = 4$).

few circuit techniques have been reported to alleviate S&H effects on high-speed interpolations [56][57][58]. The interested reader is referred to the literature for more details.

SC Decimators

As an analog sampled-data signal processing stage with an operation complementary to that of the SC interpolator, the SC decimator reduces the sampling rate from Mf_{clk} to f_{clk}, where M is an integer greater than 1 and is often called *decimation factor.* In contrast to its interpolating counterpart, the SC decimator is usually used as the input stage of an SC filter, allowing the SC filter to operate at a lower clock rate. Additionally, it can be used after an oversampled delta-sigma ADC to downsample the signals and remove the out-of-band quantization noise as well as the high-frequency replicas. However, similar to the SC interpolator, if a large M is required (e.g., for an oversampled audio ADC), a fully digital decimator is often preferred over the SC implementation.

A standard SC decimator is equivalent to a low-pass filter followed by an analog downsampler (or vice visa). A simple and power-efficient SC decimator is shown in Figure 3.10(a) [59] ($M = 4$). This decimator is basically a fast-sampling SC integrator with a slow read-out. That is, it operates like a linear-phase averaging filter and its transfer function can thus be expressed as follows [59]:

$$H(z)=\frac{V_{out}(z)}{V_{in}(z)}=\frac{C}{C_0}\cdot\sum_{i=0}^{M-1}z^{-i}\cdot z^{-1/2}=\frac{C}{C_0}\cdot\frac{1-z^{-M}}{1-z^{-1}}\cdot z^{-1/2} \tag{3.77}$$

where M is the decimation factor, and $z = \exp(-s/Mf_{clk})$. "$z^{-1/2}$" is a half-delay element and does not affect the decimation function. The transfer function derived here indicates that this SC decimator is indeed a sampled-data finite-impulse-response (FIR) filter. The Φ_3 switch provides a dc feedback from the op-amp's output to the inverting input, thereby resetting the op-amp every T ($T = 1/f_{clk}$). The ratio C/C_0 is often set to be equal to or less than $(1/M)$ in order to avoid saturating the following stages. From Equation (3.77), we can derive the magnitude response of the decimator using algebraic formulae, which is given by

$$H\left(e^{j\omega T/M}\right)=\frac{C}{C_0}\cdot\left|\frac{\sin(\omega T)}{\sin(\omega T/M)}\right| \tag{3.78}$$

The foregoing equation indicates that the decimator has a *sinc*-type magnitude response. As a result, the frequency replicas at integer multiples of f_{clk} (except for multiples of Mf_{clk}) ideally should have zero magnitudes, meaning that these high-

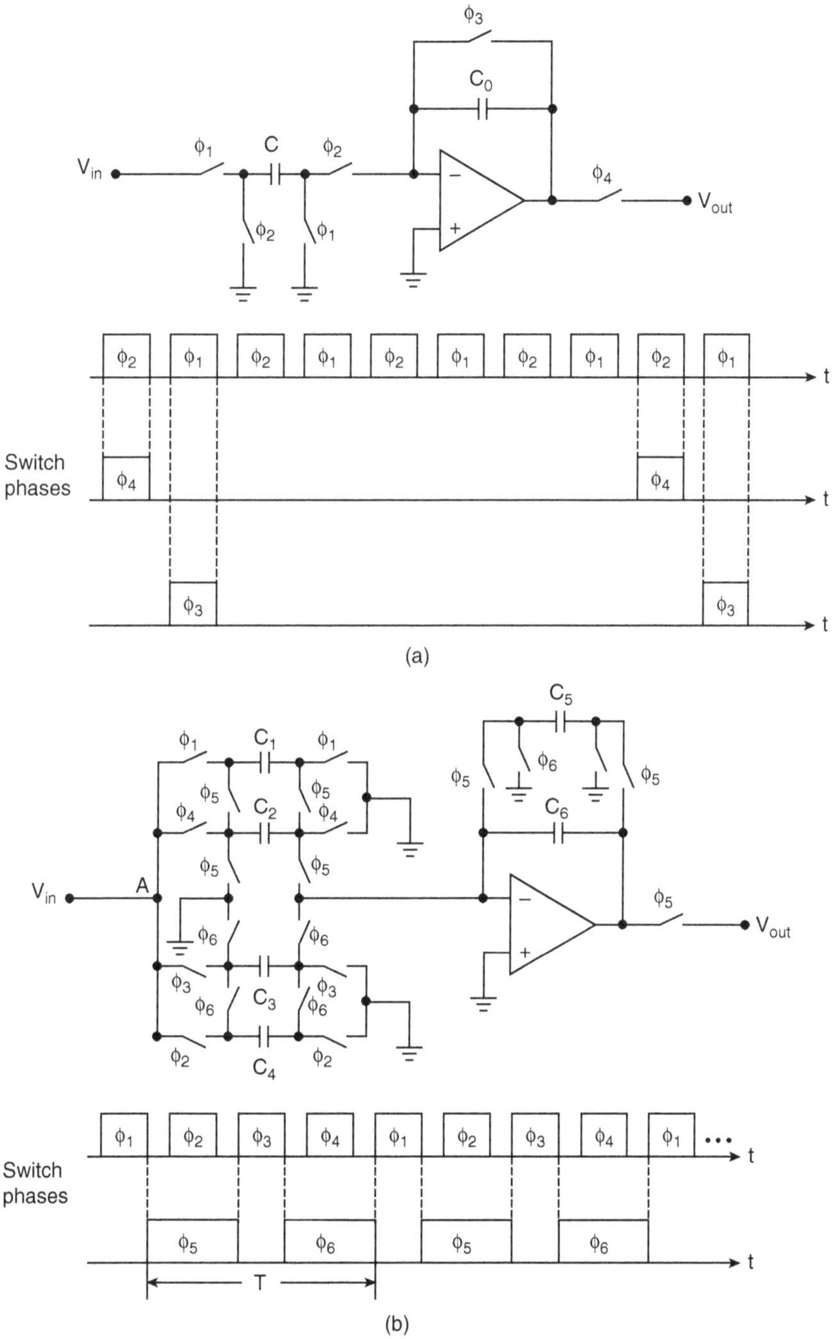

Figure 3.10 (a) FIR SC decimator. (b) IIR SC decimator ($M = 4$).

frequency images are eliminated by the decimator. However, in practice they are never removed completely due to circuit imperfections. Additionally, in this decimator, the speed of the op-amp is determined by the input sampling rate Mf_{clk}, thereby imposing a stringent limit on the maximum workable sampling frequency due to the difficulties of designing very fast and linear op-amps.

To alleviate the restrictions on the sampling frequency as well as the op-amp's settling characteristics, SC decimators based on infinite-impulse-response (IIR) transfer functions are reported [57][60]. Generally speaking, in comparison with their FIR counterparts, IIR SC decimators are more suitable for applications with higher selectivity and wider signal bandwidth.

An IIR SC decimator based on a single op-amp is shown in Figure 3.10(b) [60]. The decimation factor M is equal to 4 in the schematic. It is proved that the decimator's transfer function can be presented as follows [60]:

$$H(z) = \frac{C_1 + C_2 z^{-1} + C_3 z^{-2} + C_4 z^{-3}}{(C_5 + C_6) - C_6 z^{-4}} \tag{3.79}$$

The basic idea is to prolong the effective sampling time (i.e., to reduce the effective input sampling frequency sensed by the op-amp). In the schematic, four parallel sampling branches placed before the op-amp stage are operating at the same clock rate $4f_{clk}$, while the sampled charges accumulated on these branches are read out by the op-amp at a lower frequency f_{clk}. Also, the effective sampling function of the C_2 branch (i.e., the transmission path through the switched C_2) is delayed by z^{-1} as compared to that of the C_1 branch, whereas the C_3 and C_4 branches are delayed by z^{-2} and z^{-3}, respectively. In effect, the effective sampling frequency sensed by the op-amp and the subsequent circuitry is reduced from $4f_{clk}$ to f_{clk}, relaxing the op-amp's settling requirement.

On some occasions, parallel decimating circuits have come into use for the realizations of *subsampling* wireless receivers with a relatively high intermediate frequency (IF). An interesting IIR SC decimator was reported by Lindfors et al. [61]. The decimator was reported to be capable of operating at a maximum clock rate of 230 MHz in a 0.5-μm CMOS technology. The interested reader is referred to the reference for a complete investigation.

In summary, this section provided an overview of SC interpolators and decimators. A few design examples were presented, and their corresponding advantages and disadvantages were discussed.

3.6 Signal-Flow-Graph Analysis of Switched-Capacitor Circuits

Signal-Flow-Graph Analysis

As Section 3.3 showed, using charge equations to analyze SC circuits could be tedious, even for obtaining the transfer function of a simple SC integrator. As a result, a very useful and simple method called the *signal-flow-graph* (SFG) analysis based on the principle of superposition has been widely adopted in practice [9][10][13][19][20][21].

Consider the first-order switched capacitor filter shown in Figure 3.11(a). Based on the principle of superposition, we realize that this circuit can be decomposed into three active subcircuits: an inverting SC gain stage, a parasitic-insensitive parallel SC integrator similar to that shown in Figure 3.3(a), and a parasitic-insensitive series SC integrator similar to that in Figure 3.3(b). Moreover, utilizing some of the results derived in Section 3.3, we can write the transfer functions of these three subcircuits as follows:

$$H_{gain}(z) = -\frac{C_1}{C_0}, \quad H_{parallel}(z) = \frac{C_2}{C_0}\frac{z^{-1}}{1-z^{-1}}, \quad H_{series}(z) = -\frac{C_3}{C_0}\frac{1}{1-z^{-1}} \tag{3.80}$$

Hence, the input-output relationship of the circuit can be precisely represented by the equivalent signal flow graph shown in Figure 3.11(b). Note that in the graph, the

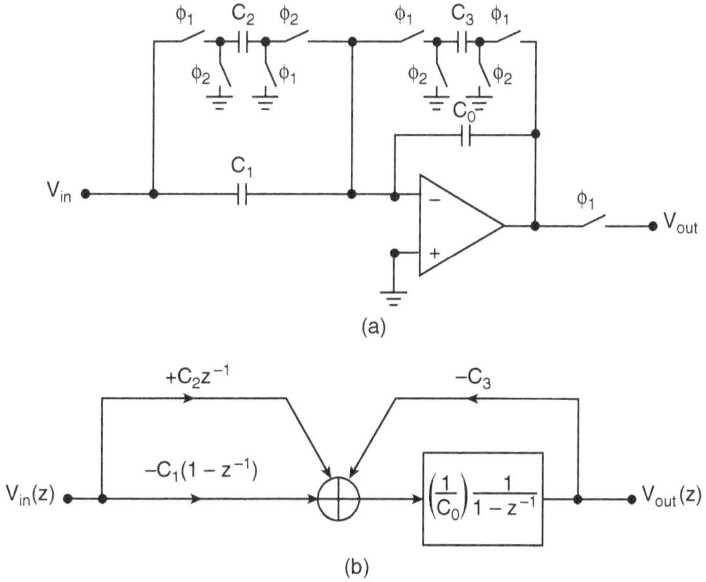

Figure 3.11 (a) Simple SC circuit. (b) Equivalent signal flow graph.

input stages (either in forward or feedback paths) are represented by three different gain factors, whereas the closed-loop op-amp is represented by an integrating function. The weighted inputs and output are summed up before entering into the op-amp. Based on the signal flow graph, we can develop the following equation:

$$\{V_{in}(z)\cdot[C_2z^{-1}-C_1(1-z^{-1})]+V_{out}\cdot(-C_3)\}\cdot\frac{1}{C_0}\cdot\frac{1}{1-z^{-1}}=V_{out}(z) \tag{3.81}$$

Thus, the overall transfer function is given by

$$H(z)=\frac{V_{out}(z)}{V_{in}(z)}=\frac{(C_1+C_2)z^{-1}-C_1}{C_0+C_3-C_0z^{-1}}=-\frac{C_1}{C_0+C_3}\cdot\frac{1-[(C_1+C_2)/C_1]z^{-1}}{1-[C_0/(C_0+C_3)]z^{-1}} \tag{3.82}$$

It can be found that the pole Z_p and zero Z_z of the previous circuit are given by the following:

$$Z_p=\frac{C_0}{C_0+C_3} \quad \text{and} \quad Z_z=1+\frac{C_2}{C_1} \tag{3.83}$$

This circuit is *always* stable since Z_p is located inside the unit circle (i.e., $Z_p < 1$). However, Z_z is located outside the unit circle ($Z_z > 1$). To implement a similar first-order transfer function with the same pole but a zero located inside the unit circle, we need to replace the noninverting and delayed input stage (i.e., C_2z^{-1}) with an inverting and delay-free (i.e., $-C_2$) one. The reader is encouraged to derive the new circuit's equivalent signal flow graph and transfer function, as well as pole/zero values.

Mason's Rule

The larger and more complex is a circuit, the more difficult it would be to quickly obtain the input-output relationship, since a large and complex circuit often consists of multiple forward paths and feedback loops. In practice, a well-known technique called the *Mason's rule* [62] is often employed to derive the transfer function of a complex circuit by simply inspecting its equivalent signal flow graph. The *z*-domain Mason's rule is stated as follows:

$$H(z)=\frac{V_{out}(z)}{V_{in}(z)}=\frac{1}{\Delta(z)}\sum_i G_i(z)\Delta_i(z) \tag{3.84}$$

$G_i(z)$ represents the transfer function of the *i*th forward path, which contains no feedback. $\Delta(z)$ is the determinant of the whole signal flow graph, which is given by

$$\Delta(z)=1-\sum_j G_{jL}+\sum_{k,l} G_{kL}G_{lL}-\sum_{m,n,p} G_{mL}G_{nL}G_{pL}+\ldots \tag{3.85}$$

Here, G_{jL} is the gain of the jth loop. $(G_{kL}G_{lL})$ is the product of the loop gains of two nontouching loops (kth and lth) if they exist. Similarly, $(G_{mL}G_{nL}G_{pL})$ is the loop-gain product of three nontouching loops (mth, nth, and pth) if they exist and so on. Lastly, $\Delta_i(z)$ is the determinant of a subgraph whose signal paths are *not* touching the ith forward path.

As a simple example, let us revisit the signal flow graph in Figure 3.11(b) and utilize the Mason's rule to obtain its transfer function. As seen in the schematic, the circuit consists of two forward paths and one feedback loop. From Equations (3.84) and (3.85), we can obtain the transfer function as follows:

$$H(z)=\frac{1}{1+\frac{C_3}{C_0}\frac{1}{1-z^{-1}}}\cdot\left[-\frac{C_1}{C_0}\cdot 1+\frac{C_2}{C_0}\frac{z^{-1}}{1-z^{-1}}\cdot 1\right]=\frac{(C_1+C_2)z^{-1}-C_1}{C_0+C_3-C_0z^{-1}} \tag{3.86}$$

As seen, this transfer function is identical to that given by Equation (3.82).

EXAMPLE 3.3

Draw an equivalent signal flow graph for the SC second-order filter (biquad) shown in Figure 3.12(a) [9][13]. Note that *switch-sharing* is not employed here for illustration purposes. Find its transfer function using the Mason's rule, and show that the dc signal level at node A is independent of the transfer function. Lastly, investigate the circuit's Q and capacitance spread.

Solution: The circuit in Figure 3.12(a) can be represented by the equivalent signal flow graph shown in Figure 3.12(b). As the graph shows, there are three forward paths and two feedback loops. Utilizing the Mason's rule, we can obtain the transfer function, which is given by

$$H(z)=\frac{V_{out}(z)}{V_{in}(z)}=-\frac{(M_2+M_3)z^2+(M_1M_5-M_2-2M_3)z+M_3}{(M_6+1)z^2+(M_4M_5-M_6-2)z+1} \tag{3.87}$$

The overall dc gain of this filter ($z = 1$) is thus given by

$$H(z)|_{z=1}=-\frac{(M_2+M_3)+(M_1M_5-M_2-2M_3)+M_3}{(M_6+1)+(M_4M_5-M_6-2)+1}=-\frac{M_1}{M_4} \tag{3.88}$$

Assuming the input dc signal level is 1 V, then ideally the output would be at $(-M_1/M_4)$ (in volts). From Figure 3.12(b), we can obtain the dc signal level at node A as follows:

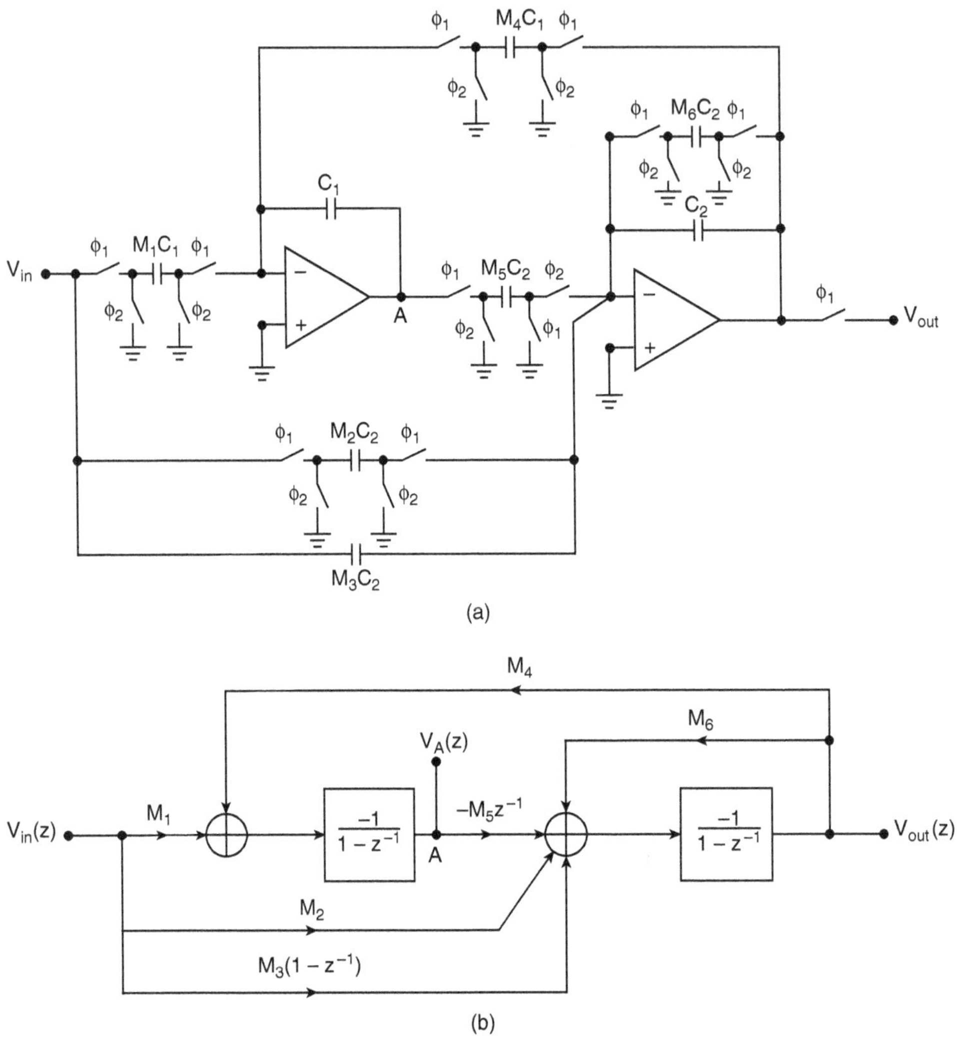

Figure 3.12 (a) SC biquad. (b) Equivalent signal flow graph.

$$V_A(z)\big|_{z=1} = \left[1 \cdot M_1 + \left(-\frac{M_1}{M_4}\right) \cdot M_4\right] \cdot \left(\frac{-1}{1-1^{-1}}\right) = \frac{0}{0} = ? \qquad (3.89)$$

The question mark in the preceding equation stands for uncertainty, which implies that the specified transfer function $H(z)$ cannot determine the dc signal level at node A. Uncertainty essentially means flexibility, therefore a certain degree of freedom in choosing the values of M_1, M_4, and M_5 is obtained, and the nominal dc signal level at node A can be set at any value regardless of the three capacitance ratios. A common and often optimal practice is to set the interstage gain M_5 to 1, equate the system coefficients of z in Equations (3.87) and (3.24), and then solve the design equations to determine all the initial capacitance ratios. The last step is to perform *dynamic range scaling* so that the peak voltage at the output is equal to that at node A [20].

Recalling Equations (3.24), (3.27), and (3.28) in Section 3.2, we find some interesting relationships among d_1, d_2, M_4, M_5, M_6, $\omega_0 T$, and Q, assuming a very fast sampling clock rate ($\omega_0 T << 1$). First, by equating the coefficients of z in Equations (3.24) and (3.87), we can write the following:

$$d_1 = \frac{M_4 M_5 - M_6 - 2}{M_6 + 1}, \quad \text{and} \quad d_2 = \frac{1}{M_6 + 1} \tag{3.90}$$

Next, making use of the flexibility in choosing the values of M_1, M_4, and M_5, we obtain the following approximations based on Equations (3.27) and (3.28) (assuming $\omega_0 T << 1$):

$$M_1 \cong \frac{c_0 + c_1 + c_2}{\omega_0 T}, \quad M_4 \cong M_5 \cong \omega_0 T, \quad M_6 \cong \frac{\omega_0 T}{Q}, \quad \text{and} \quad Q \cong \frac{\sqrt{M_4 M_5}}{M_6} \tag{3.91}$$

This indicates that the values of $\omega_0 T$ and Q can be determined by M_4, M_5, and M_6. That is, M_4, M_5, and M_6 determine the pole positions [9][13]. By contrast, it can be found that the other capacitance ratios (M_1, M_2, and M_3) are responsible for the zero(s) only.

In most cases, the largest to smallest capacitance ratio (i.e., capacitance spread) is also determined by one of M_4, M_5, and M_6 for a guaranteed stability and a convenient dynamic range scaling [13][20]. In practice, the feedback capacitors C_1 and C_2 usually have the largest capacitances, and both are normalized to unity (i.e., $C_1 = C_2 = 1$).

Based on the approximations given by Equation (3.91) we can see that when Q is high ($Q > 1$), the smallest capacitance is given by $M_6 C_2$, thereby providing a capacitance spread that approximates ($Q/\omega_0 T$), which is much larger than Q since $\omega_0 T << 1$. On the other hand, when Q is low ($Q < 1$), the smallest capacitance is given by $M_4 C_1$ or $M_5 C_2$, resulting in a capacitance spread that approximates ($1/\omega_0 T$), which is much smaller than that of the high-Q configuration. Therefore, the SC biquad shown in Figure 3.12(a) is more suitable for realizing low-Q filtering transfer functions. In Chapter 4, we will study the alternative SC biquad appropriate for building high-Q filters.

Appendix 3.1

A SWITCAP programming example is presented here. These source codes may be used to simulate the parasitic-sensitive bilinear SC integrator shown in Figure 3.2(b), and to plot the frequency responses (with and without C_{p1}, respectively) between 1 Hz and 600 kHz. Furthermore, a MATLAB program is used to verify Equation (3.50). It is assumed that the feedback capacitor $C_0 = 5\,\text{pF}$, the sampling capacitor $C = 0.5\,\text{pF}$, the parasitic capacitor $C_{p1} = 0.05\,\text{pF}$, the input signal frequency $f_{in} = 48\,\text{kHz}$, and the sampling clock frequency $f_{clk} = 1.2\,\text{MHz}$.

```
/*SWITCAP program*/
TITLE: SIMULATING PARASITIC-SENSITIVE BILINEAR INTEGRATOR


TIMING;
PERIOD 1.2E-6        /*Define the input signal frequency and clock
pulses*/
CLOCK PHI1 1 (0 0.5);
CLOCK PHI2 1 (0.5 1);
END;


CIRCUIT
C (2 3) 5E-13;
C0 (4 5) 5E-12;
CP1 (2 0) 5E-14;
VIN (1 0);
S1 (1 2) PHI1;
S2 (2 4) PHI2;
S3 (1 3) PHI2;
S4 (3 0) PHI1;
S5 (5 6) PHI2;
E1 (5 0 0 4) 1E9;        /*Define the op-amp to be an ideal VCCS*/
END;


ANALYZE SSS;        /*Select the analysis package in SWITCAP*/
INFREQ 1 6E5 LIN 100;        /*Sweep from 1 Hz to 600 kHz*/
SET VIN AC 1.0 0.0;
SAMPLE INPUT HOLD 1 3/8+;        /*Sampling the input when Φ1 → 1*/
SAMPLE OUTPUT IMPULSE 1 5/8+;        /*Sampling the output when Φ2 →
1*/
```

```
PRINT VDB(6) VP(6);
END;
END;

% MATLAB analysis after SWITCAP simulations %
CLOSE ALL;
LOAD BILINEAR.dat       %Load the response data (with parasitic
capacitance)%
FF = BILINEAR (:,1);
VDB = BILINEAR (:, 2);
VP = BILINEAR (:, 3);
NFF = FF.*(2.4E-6);
FIGURE (A1);
SUBPLOT (2,1,1);
PLOT (NFF,VDB,'*R');
TITLE ('The Frequency Response from 1 Hz to 500 kHz');
AXIS ([0,1,-100,100]);
GRID ON;
SUBPLOT (2,1,2);
PLOT (NFF, VP, 'B+');
HOLD ON;
FREQZ ([-.2], [1,-1], ((2.4E-6)*PI:0.01*PI:PI)); %Compare magnitudes and
phases%
```

References

[1] R. W. Brodersen, P. R. Gray, and D. A. Hodges, "MOS Switched-capacitor filters," *Proceedings of IEEE*, Vol. 67, pp. 61–75, January 1979.

[2] J. A. McKinney and C. A. Halijak, "The periodically reverse-switched capacitor," *IEEE Trans. on Circuits & Systems*, Vol. CT-15, No. 3, pp. 288–290, September 1968.

[3] G. C. Temes, "The derivation of switched-capacitor filters from active RC prototypes," *Electronics Letters*, Vol. 14, No. 3, pp. 361–362, June 1978.

[4] G. C. Temes, H. J. Orchard, and M. Jahanbegloo, "Switched-capacitor filter design using the bilinear *z*-transform," *IEEE Trans. on Circuits & Systems*, Vol. CAS-25, No. 12, pp. 1039–1044, December 1978.

[5] J. L. McCreary, "Matching properties, and voltage and temperature dependence of MOS capacitors," *IEEE Journal of Solid-State Circuits*, Vol. SC-16, No. 6, pp. 608–616, December 1981.

[6] D. Fried, "Analog sample-data filters," *IEEE Journal of Solid-State Circuits*, Vol. SC-7, No. 4, pp. 302–304, August 1972.

[7] G. C. Temes and I. A. Young, "An improved switched-capacitor integrator," *Electronics Letters*, Vol. 14, No. 9, pp. 287–288, April 1978.

[8] G. M. Jacobs, D. J. Allstot, R. W. Brodersen, and P. R. Gray, "MOS switched-capacitor ladder filters," *IEEE Trans. on Circuits & Systems*, Vol. CAS-25, No. 12, pp. 1014–1021, December 1978.

[9] K. Martin, "Improved circuits for the realization of switched capacitor filters," *IEEE Trans. on Circuits & Systems*, Vol. CAS-27, No. 4, pp. 237–244, April 1980.

[10] B. J. Hosticka and G. S. Moschytz, "Practical design of switched capacitor networks for integrated circuit implementation," *IEE Electronics Circuits and Systems*, Vol. 3, pp. 76–88, March 1979.

[11] E. A. Vittoz, "The design of high-performance analog circuits on digital CMOS chips," *IEEE Journal of Solid-State Circuits*, Vol. SC-20, No. 3, pp. 657–665, June 1985.

[12] A. S. Sedra and P. O. Brackett, *Filter theory and design: Active and passive*, Matrix Publishers, Beaverton, OR, 1978.

[13] P. E. Fleischer and K. R. Laker, "A family of active switched capacitor biquad building blocks," *The Bell Systems Technical Journal*, No. 58, pp. 2235–2269, April 1979.

[14] S. Signell, "On selectivity properties of discrete-time linear networks," *IEEE Trans. on Circuits & Systems*, Vol. CAS-31, No. 3, pp. 275–280, March 1984.

[15] W-H. Ki and G. C. Temes, "Optimal capacitance assignment of switched-capacitor biquads," *IEEE Trans. on Circuits & Systems I: Fundamental Theory and Applications*, Vol. CAS-42, No. 6, pp. 334–342, June 1995.

[16] D. J. Allstot, R. W. Brodersen, and P. R. Gray, "An electrically-programmable switched capacitor filter," *IEEE Journal of Solid-State Circuits*, Vol. SC-14, No. 6, pp. 1034–1041, December 1979.

[17] D. B. Cox, L. T. Lin, R. Florek, and H. F. Tseng, "A real-time programmable switched capacitor filter," *IEEE Journal of Solid-State Circuits*, Vol. SC-15, No. 6, pp. 972–977, December 1980.

[18] S. Signell, "A bilinear switched capacitor bandpass filter with independent selectivity and center frequency adjustments," *IEEE International Symposium on Circuits and Systems*, pp. 802–804, Montreal, Canada, May 1984.

[19] P. E. Allen and E. Sanchez-Sinencio, *Switched capacitor circuits*, Van Nostrand, New York, 1984.

[20] R. Gregorian and G. C. Temes, *Analog MOS integrated circuits for signal processing*, John Wiley & Sons, New York, 1986.

[21] R. Schauman, M. Ghausi, and K. Laker, *Design of analog filters passive, active RC, and switched capacitor*, Prentice Hall, Englewood Cliffs, NJ, 1990.

[22] L. T. Bruton, "Low sensitivity digital ladder filters," *IEEE Trans. on Circuits & Systems*, Vol. CAS-22, No. 3, pp. 168–176, March 1975.

[23] SWITCAP2 (v1.1) User manual. [Online]. Available at www.cisl.Columbia.edu/projects/switcap.

[24] A. Knob, "Novel stray-insensitive switched-capacitor integrator realizing the bilinear *z*-transform," *Electronics Letters*, Vol. 16, No. 5, pp. 173–174, February 1980.

[25] P. E. Fleischer, A. Ganesan, and K. Laker, "Parasitic compensated switched capacitor circuits," *Electronics Letters*, Vol. 17, No. 14, pp. 929–931, 1981.

[26] S. Eriksson and H. Akhlaghi, "Noninverting parasitic-compensated bilinear SC integrator with only one amplifier," *Electronics Letters*, Vol. 19, No. 2, pp. 450–452, April 1983.

[27] S. R. Norsworthy, R. Schreier, and G. C. Temes, *Delta-sigma data converters: Theory, design and simulation*, IEEE Press, New York, 1997.

[28] T. Brooks et al., "A cascased sigma-delta pipeline A/D converter with 1.25 MHz signal bandwidth and 89 dB SNR," *IEEE Journal of Solid-State Circuits*, Vol. 32, No. 12, pp. 1896–1906, December 1997.

[29] D. B. Ribner et al., "A third-order multistage sigma-delta modulator with reduced sensitivity to nonidealities," *IEEE Journal of Solid-State Circuits*, Vol. 26, No. 12, pp. 1764–1774, December 1991.

[30] F. Maloberti et al., "Bilinear design of fully differential switched-capacitor ladder filters," *IEE Proceeding Electronic Circuits & Systems*, No. 132, pp. 266–272, 1985.

[31] D. Senderowicz et al., "Low-voltage double-sampled ΣΔ converters," *IEEE Journal of Solid-State Circuits*, Vol. 32, No. 12, pp. 1907–1912, December 1997.

[32] K. Nagaraj, "A parasitic-insensitive area-efficient approach to realizing very large time constants in switched-capacitor circuits," *IEEE Trans. on Circuits & Systems*, Vol. CAS-36, No. 9, pp. 1210–1216, September 1989.

[33] W-H. Ki and G. C. Temes, "Area-efficient gain- and offset-compensated very-large-time-constant SC biquads," *IEEE International Symposium on Circuits and Systems*, No. 3, pp. 1187–1190, San Diego, CA, May 1992.

[34] S. Tewksbury et al., "Terminology related to the performance of S/H, A/D, and D/A circuits," *IEEE Trans. on Circuits & Systems*, Vol. CAS-25, No. 7, pp. 419–426, July 1978.

[35] M. Shinagawa et al., "Jitter analysis of high speed sampling systems," *IEEE Journal of Solid-State Circuits*, Vol. 25, No. 1, pp. 220–224, February 1990.

[36] J. Shieh, M. Patil, and B. Sheu, "Measurement and analysis of charge injection in MOS analog switches," *IEEE Journal of Solid-State Circuits*, Vol. 22, No. 2, pp. 277–281, April 1987.

[37] R. Geiger, P. Allen, and N. Strader, *VLSI: Design techniques for analog and digital circuits*, McGraw-Hill, New York, 1990.

[38] R. van de Plassche, *Integrated analog-to-digital and digital-to-analog data converters*, Kluwer Academic Publisher, Berlin, Germany, 1994.

[39] T. Kuyel, "Method and system for measuring jitter," U.S. Patent 6640193. 2003.

[40] A. Zanchi et al., "Measurement and spice prediction of sub-picosecond clock jitter in A-to-D converters," *IEEE International Symposium on Circuits and Systems*, Vol. 4, pp. 557–560, Bangkok, Thailand, May 2003.

[41] B. Razavi, *Principles of data conversion system design*, IEEE Press, Piscataway, NJ, 1995.

[42] K. Hadidi et al., "An open-loop full CMOS 103MHz −61 dB THD S&H circuit," *Proceedings of the IEEE Custom Integrated Circuits Conference*, Vol. 3, pp. 381–383, May 1998.

[43] A. Boni et al., "A 10-b 185-MS/s track-and-hold in 0.35 μm CMOS," *IEEE Journal of Solid-State Circuits*, Vol. 36, No. 2, pp. 195–203, February 2001.

[44] A. Abo and P. R. Gray, "A 1.5-V 10-bit 14.3-MS/s CMOS pipeline analog-to-digital converter," *IEEE Journal of Solid-State Circuits*, Vol. 34, No. 5, pp. 599–606, May 1999.

[45] B. Pregardier et al., "A 1.2-GS/s silicon bipolar track & hold IC," *IEEE Journal of Solid-State Circuits*, Vol. 31, No. 9, pp. 1336–1339, September 1996.

[46] T. Baumheinrich et al., "A 1-GSample/s 10-b full Nyquist silicon bipolar track and hold IC," *IEEE Journal of Solid-State Circuits*, Vol. 32, No. 12, pp. 1951–1959, December 1997.

[47] M. Nayebi and B. A. Wooley, "A 10-bit video BiCMOS track-and-hold amplifier," *IEEE Journal of Solid-State Circuits*, Vol. 24, No. 12, pp. 1507–1516, December 1989.

[48] F. Wang and G. C. Temes, "A fast offset-free S/H circuit," *IEEE Journal of Solid-State Circuits*, Vol. SC-23, No. 5, pp. 1270–1272, October 1988.

[49] P. J. Lim and B. A. Wooley, "A high-speed sample-and-hold technique using a Miller hold capacitance," *IEEE Journal of Solid-State Circuits*, Vol. 26, No. 4, pp. 643–651, April 1991.

[50] G. C. Temes et al., "A high-frequency T/H stage with offset and gain compensation," *IEEE Trans. on Circuits & Systems*, Vol. CAS-42, No. 8, pp. 559–561, August 1995.

[51] W. Yang et al., "A 3-V 340-mW 14-b 75-Msample/s CMOS ADC with 85-dB SFDR at Nyquist input," *IEEE Journal of Solid-State Circuits*, Vol. 36, No. 12, pp. 1931–1936, December 2001.

[52] J. Crols and M. Steyaert, "Switched-opamp: an approach to realize full CMOS switched-capacitor circuits at very low power supply voltages," *IEEE Journal of Solid-State Circuits*, Vol. 29, No. 8, August 1994.

[53] R. Gregorian and W. E. Nicholson, "Switched-capacitor decimation and interpolation circuits," *IEEE Trans. on Circuits & Systems*, Vol. CAS-27, No. 6, pp. 509–514, June 1980.

[54] D. Senderowicz et al., "PCM Telephony: Reduced architecture for a D/A converter and filter combination," *IEEE Journal of Solid-State Circuits*, Vol. 25, No. 8, pp. 987–995, August 1990.

[55] R. Adams et al., "A 113 dB SNR oversampling DAC with segmented noise-shaped scrambling," *IEEE Journal of Solid-State Circuits*, Vol. 33, No. 12, pp. 1871–1878, December 1998.

[56] T.-H. Hsu, *Improved design techniques for switched-capacitor ladder filters*, Ph.D. dissertation, UCLA, Los Angles, CA, 1982.

[57] R. Unbehauen and A. Cichocki, *MOS switched-capacitor and continous-time integrated circuits and systems*, Springer-Verlag, Berlin, Germany, 1989.

[58] Seng-Pan U., R. P. Martins, and J. E. Franca, "Improved switched-capacitor interpolators with reduced sample-and-hold effects," *IEEE Trans. on Circuits & Systems II*, Vol. 47, No. 8, pp. 665–684, August 2000.

[59] D. von Grunigen et al., "Integrated switched-capacitor low-pass filter with combined anti-aliasing decimation filter for low frequencies," *IEEE Journal of Solid-State Circuits*, Vol. SC–17, No. 6, pp. 1024–1029, December 1982.

[60] J. E. Franca and R. P. Martins, "IIR switched-capacitor decimator building blocks with optimum implementation," *IEEE Trans. on Circuits and Systems*, Vol. 37, No. 1, pp. 81–90, January 1990.

[61] S. Lindfors et al., "A 3-V 230-MHz CMOS decimation subsampler," *IEEE Trans. on Circuits and Systems II*, Vol. 50, No. 3, pp. 105–117, March 2003.

[62] S. Mason and H. Zimmermann, *Electronic circuits, signals and systems*, John Wiley & Sons, New York, 1960.

CHAPTER 4

Switched-Capacitor Filters

4.1 Introduction

In the 1970s, a surging demand for high-quality monolithic MOSFET active filters in the fields of voice/data communications and instrumentations stimulated tremendous research-and-development (R&D) efforts of switched-capacitor filters (SCF) [1][2][3]. Figure 4.1 shows a generic block diagram of an SCF-based sampled-data filtering system.

As shown in the diagram, the analog input signal is preprocessed by an antialiasing filter (AAF) to eliminate the unwanted signals located beyond half the clock sampling frequency. The input sample-and-hold (S&H) stage samples the analog input signal and sends a sampled-data signal to the subsequent SCF. Depending on the application, an SC decimator may be incorporated into the input S&H stage so that the signal frequency can be reduced from Mf to f ($M > 1$). The output from the SCF is sent to a second S&H stage, which is typically built from a sampled-data-to-continuous-time voltage buffer (see Chapter 5). In some cases, an interpolator is employed in this stage to upsample the signal from f to Nf ($N > 1$). The last stage is a reconstruction filter that is typically used for smoothing the output waveform.

Chapter Outline

This chapter is organized as follows. Section 4.2 describes the fundamental aspects of first-order and second-order (biquad) active SC filters (SCF). Section 4.3 discusses the design principles of high-order SCFs. A step-by-step design example of a sixth-order elliptical low-pass SCF is provided in this section (the relevant computer simulation source codes are included in Appendix 4.1). Finally, Section 4.4 includes a brief introduction to high-frequency CMOS SCFs.

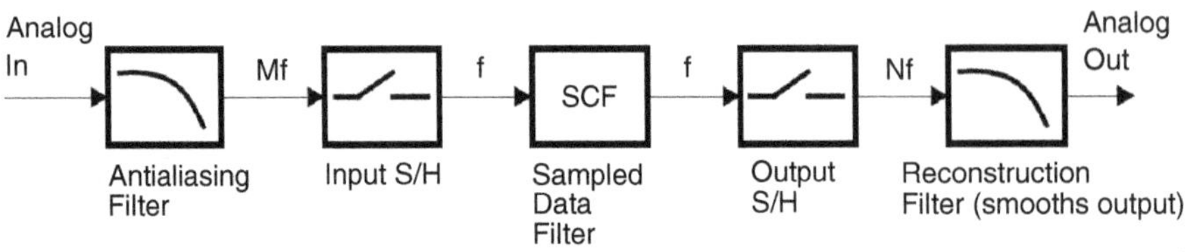

Figure 4.1 Block diagram of a sampled-data SC filtering system.

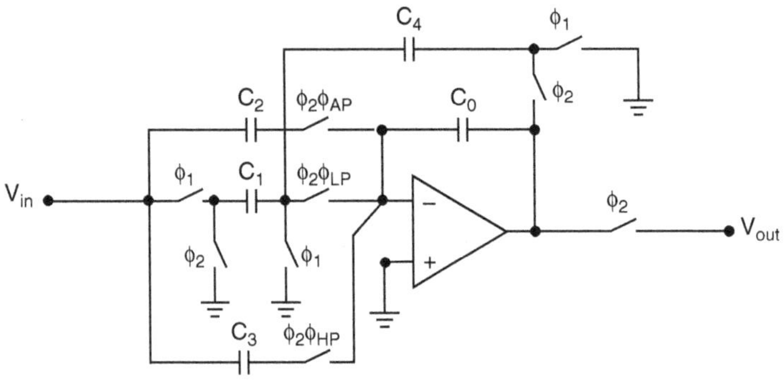

Figure 4.2 First-order SC filter.

4.2 Low-Order Switched-Capacitor Filters

First-Order SC Filters

A generic first-order active SCF is shown in Figure 4.2. This is a *three-in-one* filtering circuit that incorporates three different types of SCF: low-pass, all-pass, and high-pass. Note that there are three specially labeled switches. For instance, the switch labeled $\Phi_2\Phi_{LP}$ will be turned on when both Φ_2 *and* Φ_{LP} go to 1. Here, *LP*, *AP*, and *HP* stand for low-pass, all-pass, and high-pass, respectively.

As the clocking scheme for this filter, Φ_{AP} takes up a portion of Φ_{LP}, whereas Φ_{HP} is not overlapping with Φ_{LP}. In other words, when the circuit is employed to realize a first-order all-pass filter, both C_1 and C_2 branches are activated. By contrast, when the circuit is used as a first-order high-pass filter, only the C_3 branch is activated. And when the circuit is used as a first-order low-pass filter, only the C_1 branch is activated. Utilizing the signal-flow-graph (SFG) technique introduced in Chapter 3, we can find these three filters' transfer functions, which are given by

$$\begin{cases} H_{lp}(z) = \dfrac{\dfrac{C_1}{C_0 + C_4}}{z - \dfrac{C_0}{C_0 + C_4}}, (low\text{-}pass) \\ H_{ap}(z) = \dfrac{-\dfrac{C_2}{C_0 + C_4}\left(z - \dfrac{C_1 + C_2}{C_2}\right)}{z - \dfrac{C_0}{C_0 + C_4}}, (all\text{-}pass) \\ H_{hp}(z) = \dfrac{-\dfrac{C_3}{C_0 + C_4}(z-1)}{z - \dfrac{C_0}{C_0 + C_4}}, (high\text{-}pass) \end{cases} \tag{4.1}$$

Note that these three functions have identical denominators, which implies that they can be transformed from the same generic first-order transfer function. Specifically, consider a generic first-order sampled-data transfer function given by the following expression:

$$H(z) = \frac{c_1 z \pm c_0}{z - d_0} \tag{4.2}$$

First, to realize a first-order low-pass filter using the circuit in Figure 4.2, the capacitors should have the following relationships:

$$C_1 = \frac{c_0}{d_0} C_0, \quad \text{and} \quad C_4 = \frac{1 - d_0}{d_0} C_0 \tag{4.3}$$

Second, to realize a first-order all-pass filter, the capacitors in the circuit should have the following relationships:

$$C_1 = C_4 = \frac{1 - d_0}{d_0} C_0, \quad \text{and} \quad C_2 = C_0 \; (\mathit{iff}\; c_1 \equiv c_0 d_0) \tag{4.4}$$

Finally, to realize a first-order high-pass filter, the capacitors in the circuit should relate to one another in the following way:

$$C_3 = \frac{c_1}{d_0} C_0, \quad \text{and} \quad C_4 = \frac{1 - d_0}{d_0} C_0 \; (\mathit{iff}\; c_1 \equiv c_0) \tag{4.5}$$

Second-Order SC Filters

A second-order SC filter (or *biquad*) may be realized using one of many approaches, depending on the application's requirements and the arrangement of the switches. Recall that in Chapter 3 we investigated an SC biquad, which is shown in Figure

3.12 [3][4]. That biquad is also a three-in-one filtering system capable of realizing second-order low-pass, (all-pole) band-pass, and high-pass filters. Specifically, the input should be sent through the M_1C_1, M_2C_2, and M_3C_2 signal paths into the core circuitry when realizing low-pass, band-pass, and high-pass filters, respectively. However, as mentioned in Chapter 3, the biquad shown in Figure 3.12 is not suitable to realize high-Q filters because when Q is high ($Q > 1$), its capacitance spread approximates ($Q/\omega_0 T$), which is quite large since $\omega_0 T \ll 1$.

A well-known SC biquad appropriate for high-Q filtering applications is shown in Figure 4.3 [5]. Similar to the low-Q case, the input signal is passed through the M_1C_1, M_2C_1, and M_3C_2 signal paths to realize low-pass, band-pass, and high-pass filters, respectively.

Following the same process introduced in Chapter 3 (i.e., draw the signal flow graph and then apply the Mason's rule), we can obtain the biquad's transfer function, which is given by

$$H(z)=\frac{V_{out}(z)}{V_{in}(z)}=-\frac{M_3z^2+(M_1M_5+M_2M_5-2M_3)z+M_3-M_2M_5}{z^2+(M_4M_5+M_5M_6-2)z+1-M_5M_6} \tag{4.6}$$

The overall dc gain of this filter ($z = 1$) is then given by

$$H(z)|_{z=1}=-\frac{M_1M_5}{M_4M_5}=-\frac{M_1}{M_4} \tag{4.7}$$

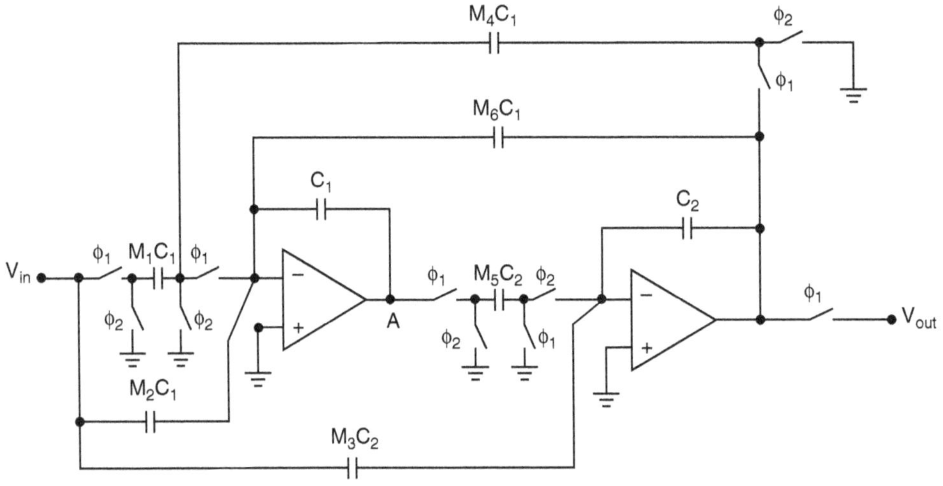

Figure 4.3 SC biquad for high-Q applications.

Interestingly, this dc gain is identical to that given by Equation (3.88). In addition, it can be found that the signal at node A is expressed as

$$V_A(z) = -\frac{M_1 + M_2(1-z^{-1}) + H(z)\cdot[M_4 + M_6(1-z^{-1})]}{1-z^{-1}} \cdot V_{in}(z) \tag{4.8}$$

Similar to what we saw in Chapter 3, it can be proved that the dc signal level at node A is independent of $H(z)$, hence there is some freedom in deciding the values of M_1, M_4, M_5, and M_6. By equating the corresponding system coefficients in Equations (3.24) and (4.6), we can write the following:

$$d_1 = M_4M_5 + M_5M_6 - 2, \text{ and } d_2 = 1 - M_5M_6 \tag{4.9}$$

Utilizing the freedom of choosing the values of M_4, M_5, and M_6, we obtain the following approximations based on Equations (3.27) and (3.28) (assuming $\omega_0 T \ll 1$)

$$M_4 \cong M_5 \cong \omega_0 T,\ M_6 \cong \frac{1}{Q}, \text{ and } Q \cong \frac{\sqrt{M_4M_5}}{M_5M_6} \tag{4.10}$$

For standard high-Q SC applications ($1 < Q < 2\pi$), Q is typically smaller than $(\omega_0 T)^{-1}$ in the presence of a high sampling clock frequency (i.e., $f_{clk} > 40f_0$, or, equivalently, $\omega_0 T < 0.05\pi$). Thus, the smallest capacitance in the circuit is equal to $(\omega_0 T)$, and the resultant capacitance spread is equal to $(\omega_0 T)^{-1}$. However, for applications of higher Q ($Q > 2\pi$) and larger signal bandwidth ($4f_0 < f_{clk} < 40f_0$), Q is usually larger than $(\omega_0 T)^{-1}$, and as a result, the capacitance spread is equal to Q.

By contrast, for low-Q SC applications ($Q < 1$), the largest capacitance in the circuit is equal to $(Q)^{-1}$, hence the capacitance spread is equal to $(\omega_0 TQ)^{-1}$, which is larger than both $(\omega_0 T)^{-1}$ and Q (assuming $\omega_0 T \ll 1$). Thus, the biquad shown in Figure 4.3 is more suitable for high-Q than for low-Q applications.

There are many other SC biquad realizations beyond the two mentioned in this section. For a fruitful analysis of biquad construction, the reader is referred to Section 8-7-2 of Laker and Sansen [6]. In addition, fully differential configurations are usually adopted in practice to optimize the common noise rejection performance. A comprehensive comparison between different fully differential SC biquads can be found in [7].

Area-Efficient High-*Q* SC Filters

The aforementioned development indicates that the capacitance spread can be large, particularly in a realized high-Q SC filter. It is well known that a custom capacitor in the modern integrated circuit is typically built from a number of equal-sized unit capacitors or so-called unit elements. To realize the desired capacitance with adequate accuracy, the size of each unit element should not be too small, otherwise its desired value would be submerged by normal fabrication tolerances. In such a case, a large capacitance ratio essentially means a large total capacitance, which results in large chip area and high power consumption.

A number of practical *area-efficient* approaches are capable of reducing the capacitance spread of a high-Q SC filter. Among them, the *T-network* approach [8] is the simplest. Figure 4.4 shows the procedure of replacing the smallest capacitor in the circuit, namely C, with a T-shape capacitor network consisting of C_1, C_2, and C_3. Here, the basic idea is to make use of a T-network to simulate a very small capacitor without actually implementing it, thereby considerably reducing the physical capacitance spread. It can be proved based on the charge reservation principle that for both of the circuits shown in Figure 4.4, the equivalent capacitance realized by the T-network is given by

$$C_{equ} = \frac{C_1 C_2}{C_1 + C_2 + C_3} \tag{4.11}$$

For example, if it is assumed that $C_1 = C_2 = C$ and $C_3 = 6C$, then from Equation (4.11) we can find that the equivalent smallest capacitance is equal to $0.125C$. If the original capacitance spread is assumed to be 64 (i.e., the maximum capacitance is equal to $8C$), then the new capacitance spread after applying the T-network scheme is the result of dividing $8C$ by C (instead of $0.125C$), which is equal to 8.

Generally speaking, if we assume that the original capacitance spread is A, then the new capacitance spread after applying the T-network scheme is given by

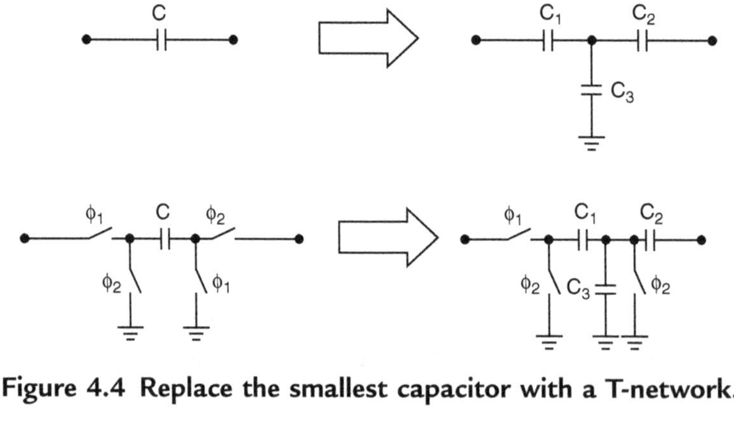

Figure 4.4 Replace the smallest capacitor with a T-network.

$$A_{new} = A \cdot \frac{C_1 C_2}{(C_1 + C_2 + C_3)C} \tag{4.12}$$

Next, let us look into this feature from another perspective. If we assume that the initial input and feedback capacitors of an SC amplifier are equal to C and $64C$, respectively, then the dc gain of the amplifier is equal to 1/64. After replacing the input capacitor C with the T-network, the resultant equivalent input capacitance is now $0.125C$. To maintain the same amplifier gain of 1/64, the feedback capacitor must be changed to $8C$. Thus, the total capacitance and hence the silicon area is drastically decreased, and the *physical* capacitance spread is reduced from 64 to 8.

However, note that the T-network is sensitive to the parasitic capacitance between the top-plate of C_3 and ground, thereby requiring careful layout to reduce the parasitic effect. In addition, there is a limit on the achievable factor of capacitance spread reduction. For instance, if we assume $A = 64$, $C_1 = C_2 = C$, and $C_3 = 14C$, then from Equation (4.12) we calculate that the new capacitance spread would be equal to 4, but this is not true because $C_3 = 14C$ and the capacitance spread is indeed equal to 14. It can be shown that theoretically the lowest possible capacitance spread after being modified by the T-network is equal to the square root of that before the modification [8].

Huang [9] proposed an alternative parasitic-insensitive and what's more, a more effective approach. The key idea of this approach is to make use of the filter's idle phase (in this case when $\Phi_2 = 1$) to reduce the capacitance spread.

Consider the first integrator of the high-Q biquad shown in Figure 4.3. Modify it using Huang's approach, and the resultant schematic is shown in Figure 4.5. The

Figure 4.5 Modified first integrator in the SC biquad.

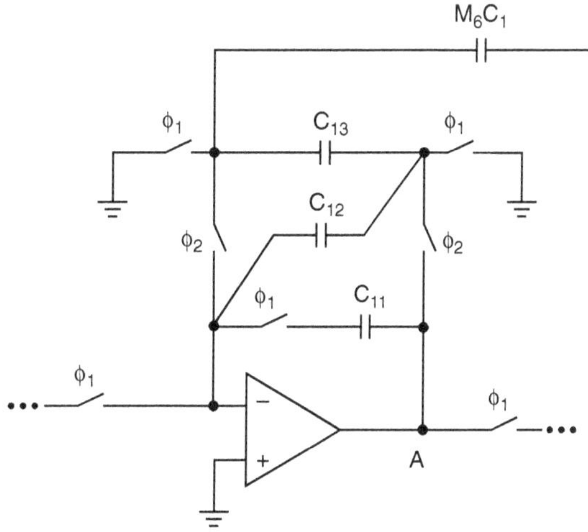

schematic shows that the integrating capacitor of the first integrator, C_1, is split into three capacitors, namely C_{11}, C_{12}, and C_{13}. Let us assume that the input signal comes into the circuit through the M_1C_1 path—that is, the SC circuit realizes a low-pass filter. During the idle case ($\Phi_2 = 1$), the op-amp's output at node A is given by

$$V_A(z)|_{idle} = -\frac{M_1C_1}{C_{12}+C_{13}}V_{in}(z) \tag{4.13}$$

Next, $\Phi_1 = 1$, and the capacitor C_{13} is discharged to ground while the stored charge on C_{12} is transferred to C_{11}. Based on the charge reservation rule, we can write the following expression of $V_A(z)$:

$$V_A(z)|_{active} = -\frac{M_1C_1C_{12}}{(C_{12}+C_{13})C_{11}}\cdot\left(\frac{1}{z-1}\right)V_{in}(z) \equiv -M_1\left(\frac{1}{z-1}\right)V_{in}(z) \tag{4.14}$$

As shown here, to maintain the same op-amp voltage gain, the following condition must be met:

$$C_1 = \frac{(C_{12}+C_{13})C_{11}}{C_{12}} \tag{4.15}$$

One of many scenarios able to satisfy the foregoing condition is described as follows: $C_{11} = 0.125C_1$, $C_{12} = 0.125C_{11}$, and $C_{13} = 0.875C_{11}$. In such a case, the largest capacitance is equal to only $0.125C_1$ rather than C_1; thus, both the total capacitance and capacitance spread are reduced. Specifically, the capacitance spread is reduced by a factor of 8. And it can be shown that this first-integrator is insensitive to op-amp's offset and finite gain errors since it adopts the compensation design method reported in [10] (see also Chapter 7). Furthermore, it is of practical interest to find that this approach can reduce the capacitance spread to less than the square root of its original value (as compared to the T-network case). The proof is left as an exercise for the reader.

These area-efficient approaches are among the most widely adopted in practice. There may be as many capacitance spread reduction techniques as different realizations of an SC biquad. It is not the author's intention to be exhaustive on this subject, and the interested reader is referred to the literature for more information.

4.3 High-Order Switched-Capacitor Filters

Realizations of SC Filters

The realizations of SCFs can be categorized into three basic groups. The first group is called *continuous-time filter emulations*. In such cases, SC passive elements are used to replace the resistors in a classical continuous-time active RC filter (e.g., *Tow-Thomas* or *KHN biquad*). The major drawbacks include the limited number of filter types and inappropriateness for high-Q applications.

The second group is called *SC ladder filters*. The idea is to use SC circuits to simulate the low-sensitivity response of a high-Q doubly terminated passive reactance two-port, or, equivalently, an RLC ladder network for realizing a desired high-order transfer function with high-Q poles [11]. There are two different approaches to the realization of an SC ladder filter. One is called *ladder component substitution*, in which case an equivalent SC element is used to substitute each R, L, and C component of the original continuous-time ladder network. The advantage of this approach is that the original filter's low sensitivity to component variations is perfectly maintained. The *voltage-inverter-switch* (VIS) [12] is a representative of this approach, but it has been disregarded by circuit designers due to its inherent drawbacks such as *not* being fully insensitive to the top-plate parasitic of the capacitors, and requiring more than two clock phases to operate. Although a modified VIS-SCF requiring only two-phase clocking was proposed in [13], the yield is still too low for economical production in the existing CMOS technologies.

The other approach is called the *SFG-based ladder realization*. The basic idea is to convert the s-domain transfer function of the RLC ladder filter into its z-domain counterpart using either LDI (i.e., *approximate design*) or bilinear (i.e., *exact design*) s-to-z transformations, depending on the accuracy requirements and the ratio of clock frequency over signal bandwidth (i.e., f_{clk}/f_0). Once the conversion is decided, a signal flow graph (SFG) consisting of integrators, delays, and summing stages is developed from the resultant z-domain transfer function, and the SCF circuitry is implemented based on the SFG. Generally speaking, it is much easier to build a parasitic-insensitive SFG-based SC ladder filter than it is to use the VIS approach since it makes use of integrators rather than voltage inverters.

The third group of SCF realizations is somewhat similar to the aforementioned SFG-based ladder filter. It is called the *cascade SCF realizations*. This type of realization also adopts a direct building block approach to determine the filter's transfer function in the z-domain. However, here the main transfer function is broken into products of first- and second-order terms. In other words, the numerator and denominator of a high-order ($L \geq 3$) transfer function are factored into first- and second-order subfunctions. Each subfunction can be realized by one of the low-order (first-order or biquad) filters mentioned earlier. Since each of the low-order filters is individually buffered and capable of functioning independently, cascading them together will not affect their own transfer functions [14], thereby enabling a straightforward and easy-to-troubleshoot design. In practice, the cascade configuration is usually adopted to realize medium-to-high-order ($3 \leq L \leq 12$) SCFs that provide a medium-to-high selectivity ($1 \leq Q \leq 30$).

However, when it comes to even higher-order and more stringent selectivity requirements, the cascade SCF realizations are inappropriate because the resulting high-order transfer functions typically have high-Q poles (i.e., poles that are located very close to the unit circle in the z-plane). In effect, the value of each system coefficient, or, equivalently, capacitance ratio has a long string of digits after the decimal point, resulting in a frequency response so sensitive to the component accuracy that it is impractical to fabricate the SCF using any existing CMOS technologies.

By contrast, an SFG-based SC ladder filter is typically less sensitive to component variations because it is built from the doubly terminated LC two-port circuit, whose frequency response normally has a very low sensitivity to component variations. This implies that the passband response of a realized high-order SC ladder filter is flatter (i.e., with smaller ripples) as compared to that of its cascade counterpart.

Despite its superior sensitivity performance, an SC ladder filter typically requires a more sophisticated design strategy and advanced mathematical discussions, which involve multiple-parameter sensitivity analyses and optimal signal terminations. As a result, it is more difficult to troubleshoot a high-order SC ladder filter than it is to troubleshoot a cascade SCF. Moreover, in an LDI-transformed or an approximate SC ladder filter design, many feedback capacitors with a large capacitance spread are required, limiting the clock sampling frequency and occupying a large silicon area [11][14]. Due to the space constraints of this book, we will investigate the cascade SCF approach only (through a low-pass SCF design example). For more information on the ladder approach, the reader is referred to the literature.

Biquad Ordering and Dynamic Range Scaling

As mentioned earlier, multiple SC biquads can be cascaded to realize a high-order SCF. We have seen the biquad suitable for realizing a low-Q second-order transfer function (namely, *Type-1 biquad* as shown in Figure 3.12), and the other for realizing a high-Q function (namely, *Type-2 biquad* as shown in Figure 4.3). They are widely used as the building blocks of a high-order SCF.

The next logical step is to determine the ordering of biquads when more than two biquads are required to realize an SCF (i.e., $L \geq 6$). Apparently, it is impractical to use either only low-Q or only high-Q biquads to realize the SCF, because the former cannot provide the adequate stopband attenuation, while the latter will cause the output to have a significant peaking near the corner frequency (i.e., ω_0) and the resulting SCF may not be stable (see Chapter 3).

To simplify the analysis, we assume that three biquads (namely, Q_1, Q_2, and Q_3) are needed to build a sixth-order SCF, with their quality factors differing from each other such that $Q_1 > Q_2 > Q_3$. Then we have six ordering options as follows: $Q_1 \rightarrow Q_2 \rightarrow Q_3$, $Q_3 \rightarrow Q_2 \rightarrow Q_1$, $Q_1 \rightarrow Q_3 \rightarrow Q_2$, $Q_2 \rightarrow Q_3 \rightarrow Q_1$, $Q_3 \rightarrow Q_1 \rightarrow Q_2$, and $Q_2 \rightarrow Q_1 \rightarrow Q_3$.

The $Q_1 \rightarrow Q_2 \rightarrow Q_3$ configuration allows the last two stages to attenuate the peaking near the corner frequency. However, it can be shown that a high-Q first stage typically results in large capacitance spread and high sensitivity because the most sensitive pole (i.e., the pole closest to the unit circle) is not *paired with* a nearby zero [15]. In comparison, the $Q_3 \rightarrow Q_2 \rightarrow Q_1$ configuration generally results in a smoother passband with smaller ripples, but the high-Q final stage still causes some peaking near the corner frequency since the pole-zero pairing operation is not optimized. The $Q_1 \rightarrow Q_3 \rightarrow Q_2$ and $Q_2 \rightarrow Q_3 \rightarrow Q_1$ configurations can be used to achieve similar (but not much better) capacitance spreads as compared to the aforementioned two.

In practice, the last two configurations, $Q_3 \rightarrow Q_1 \rightarrow Q_2$ and $Q_2 \rightarrow Q_1 \rightarrow Q_3$, are more widely adopted since either of them can be used to realize the cascade-type SCF of the smallest possible capacitance spread and the lowest possible sensitivity to component variations. Which one to choose is highly dependent on the application's requirements, and it is always worthwhile to test both configurations using a computer-based simulation before making the final decision.

Having determined the biquad ordering, we shall work on the *capacitance assignment*. The common practice is as follows. In each biquad stage, both op-amps' feedback capacitors (which usually have the largest capacitance values in the circuitry) are normalized to 1, and substituted into the design equations generated from the z-domain transfer function of the biquad to determine the remaining initial capacitance values. The next important step is called *dynamic range scaling*. The basic idea of dynamic range scaling is to scale the magnitude of each op-amp output voltage such that all op-amps saturate for the same input level, enabling the SCF to work with the largest possible input dynamic range.

The typical implementation of dynamic range scaling is as follows [14]. First, increase the input voltage V_{in} till the filter's output is on the edge of being saturated (not saturated yet). Calculate (usually by computer simulation) the voltage value of each internal op-amp's output at this saturation point. Then multiply all the capacitors connected to the output node of each op-amp by its output voltage value. Next, *capacitance minimization* or *scaling* is carried out. Scale all the capacitors connected to the input of each op-amp such that the smallest (nonzero) capacitor among them is normalized to 1. Start from the first op-amp in the first biquad stage, and repeat the process for all the succeeding op-amps in the system.

EXAMPLE 4.1

Assume that the low-Q biquad shown in Figure 3.12 is the last stage of a high-order cascade SCF. Also assume that initially $C_1 = C_2 = 1$, $M_2 = 0$, $M_4 = 0.1256$, $M_5 = 0.1558$, $M_6 = 0.1103$, $M_1 = 42.568$, and $M_3 = 59.965$ (the last two items are the results of the preceding stage's scaling operation). Before scaling this stage, the maximum values of the first and second op-amp output voltages are 500 and 250, respectively. Perform dynamic range scaling on the capacitance values. The desired output voltage value of each op-amp should be normalized to 1 after scaling.

Solution: We start with the first op-amp. First, let us multiply the capacitors connected to its output terminal by 500, and we get

$$C_1' = 500 \quad \text{and} \quad M_5' = 77.9 \tag{4.16}$$

Then we perform the same operation for the second op-amp, and we have

$$C_2' = 250,\ M_4' = 31.4, \quad \text{and} \quad M_6' = 27.575 \tag{4.17}$$

Next, scale all the capacitors connected to the input of each op-amp such that the smallest capacitor among them (in this case it is M_4') is normalized to 1. We can write

$$C_1'' \approx 7.962,\ M_4'' = 1, \quad \text{and} \quad M_1'' \approx 1.356 \tag{4.18}$$

For the second op-amp, we carry out the calculations in a similar manner:

$$C_2'' \approx 9.066,\ M_6'' = 1,\ M_5'' \approx 2.825, \quad \text{and} \quad M_3'' \approx 2.175 \tag{4.19}$$

The capacitance values given by Equations (4.18) and (4.19) are the scaled results. If we assume that the smallest capacitor—that is, the unit capacitor is equal to C, then the final capacitance values are obtained by multiplying these results with C. The size of C is determined by two main factors: One is the minimum-size limit due to fabrication tolerances, and the other is the noise (typically kT/C or *sampling noise*) requirement [14][16].

An interesting question with respect to dynamic range scaling is often raised: Does the operation of dynamic range scaling affect the SC filter's frequency response? As we learned in Chapter 3, in a typical SC circuit, an SC passive element is used to simulate a physical resistor. Thus, if we assume that the SCF, which undergoes dynamic range scaling, can be realized by replacing all physical resistors in a certain RC filter with SC elements, then it is easy to find that in essence the operation of dynamic range scaling in the SCF is equivalent to multiplying all resistance values by a scaling factor S and all capacitance values by $(1/S)$ in that RC filter. Therefore, theoretically speaking, as far as the overall time constant (either the true or the SC-simulated RC product) is concerned, no significant change is made, and the frequency response of the SCF should not be influenced by dynamic range scaling, except that the maximum output voltage magnitude may be altered.

Finally, although adequate to many practical SCFs' specifications, the aforementioned capacitance assignment process cannot guarantee to achieve the smallest possible total capacitance or capacitance spread. In other words, it is not the *optimal capacitance assignment.* This is due to the inherent assumption made for the preceding analysis that dynamic range scaling and capacitance minimization are independent of each other and can be performed separately [16]. The reader is referred to [16] and the literature for more information about how to fulfill the optimal capacitance assignment.

Design Example: An Elliptical Low-Pass SC Filter

This subsection details the design of a high-order low-pass SCF. Techniques such as dynamic range scaling and capacitance minimization are utilized in this example. Before proceeding, note that the goal of presenting this work is not to report an optimal result but rather to demonstrate the top-down design procedure in the form of a step-by-step tutorial.

The target specifications of the SCF are listed in Table 4.1. Based on these specifications, the filter's z-domain transfer function can be simulated and determined by using MATLAB. The source codes are included in Appendix 4.1. In such a system-level simulation, we assume that the maximum allowable passband ripple is equal to 0.1 dB, and that the stopband attenuation is at least 75 dB to provide some design margin for the circuit-level implementation. Next, let us go through the design.

To determine the filter topology, four different filter types (*Butterworth*, *Chebyshev I*, *Chebyshev II*, and *Elliptical*) are simulated based on the specifications in Table 4.1, and their simulated (ideal) frequency responses are shown in Figure 4.6 (the

Table 4.1 Target specifications of the low-pass SCF.

Parameter	Value	Units
Sampling frequency	30	MHz
DC gain in passband	0	dB
Passband	0–1.5	MHz
Ripple in passband	<0.3	dB
Stopband	3–9	MHz
Attenuation in stopband	65	dB
Minimum capacitor size	0.5	pF
Power supply voltage	3.3	V

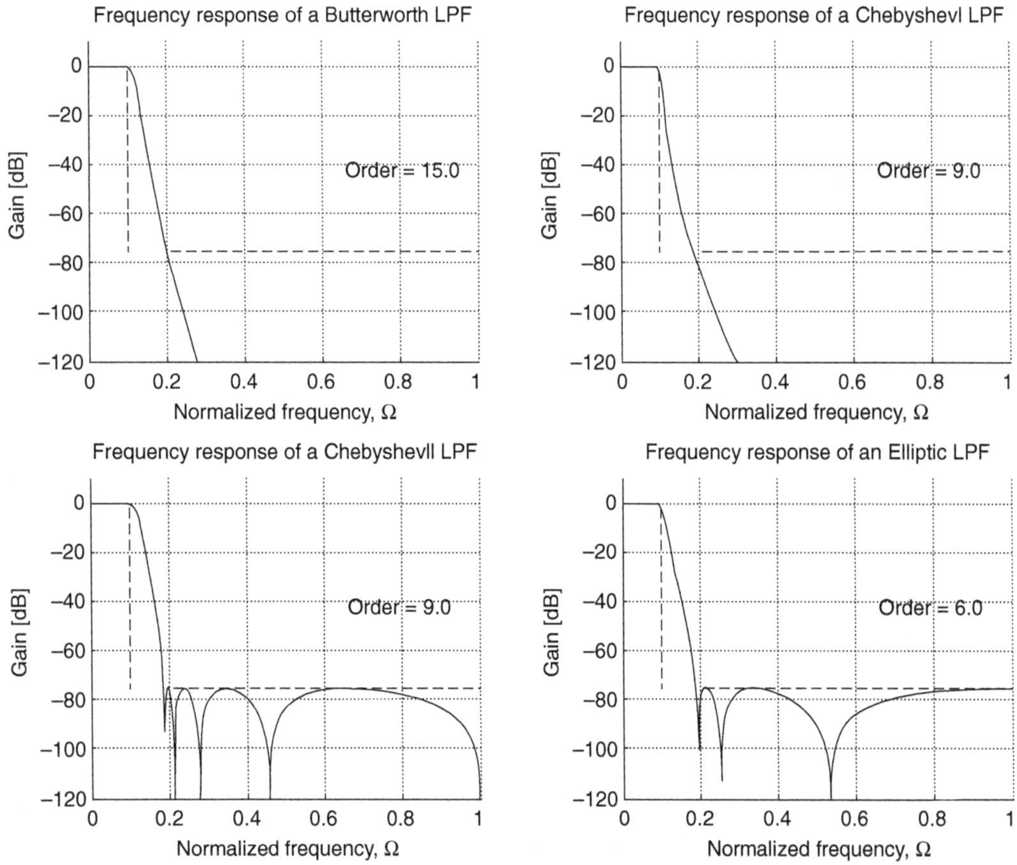

Figure 4.6 Frequency responses of four filters.

dot-and-dash lines stand for the passband and stopband requirements). As Figure 4.6 shows, the elliptical low-pass filter is the best choice out of the four, because it requires only six op-amps (i.e., it is a sixth-order filter). From computer simulation, we obtain the ideal transfer function of this sixth-order elliptical low-pass filter, which is given by

$$H(z)=\frac{G\cdot(z^2-1.63z+1)\cdot(z^2-1.394z+1)\cdot(z^2+0.2202z+1)}{(z^2-1.721z+0.7479)\cdot(z^2-1.764z+0.8316)\cdot(z^2-1.837z+0.9427)} \tag{4.20}$$

Here, the constant G is equal to 0.00038221. From simulation, we can find the zeros and poles of this transfer function. The zeros are as follows:

$$Zeros=\begin{cases}-0.1101+j0.9939\\-0.1101-j0.9939\\0.8150+j0.5795\\0.8150-j0.5795\\0.6970+j0.7171\\0.6970-j0.7171\end{cases} \tag{4.21}$$

And the poles are

$$Poles=\begin{cases}0.9185+j0.3147\\0.9185-j0.3147\\0.8820+j0.2317\\0.8820-j0.2317\\0.8605+j0.0863\\0.8605-j0.0863\end{cases} \tag{4.22}$$

From these results, we can obtain the approximate value of the highest pole-Q for each biquad. A more accurate approach of finding pole-Q can be found in [16].

The next step is to pair the poles with the nearby zeros. As a general rule of thumb, the most sensitive pole (the pole closest to the unit circle in the z-plane) should have the priority to pair with the zero closest to it, the second most sensitive pole will be the next to pair with a nearby zero, and so on [15]. Following this rule, we pair the poles and zeros, and the three biquadratic transfer functions are found as follows:

$$\begin{cases}H_1(z)=C_1\dfrac{z^2+0.2202z+1}{z^2-1.721z+0.7479},\ (Q_1\cong 8.071)\\H_2(z)=C_2\dfrac{z^2-1.63z+1}{z^2-1.837z+0.9427},\ (Q_2\cong 9.738)\\H_3(z)=C_3\dfrac{z^2-1.394z+1}{z^2-1.764z+0.8316},\ (Q_3\cong 3.311)\end{cases} \tag{4.23}$$

Note that $C_1 \cdot C_2 \cdot C_3 = G = 0.00038221$.

As mentioned earlier, experience shows that placing the higher-Q biquad in the middle and the lower-Q biquads in the first and last stages tends to result in a lower sensitivity and a lower capacitance spread. Thus, we here decide to place the high-Q $H_2(z)$ in the middle.

Now, there are two possible ways to place the remaining two biquads: place $H_1(z)$ in the first and $H_3(z)$ in the third stage or the other way around. Computer simulation shows that in this particular case, the former is a better configuration than the latter because it results in a lower capacitance spread. In fact, after dynamic range scaling, the first configuration provides a capacitance spread of 16.395, while the second provides 24.221. Therefore, in this example, the chosen biquad ordering is $H_1(z) \rightarrow H_2(z) \rightarrow H_3(z)$. The frequency response of each individual biquad is shown in Figure 4.7.

The complete SCF is shown in Figure 4.8. As the schematic shows, the first and third stages are realized based on the *Type-1 biquad* presented in Figure 3.12 (note

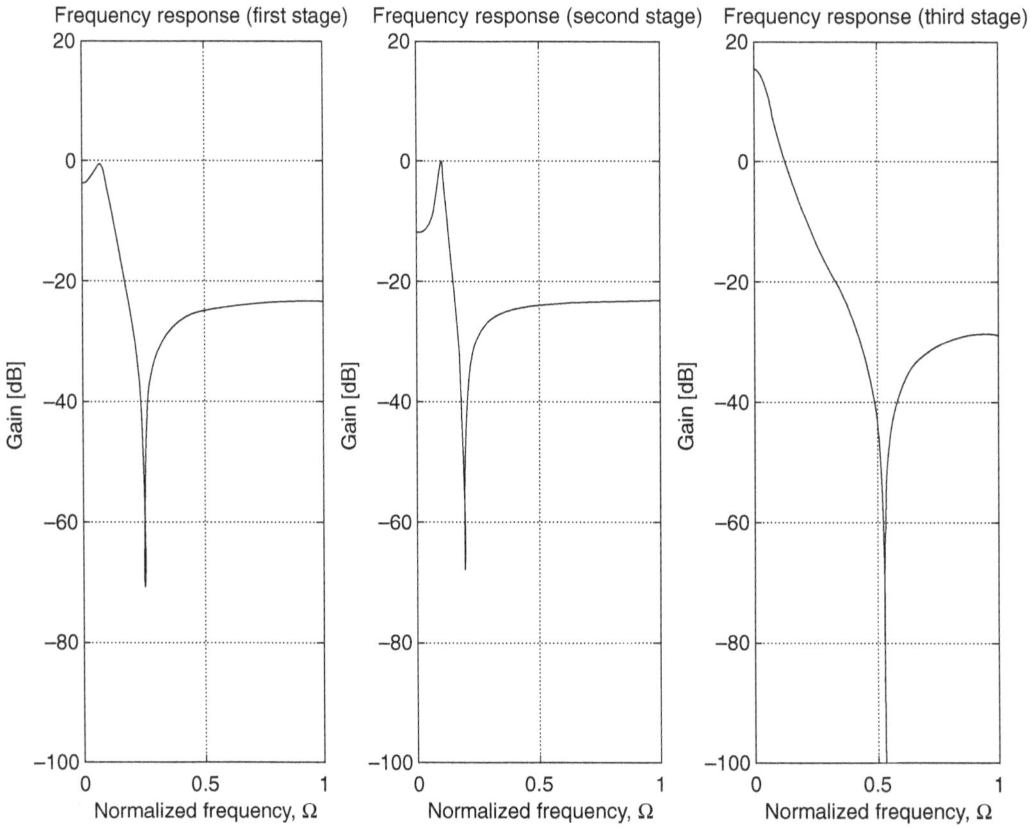

Figure 4.7 Frequency responses of three individual biquads.

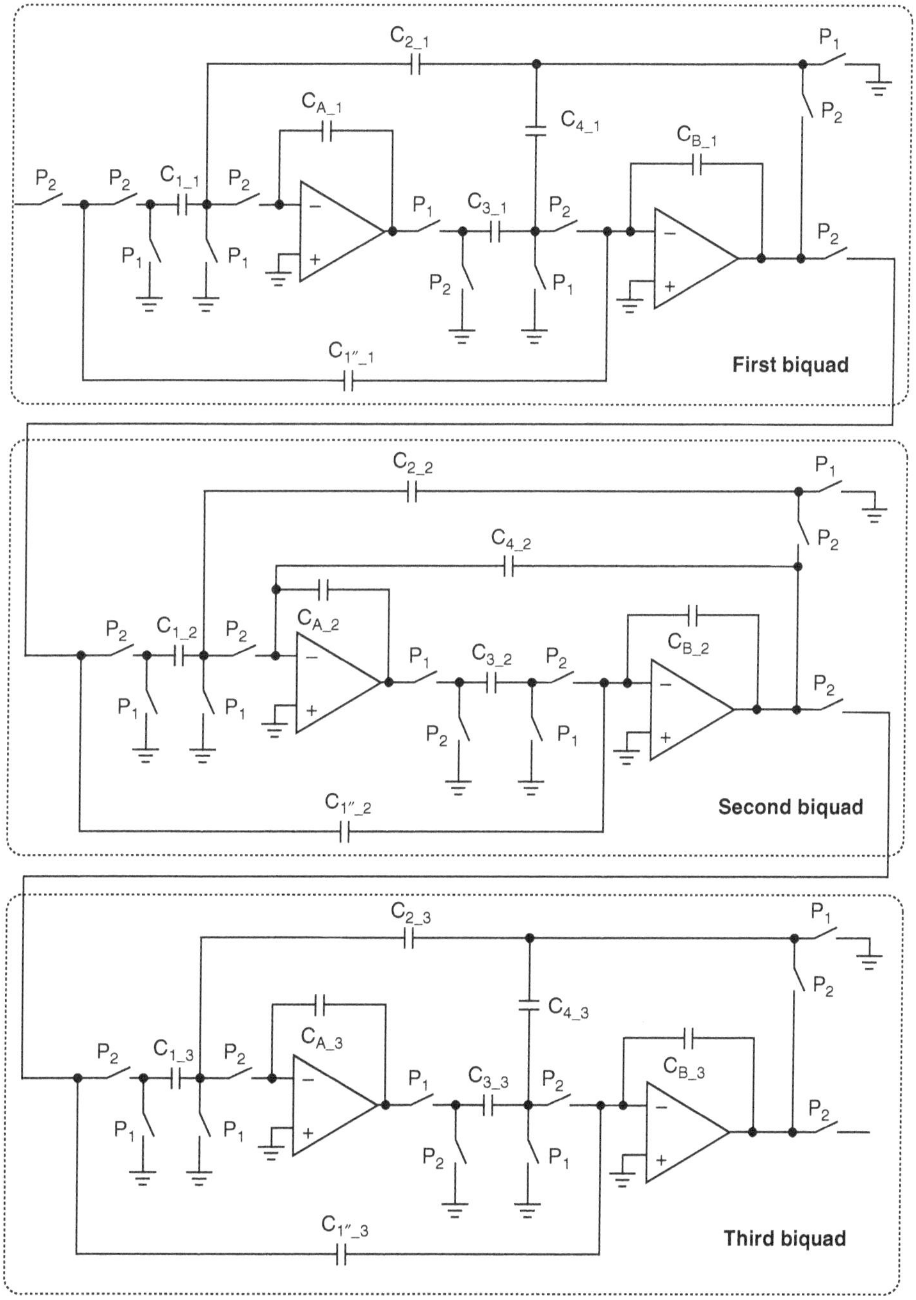

Figure 4.8 Sixth-order elliptical low-pass SCF.

that switch sharing is employed), whereas the second stage is realized based on the *Type-2 biquad* shown in Figure 4.3. Additionally, note that the filter output jumps only during the P_2 phase.

The next step is to use dynamic range scaling to scale the peak output voltage of each op-amp. Figures 4.9(a) and (b) illustrate the output of each op-amp before and after scaling, respectively. The peak op-amp output voltages before scaling shown in Figure 4.9(a) are obtained from SWITCAP (see Chapter 3) simulation based on the circuit shown in Figure 4.8, using the initial capacitance values calculated by

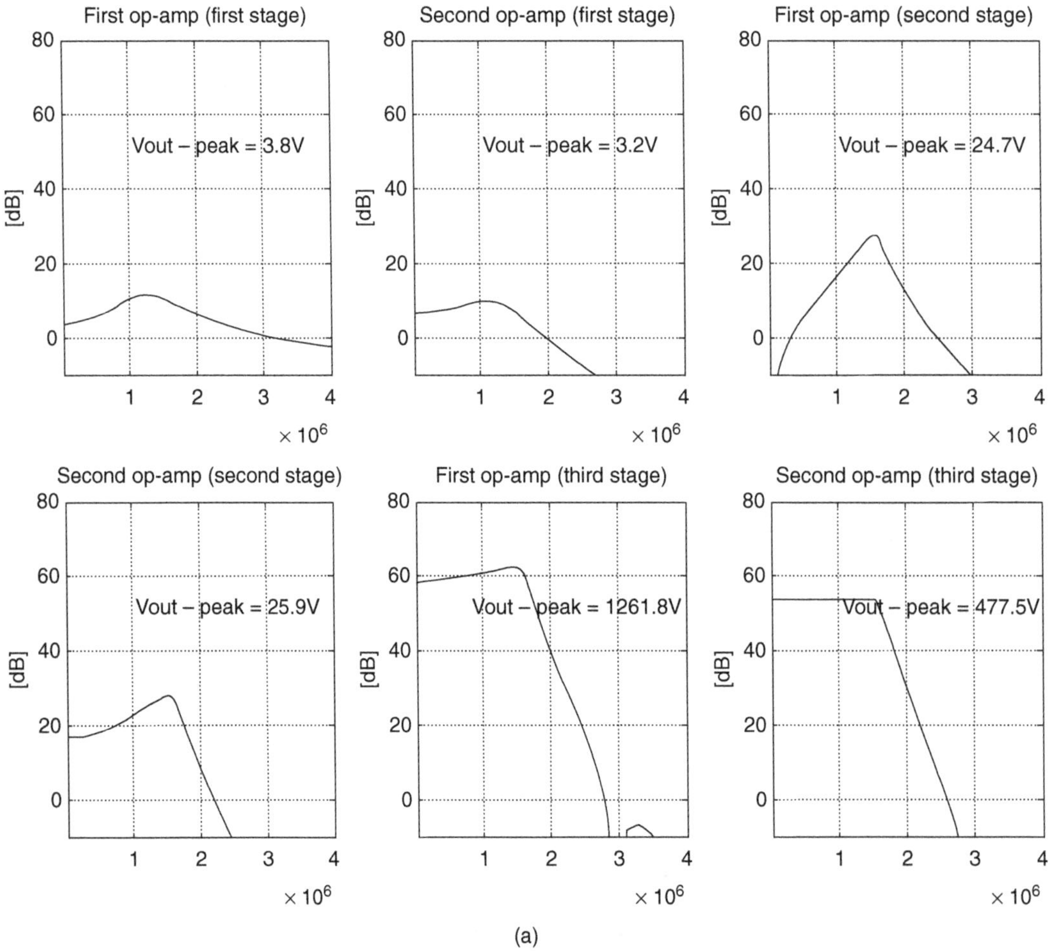

Figure 4.9 Peak op-amp output voltages (a) before scaling and (b) after scaling.

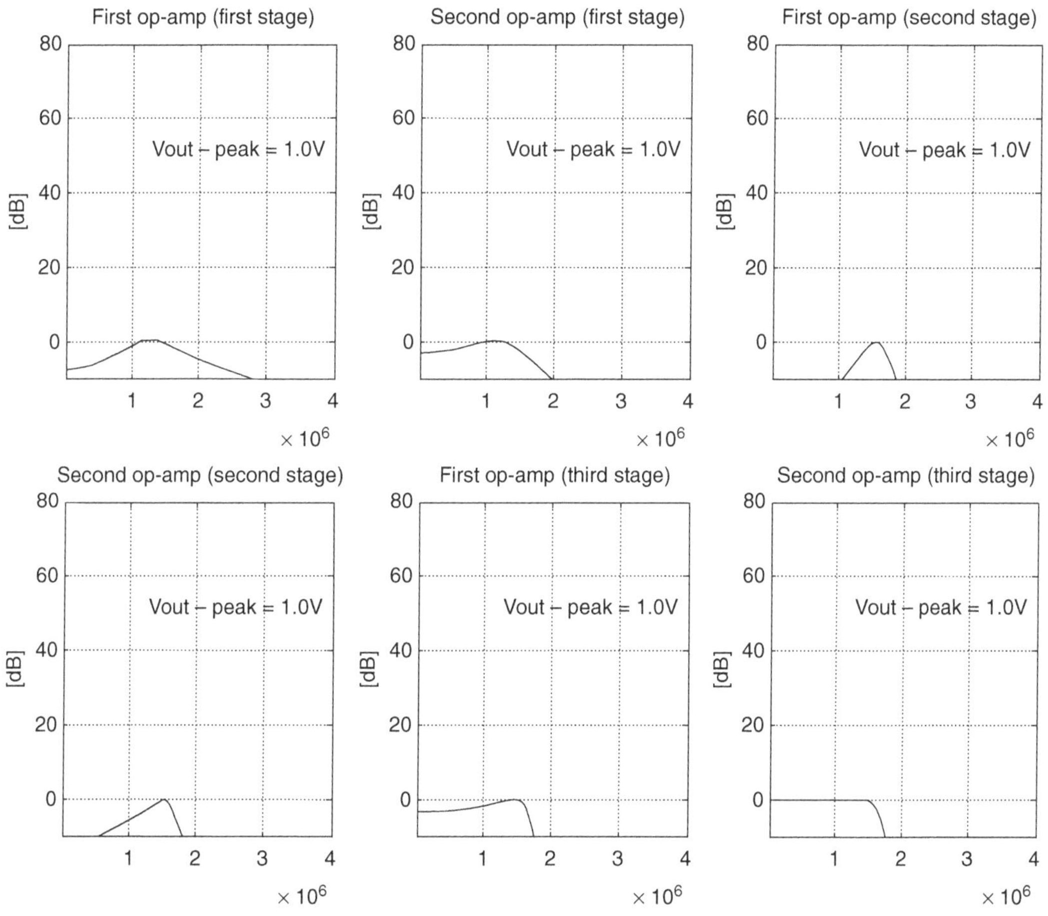

(b)

Figure 4.9 ***Continuted***

MATLAB simulation. After dynamic range scaling is applied to the initial capacitance values using MATLAB, the resulting scaled capacitance values are used to replace the initial ones in the circuit presented in Figure 4.8, and the circuit simulation in SWITCAP is carried out again. The new peak op-amp output voltages resulting from SWITCAP simulation are shown in Figure 4.9(b). The reader is referred to Appendix 4.1 for more simulation details.

To minimize the total capacitance, these scaled capacitances are normalized with respect to the smallest capacitance (0.5 pF in this case). The resultant capacitance values are listed in Table 4.2. The resulting capacitance spread is about 16.4, and the nominal total capacitance value is equal to 50.95 pF.

Table 4.2 Capacitance values before and after scaling.

Biquad stage	Capacitor	Initial capacitances before scaling (pF)	Scaled capacitances (pF)
	C1_1	2.1262	0.5
	C2_1	0.2848	0.71676
	C3_1	0.2848	1.8515
First stage	C4_1	0.2025	1.0835
	CA_1	1	3.0571
	CB_1	1	5.3502
	C1″_1	1	0.5
	C1_2	1.0732	0.5
	C2_2	0.3248	1.3220
	C3_2	0.3248	1.4333
Second stage	C4_2	0.1765	0.71859
	CA_2	1	3.8760
	CB_2	1	4.6341
	C1″_2	0.9427	0.5
	C1_3	11.6834	1.8749
	C2_3	0.1900	0.5
	C3_3	0.1900	4.1171
Third stage	C4_3	0.3370	2.7630
	CA_3	1	6.9535
	CB_3	1	8.1982
	C1″_3	1	0.5

Having determined the capacitance values, we should investigate the issues involved in the circuit implementation. Essentially, we must specify three important parameters with respect to the op-amps used in the SCF: dc gain, unity-gain-bandwidth (UGBW), and slew rate.

Assume that the average dc gain of the op-amps in a SCF is given by A_0 (i.e., A_0 is the finite op-amp gain), and as the result, the corner frequency of the filter will be shifted from ω_0 to $A_0\omega_0/(A_0 - 1)$ [17]. If the average op-amp gain is no less than 1000 V/V, or, equivalently, 60 dB, this deviation of the corner frequency is negligible. However, the phase error due to the finite op-amp gain, which often appears in the form of *pole-Q shift*, cannot be neglected and is indeed rather problematic particularly for high-Q biquads. It can be found that the ith pole-Q in the presence of a finite op-amp gain A_0 is given by [17][18]

$$\frac{1}{Q_i} = \frac{1}{Q_{i0}} + \frac{2}{A_0} \tag{4.24}$$

where Q_{i0} is the desired value of the ith pole-Q, and the factor 2 means that two op-amps are used in a SC biquad. The pole-Q shift contributes to an undesirable change in magnitude near the peak of the gain response (or the *passband ripple*), whose maximum value (in decibels) is given by

$$\Delta G = 20\lg\left(1 + \frac{2Q_{\max}}{A_0}\right) \tag{4.25}$$

Recall from previous simulation results that the largest pole-Q in this low-pass SCF is equal to 9.738. If we assume that $\Delta G = 0.1$ dB, then from the foregoing formula we can calculate the minimum op-amp dc gain A_0 required, and the result is about 1682 V/V or 64.5 dB. To provide a sufficient design margin, here we specify the minimum op-amp dc gain to be 2500 V/V or 68 dB.

The effect of finite op-amp bandwidth on the SCF is typically analyzed based on the circuit's linear settling behavior. Specifically, we should focus on the time taken for each op-amp to reach a specified percentage of its final value when the input's step size is still relatively small, such that the slew rate (which governs the nonlinear settling performance) limit is not yet a concern.

A first-order closed-loop op-amp model that has a 90-degree phase margin (PM) is often employed to simulate and analyze the linear settling characteristics of an op-amp. In this model, it is assumed that the poles are widely separated from each other, and that the op-amp frequency response is dominated by only one pole, hence it is also called the *single-pole op-amp model* [17]. The generic closed-loop op-amp gain function can then be written as

$$A_{cl}(s) = \frac{1}{\beta + \dfrac{s}{\omega_0}} \tag{4.26}$$

where A_{cl} is the closed-loop op-amp dc gain, β is the feedback gain or the feedback factor, and ω_0 is the –3 dB frequency of the closed-loop op-amp's frequency response. Since a *perfectly compensated* (i.e., with a 90-degree phase margin) closed-loop op-amp model is employed here, we can approximate ω_0 to the product of the unity-gain bandwidth (UGBW) of the *open-loop* op-amp ω_{ol}, and β [19].

Also, it can be found that the time constant τ of the closed-loop op-amp model is given by

$$\tau = \frac{1}{\omega_0} \cong \frac{1}{\beta\omega_{ol}} \tag{4.27}$$

In the presence of a small step input, the time constant τ dictates the settling characteristic of the op-amp. From signals and systems theories we know that the step response of a linear feedback system is given by

$$V_{out}(t) = V_{step}\left(1 - e^{-t/\tau}\right) \tag{4.28}$$

Through calculations, we find that the time needed for a closed-loop single-pole op-amp's output voltage (V_{out}) to settle within 0.1% of its final value (V_{step}) is about 7 τ. On the other hand, it is the golden rule of implementing any SC circuit that the output voltage must settle down satisfactorily within half the clock period, otherwise the accuracy of the SC circuit will be corrupted. Therefore, the settling time (7 τ) must be kept less than or equal to $T/2$. Here, we set the requisite (to leave some design margin) as follows:

$$7\tau \le \frac{3}{8}T \tag{4.29}$$

Apply Equation (4.27) to Equation (4.29), and we have

$$\omega_{ol} \ge \frac{56}{3\beta} \cdot f_{clk} \tag{4.30}$$

Note that the clock sample frequency f_{clk} is equal to 30 MHz here. The next step is to find the smallest possible value of β, so that the open-loop UGBW (ω_{ol}) required in the worst-case scenario can be decided. For this reason, the feedback factor of each SC integrator in the filter is calculated. The results are listed in Table 4.3.

Table 4.3 Feedback factors of the integrators in the SCF.

Feedback factor (i = 1, 2, and 3)	First stage	Second stage	Third stage
β (of the first integrator) $= C_{A\_i}/(C_{A\_i} + C_{1\_i})$	0.860	0.890	0.665
β (of the second integrator) $= C_{B\_i}/(C_{B\_i} + C_{3\_i})$	0.743	0.764	0.788

Note that the smallest feedback factor is equal to 0.665. From Equation (4.30), we can thus find that ω_{ol} should not be smaller than 842.11 Mrad/s, or, equivalently, 134.026 MHz. For the sake of design margin, we specify that each op-amp in the filter have a UGBW of no less than 140 MHz.

The third important op-amp parameter to be determined is the slew rate. Slew rate is known as the rate at which the output changes when the input signal has large steps. The slew rate of an op-amp can also be defined as the rate at which the op-amp is able to charge or discharge the capacitive load (C_L) at its output terminal [19]. A precise slew-rate analysis should not ignore the correlation between the slew rate (i.e., nonlinear settling) and the finite op-amp bandwidth (i.e., linear settling). As we know from the small-signal model of an op-amp, the transconductance (g_m) of the op-amp's input device is *not* independent of the current (I_c) that flows through it. As the result, computer simulation is usually adopted to generate contours to investigate the exact relationship between the UGBW (which is generically represented by g_m/C_L), the slew rate (which is generically represented by I_c/C_L), and the capacitive load C_L.

As an alternative to the preceding method of calculating slew rate, a simple but slightly pessimistic estimation can be used to obtain the minimum value of the slew rate required to maintain the accuracy of the filter. First, express the output slew time of an op-amp as

$$t_{SR} = \frac{\Delta V_{out}}{SR} = \frac{\Delta V_{out}}{\left|\frac{\partial V_{out}}{\partial t}\right|} \tag{4.31}$$

where *SR* is short for slew rate. Assume a sinusoidal input $v_{in}(t) = V_{\max} \sin \omega_a t$ is applied to the filter. To consider the worst-case scenario, we go on to assume that ω_a is the highest frequency within the passband of the SCF (it is equal to $2\pi \times$ 1.5 Mrad/s in this case), and that V_{max} is the largest possible op-amp voltage swing (use the power supply voltage $V_{dd} = 3.3$ V). The maximum input slope is then given by

$$\left|\frac{\partial V_{in}}{\partial t}\right|_{\max} = \omega_a V_{\max} \tag{4.32}$$

To keep up with the input change during the sampling phase (i.e., within half the clock period), the output step ΔV_{out} should reach its peak within the slew time. Assume that the output slew time takes up a portion of the clock period—that is,

$$t_{SR} = aT, \quad 0 < a \leq 0.5 \tag{4.33}$$

Combining Equations (4.31) to (4.33), we obtain the expression of the minimum output slew rate as follows:

$$SR_{\min} = \frac{\omega_a V_{\max} \cdot \frac{T}{2}}{aT} = \frac{1}{2a}\omega_a V_{\max} \tag{4.34}$$

In practice, the value of a typically ranges between 0.25 and 0.75; however, to leave some design margin for the circuit implementation, here we assume $a = 0.1$, which is smaller than most practical circuits would allow, and we can specify the minimum op-amp slew rate, which is about 155.51 V/μs.

Having determined the op-amp design parameters, we should consider the realization of switches using MOSFET devices. As we learned in Chapter 1, on-chip switches built of MOSFET devices suffer from several nonidealities, one of which is the nonzero on-resistance. When a MOS transistor is operated in the linear or triode region, it behaves like a resistor whose resistance value is given by

$$R_{on} = \frac{1}{2K \cdot (V_{gs} - V_{th})} \tag{4.35}$$

where K is a constant for a given process, and $(V_{gs} - V_{th})$ is the effective overdrive voltage. In an SC circuit, the MOSFET switch is typically used to charge a capacitor to a specific voltage (e.g., V_{in}). Also, the on-resistance R_{on} of the switch directly affects the time constant τ, hence the linear settling time of the op-amp.

From Table 4.2 we can calculate and conclude that the largest time constant in the low-pass SCF under design is due to the SC combination located around the second op-amp of the third biquad stage; this is because C_{B_3} has the largest capacitance value, which is equal to 8.1982 pF. From Equation (4.29), we can calculate the maximum value of R_{on} that the SCF permits, which is given by

$$R_{on\_\max} = \frac{3T}{56C_{B\_3}} \approx 218\,\Omega \tag{4.36}$$

Insert this value into Equation (4.35), and assume that $\mu_n C_{ox} = 60\,\mu\text{A/V}^2$, $\mu_p C_{ox} = 20\,\mu\text{A/V}^2$, $(V_{gs} - V_{th})_{\max} = 3.0$ V, then we have

$$\left(\frac{W}{L}\right)_{n\_\min} = \frac{1}{3}\left(\frac{W}{L}\right)_{p\_\min} = \frac{K_{\min}}{\mu_n C_{ox}} = \frac{1}{\mu_n C_{ox} \cdot 2R_{on\_\max} \cdot (V_{gs} - V_{th})_{\max}} \approx 12.742 \tag{4.37}$$

To leave some room for adjustment, we can choose the minimum transistor sizing ratio of 15/1 for NMOS and 45/1 for PMOS.

CMOS transmission gates can be used to construct the critical switches in the circuit to alleviate the charge injection errors. In this SCF, the input branches of the first and second biquad stages as well as the C_{2_1}, C_{4_2}, and C''_{1_3} branches can be built using transmission gates. Additionally, the fully differential configuration can be employed to optimize the SCF's common-mode noise rejection performance.

Henceforth, the circuit implementation of the SCF is fairly straightforward—that is, it is all about the op-amp. A folded-cascode op-amp is adequate to realize the parameter values that we specified earlier. The postlayout simulation results are shown in Figure 4.10. The circuit layout is done using a 3.3-V 0.5-μm CMOS process. The frequency response of the low-pass SCF is shown in Figure 4.10(a), while the passband ripples are shown in Figure 4.10(b).

As a final note, it is worthwhile to investigate the effect of capacitor mismatch on the SCF's frequency response. A random capacitor-mismatch sequence with a

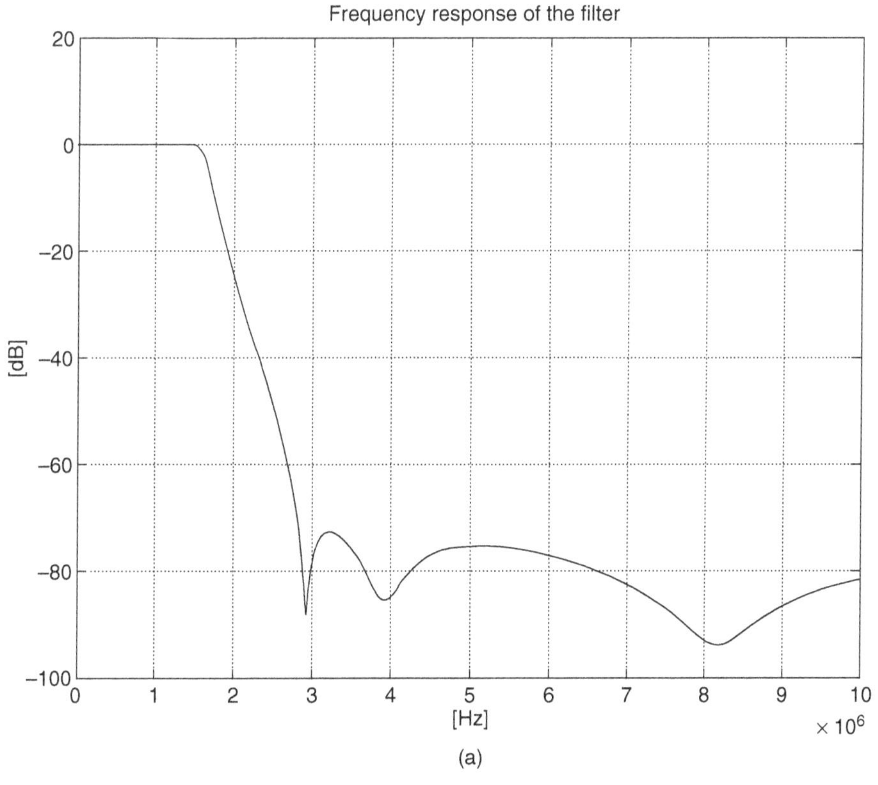

Figure 4.10 Postlayout simulation results: (a) frequency response, (b) passband ripples.

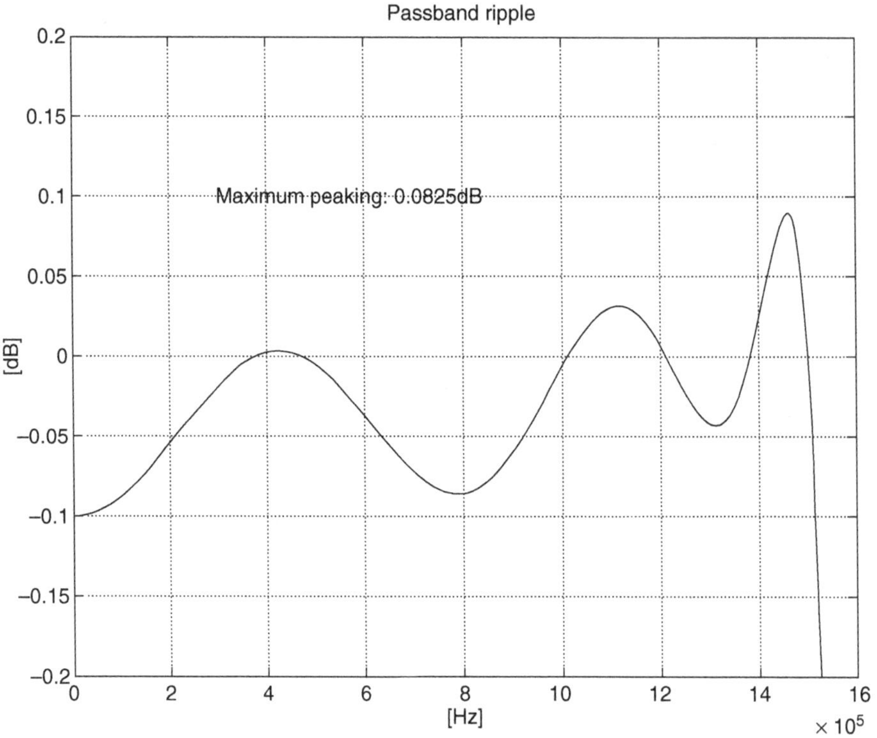

(b)

Figure 4.10 ***Continued***

standard deviation of 0.1% is generated using MATLAB. The new capacitor values with the mismatch elements are used to replace the ideal ones in the SWITCAP netlist. Then the effect of capacitor mismatch on the filter's passband performance is simulated in a SWITCAP environment, and the result is shown in Figure 4.11.

Note that the capacitor mismatch has caused the peaking in the passband to exceed 0.1 dB, which is not trivial as compared to the 0.1-dB limit that we specified earlier. Nevertheless, the peaking is still below the original specification in Table 4.1 (i.e., 0.3 dB).

4.4 High-Frequency CMOS Switched-Capacitor Filters

In comparison with active RC and Gm-C filters, active SC filters are capable of providing highly precise and programmable frequency responses, without the helps of dedicated on-chip tuning devices. Very accurate CMOS SC filters that have a maximum signal-to-noise-plus-distortion-ratio (SNDR) of nearly 96 dB have been

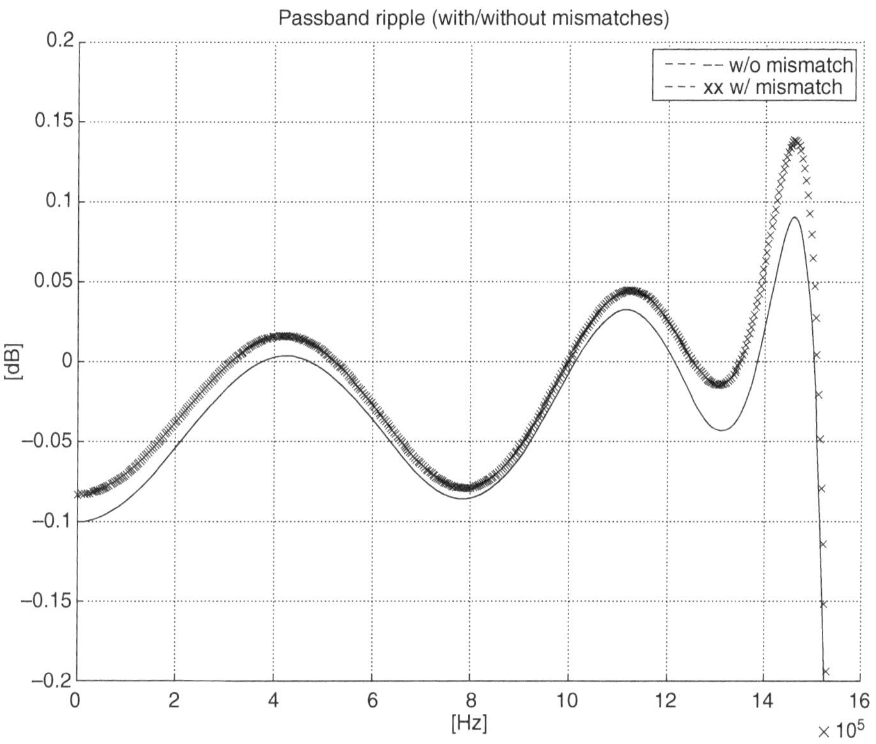

Figure 4.11 Passband ripples (with and without mismatch).

realized, most of which are mainly adopted to meet the needs of high-quality audio and sensor applications. Also, it is easier to build a parasitic-insensitive filter using SC than it is to use active RC or Gm-C techniques.

However, as we know, an SCF is a sampled-data system that must satisfy the Nyquist-rate requirement. That is, the clock sampling frequency must be at least twice the highest signal frequency to avoid aliasing. In practice, the clock frequency is several times higher than the maximum signal frequency to reduce the selectivity requirements for the antialiasing filter. This imposes an inherent limitation on the capability of the SCF to process high-frequency signals [20][21].

By contrast, active RC and Gm-C filters are continuous-time (CT) systems with no inherent limitations due to the Nyquist-rate sampling requirement. Furthermore, Gm-C filters often make use of open-loop rather than closed-loop operational transconductance amplifiers (OTAs), and as a result, Gm-C filters typically have a speed advantage over both SC and RC filters, particularly for applications in the hundreds of megahertz range. Nevertheless, as mentioned earlier, continuous-time Gm-C filters cannot provide accurate system coefficients, which are primarily determined by the products of resistors and capacitors, without using the sophisticated and power-hungry tuning circuitry. Consequently, the achievable SNDR performance of an active Gm-C filter is typically worse than 65 dB. A few Gm-C filters that are able to provide SNDRs within the range between 65 dB and 75 dB using BiCMOS technologies (i.e., integrating bipolar junction transistors and CMOS transistors into a single device) have been reported recently, in order to meet the needs of wireless communications and TV tuning applications. However, most of them are still inadequate to satisfy the accuracy requirements placed by applications such as instrumentations and analog video processing. Moreover, in these applications, the requirements of operating speed are, though still often on the megahertz level, not as stringent as that required by the aforementioned wireless communications. Therefore, it is highly anticipated that SCFs may take over these territories, and the design of SCFs for high-speed applications requiring medium-to-high accuracy has been a prevalent research topic in the area of analog circuits and systems.

Although high-frequency SCFs implemented in GaAs MESFET (i.e., Gallium arsenide metal-semiconductor field-effect transistor) [22] have been exploited, the majority of research activities so far have emphasized the practical implementation of high-frequency SCFs in standard CMOS technologies for the sakes of cost and adaptability to the system-on-a-chip (SoC) environment, where analog and digital circuitries are integrated onto a single chip.

It can be shown that the major bottleneck of realizing a high-frequency CMOS SCF is the requirement for high-gain and large-bandwidth op-amps, which provide virtual grounds for accurate charge transfer [17][18]. Although it is possible to realize CMOS op-amps that can provide the required gain and bandwidth values, the resultant high power consumption often hinders the practical realization [23][24].

Besides the large-gain-bandwidth approach, a few interesting techniques have been reported to enhance a CMOS SCF's ability to operate in the hundred-megahertz range. They can be roughly classified into two groups: op-amp based and unity-gain-buffer based. The former keeps the op-amp in the SCF, while the latter replaces it with a unity-gain buffer (UGB) [20].

It is well known that large-bandwidth or high-speed op-amps tend to have low dc gains due to the fundamental tradeoff between RC time constant and $g_m R_{out}$. A low op-amp dc gain tends to introduce nonlinearity errors to the SC integrator's output, hence compromising the filter's accuracy performance. As a response to this problem, op-amp-based techniques typically emphasize modifying the traditional op-amp structures so that the resultant new op-amps are capable of fulfilling both the speed and accuracy requirements of a high-frequency SCF, despite the fact that they typically have low dc gains [10]. An op-amp-based SC biquad that operates at a clock sampling rate of 200 Msamples/s (with the double sampling configuration) and consumes 10 mW in a 3.3-V 0.5-μm CMOS technology has been reported by Baschirotto et al. [25].

The biquad reported by Baschirotto et al. is claimed to achieve a dynamic range of 62 dB with a 1% total-harmonic-distortion (THD) while using op-amps with an average dc gain of as low as 80 V/V, or equivalently, 38 dB. The circuit is realized based on a technique called *gain regulating*, or equivalently, the precise op-amp gain design approach as named in [25]. The basic idea is to precisely control (or regulate) the low dc gain of each large-bandwidth op-amp in the high-frequency SCF and use the regulated gain value as a reference to scale the capacitances in the circuit, so that the effective error in the SCF's frequency response due to the regulated op-amp dc gain (A) is equivalent to the error that would have occurred if the op-amp had an *effective* op-amp gain of (A/ε) but the same large UGBW (apparently, it is unlikely that this type of op-amp would exist physically), where ε is the maximum variance of the regulated op-amp dc gain A [25]. Therefore, as far as the accuracy of the biquad is concerned, the effective op-amp gain is approximately (A/ε). Use the proposed biquad as an example: each op-amp's gain is regulated by a control-loop composed of a replica op-amp and an integrator so that the value of ε is limited to about 2%, hence the effective op-amp gain can be obtained by dividing 80 V/V by 0.02, and the result is 4000 V/V, or equivalently, 72 dB.

As the alternative to the op-amp-based approach, the unity-gain-buffer-based technique makes use of unity-gain buffers to build SC integrators [20][26][27][28]. A typical unity-gain buffer (UGB) is able to work over a much wider signal bandwidth than the conventional op-amp. In addition, a UGB can be realized using simpler circuitry; as a result, it occupies less silicon area and consumes less power than a conventional large-gain-bandwidth op-amp does. Thus, SCFs built on UGBs are particularly attractive to small-form-factor portable devices that require a low power consumption to maintain a long battery life.

However, UGB-based SC integrators suffer from parasitic capacitances, which are mainly caused by the source-gate diffusions in the UGB's input transistors, whose values tend to vary with process and temperature. Moreover, the parasitic-insensitive techniques introduced in Chapter 3 cannot be used on UGB-based SC integrators, because the low gain (nominally equal to 1) UGB does not have a virtual ground that is essential for accurate charge transfer. For the same reason, the offset voltage at the input terminal of the UGB is equally problematic. At the time of this writing, the appropriate solution to the preceding issues with respect to the UGB-based SC circuit is not yet available, but the reader can find a few useful discussions in [26][28].

Appendix 4.1

Part I: Filter Specification

```
% ----- original specifications of the filter ------
clear all;
close all;
Fp = 1.5e6;    % passband edge frequency [Hz]
Fs = 3.0e6;    % stopband edge frequency [Hz]
Fc = 3.0e7;    % sampling frequency [Hz]
Wp = 2*Fp/Fc;  % passband normalized frequency
Ws = 2*Fs/Fc;  % stopband normalized frequency
Wc = 2;        % sampling normalized frequency
Rp = 0.10;     % maximum passband ripple [dB]
Rs = 75;       % minimum stopband attenuation [dB]
nfig0=1;
Wa_max=1;
Wa_nr=2^10;
Wa=linspace(0,Wa_max,Wa_nr); % frequency range

% --- Locations of zeros and poles of the Elliptic filter ----

  figure(nfig0+2);
  zplane(z,p); hold on;
  title('Locations of zeros and poles of the Elliptic filter');
  xlabel('real part');
  ylabel('imag part');
  grid on;

% ---- Decomposition of H(z) into three sections ------

  figure(nfig0+3);
  H1_a3=(real(p(3)))^2+(imag(p(3)))^2;
```

```
H1=tf([1 -2*real(z(5)) 1],[1 -2*real(p(3)) H1_a3],1/Fc)
[z1,p1,k1]=tf2zp([1 -2*real(z(5)) 1],[1 -2*real(p(3)) H1_a3]);
Q1=abs(atan(imag(p1)./real(p1))./2./(1-sqrt((real(p1).^2+imag(p1).^2))))
% Q1 is an approximate value.
alpha1=1-2*sqrt(H1_a3)*cos(atan(imag(p(3))/real(p(3))))+(H1_a3);
beta1=1-H1_a3;
Q1_accurate=sqrt(alpha1*(1-0.25*alpha1-0.5*beta1))/beta1
% The first biquad is a low_Q one.
omega1=Fc*(sqrt(alpha1/(1-0.25*alpha1-0.5*beta1)))

H2_a3=(real(p(1)))^2+(imag(p(1)))^2;
H2=tf([1 -2*real(z(3)) 1],[1 -2*real(p(1)) H2_a3],1/Fc)
[z2,p2,k2]=tf2zp([1 -2*real(z(3)) 1],[1 -2*real(p(1)) H2_a3]);
Q2=abs(atan(imag(p2)./real(p2))./2./(1-sqrt((real(p2).^2+imag(p2).^2))))
% Q2 is an approximate value.
alpha2=1-2*sqrt(H2_a3)*cos(atan(imag(p(1))/real(p(1))))+(H2_a3);
beta2=1-H2_a3;
Q2_accurate=sqrt(alpha2*(1-0.25*alpha2-0.5*beta2))/beta2
% The second biquad is a high_Q one.
omega2=Fc*(sqrt(alpha2/(1-0.25*alpha2-0.5*beta2)))

H0_a3=(real(p(5)))^2+(imag(p(5)))^2;
H0=tf([1 -2*real(z(1)) 1],[1 -2*real(p(5)) H0_a3],1/Fc)
[z0,p0,k0]=tf2zp([1 -2*real(z(1)) 1],[1 -2*real(p(5)) H0_a3]);
Q0=abs(atan(imag(p0)./real(p0))./2./(1-sqrt((real(p0).^2+imag(p0).^2))))
% Q0 is an approximate value.
alpha0=1-2*sqrt(H0_a3)*cos(atan(imag(p(5))/real(p(5))))+(H0_a3);
beta0=1-H0_a3;
Q0_accurate=sqrt(alpha0*(1-0.25*alpha0-0.5*beta0))/beta0
% The third biquad is a low_Q one.
omega0=Fc*(sqrt(alpha0/(1-0.25*alpha0-0.5*beta0)))

k_decomp=k^(1/3); % Decompose the gain.

subplot(131);
HH1=freqz(k_decomp*[1 -2*real(z(5)) 1],[1 -2*real(p(3)) H1_a3],Wa_nr);
plot(Wa, 20*log10(abs(HH1)),'k-'); hold on;
axis([0 Wa_max -100 20]);
```

```
title('frequency response(1st stage)');
ylabel('Gain[dB]');
xlabel('Normalized frequency, \Omega');
grid on;

subplot(132);
HH2=freqz(k_decomp*[1 -2*real(z(3)) 1],[1 -2*real(p(1)) H2_a3],Wa_nr);
plot(Wa, 20*log10(abs(HH2)),'k-'); hold on;
axis([0 Wa_max -100 20]);
title('frequency response(2nd stage)');
ylabel('Gain[dB]');
xlabel('Normalized frequency, \Omega');
grid on;

subplot(133);
HH0=freqz(k_decomp*[1 -2*real(z(1)) 1],[1 -2*real(p(5)) H0_a3],Wa_nr);
plot(Wa, 20*log10(abs(HH0)),'k-'); hold on;
axis([0 Wa_max -100 20]);
title('frequency response(3rd stage)');
ylabel('Gain[dB]');
xlabel('Normalized frequency, \Omega');
grid on;
```

Part II: Dynamic Range Scaling and Capacitance Assignment

```
% ----- The capacitance values before scaling ------

% -------- The first stage (low-Q) ---------

 H1=tf([1 -2*real(z(5)) 1],[1 -2*real(p(3)) H1_a3],1/Fc)
 % A reminder. Next, format the transfer function.
 n1_reformed=[1 -2*real(z(5)) 1];
 d1_reformed=[1/(H1_a3) -2*real(p(3))/(H1_a3) 1];
 C1_pipi_first=1; % C1''
 C1_pi_first=0; % C1'
 C2_first=sqrt(d1_reformed(1)+d1_reformed(2)+1)
 C3_first=C2_first
 C1_first=(n1_reformed(1)+n1_reformed(2)+n1_reformed(3))/C3_first
```

```
  C4_first=d1_reformed(1)-1
  Ca_first=1
  Cb_first=1

% ------- The second stage (high-Q) --------

  H2=tf([1 -2*real(z(3)) 1],[1 -2*real(p(1)) H2_a3],1/Fc)
  % A reminder. Next, format the transfer function.
  n2_reformed=[1 -2*real(z(3)) 1];
  d2_reformed=[1/(H2_a3) -2*real(p(1))/(H2_a3) 1];
  C1_pipi_second=n2_reformed(1)/d2_reformed(1) % C1''
  C2_second=sqrt((1+d2_reformed(1)+d2_reformed(2))/d2_reformed(1))
  C3_second=C2_second
  C4_second=(1-1/d2_reformed(1))/C3_second
  C1_second=sum(n2_reformed)/(d2_reformed(1)*C3_second)
  C1_pi_second=(n2_reformed(2)-n2_reformed(3))/(d2_reformed(1)*C3_second);
  if C1_pi_second <= 0
    C1_pi_second = 0
  end
  Ca_second=1
  Cb_second=1

% -------- The third stage (low-Q) ---------

  H0=tf([1 -2*real(z(1)) 1],[1 -2*real(p(5)) H0_a3],1/Fc)
  % A reminder. Next, format the transfer function.
  n0_reformed=[1 -2*real(z(1)) 1];
  d0_reformed=[1/(H0_a3) -2*real(p(5))/(H0_a3) 1];
  C1_pipi_third=1; % C1''
  C1_pi_third=0; % C1'
  C2_third=sqrt(d0_reformed(1)+d0_reformed(2)+1)
  C3_third=C2_third
  C1_third=sum(n0_reformed)/C3_third
  C4_third=d0_reformed(1)-1
  Ca_third=1
  Cb_third=1

% --------- Dynamic Range Scaling ---------------
```

```
% Before scaling, use SWITCAP to simulate the peak output voltage
of each op-amp.
% The SWITCAP code is in biquad_6thorder_ideal.in. Its results are
% stored in the *.dat files, which will be loaded in the following.

load biquad_6thorder_bfscl_v2_1.dat;
ff=biquad_6thorder_bfscl_v2_1(:,1);
vdb2=biquad_6thorder_bfscl_v2_1(:,2);
vdb3=biquad_6thorder_bfscl_v2_1(:,3);
vdb4=biquad_6thorder_bfscl_v2_1(:,4);
vdb5=biquad_6thorder_bfscl_v2_1(:,5);
load biquad_6thorder_bfscl_v2_2.dat;
vdb6=biquad_6thorder_bfscl_v2_2(:,2);
vdb7=biquad_6thorder_bfscl_v2_2(:,3);

figure(nfig0+5);
subplot(231);
max_vdb2=max(vdb2);
max_vp2=10^(max_vdb2/20);
plot(ff,vdb2); hold on;
s=sprintf('Vout-peak=%4.1fV\n',max_vp2);
text(1.5e6,48,s);
title('1st opamp(1st stage)');
ylabel('[dB]');
axis([1,4e6,-10,80]);
grid on;
subplot(232);
max_vdb3=max(vdb3);
max_vp3=10^(max_vdb3/20);
plot(ff,vdb3); hold on;
s=sprintf('Vout-peak=%4.1fV\n',max_vp3);
text(1.5e6,48,s);
title('2nd opamp(1st stage)');
ylabel('[dB]');
axis([1,4e6,-10,80]);
grid on;
subplot(233);
max_vdb4=max(vdb4);
```

```
max_vp4=10^(max_vdb4/20);
plot(ff,vdb4); hold on;
s=sprintf('Vout-peak=%4.1fV\n',max_vp4);
text(1.5e6,48,s);
title('1st opamp(2nd stage)');
ylabel('[dB]');
axis([1,4e6,-10,80]);
grid on;

subplot(234);
max_vdb5=max(vdb5);
max_vp5=10^(max_vdb5/20);
plot(ff,vdb5); hold on;
s=sprintf('Vout-peak=%4.1fV\n',max_vp5);
text(1.5e6,48,s);
title('2nd opamp(2nd stage)');
ylabel('[dB]');
axis([1,4e6,-10,80]);
grid on;

subplot(235);
max_vdb6=max(vdb6);
max_vp6=10^(max_vdb6/20);
plot(ff,vdb6); hold on;
s=sprintf('Vout-peak=%4.1fV\n',max_vp6);
text(1.5e6,48,s);
title('1st opamp(3rd stage)');
ylabel('[dB]');
axis([1,4e6,-10,80]);
grid on;

subplot(236);
max_vdb7=max(vdb7);
max_vp7=10^(max_vdb7/20);
plot(ff,vdb7); hold on;
s=sprintf('Vout-peak=%4.1fV\n',max_vp7);
text(1.5e6,48,s);
title('2nd opamp(3rd stage)');
```

```
ylabel('[dB]');
axis([1,4e6,-10,80]);
grid on;

% Now, all the peak output voltages are known. Next, we start the
scaling.

% ---- Scale the first stage after SWITCAP simulations --------

  Ca_first_scaled=Ca_first*max_vp2;
  C3_first_scaled=C3_first*max_vp2;
  Cb_first_scaled=Cb_first*max_vp3;
  C2_first_scaled=C2_first*max_vp3;
  C4_first_scaled=C4_first*max_vp3;
  C1_second_scaled=C1_second*max_vp3;
  C1_pipi_second_scaled=C1_pipi_second*max_vp3;
% ---- Obtain minimum total capacitance in the first stage ---

% Among C1_first, C2, and Ca(_first_scaled), divide all by the
smallest.

% Among C1_pipi_first, C3, C4 and Cb(_first_scaled), divide by the
smallest.

% ---- Scale the second stage after SWITCAP simulations ------

  Ca_second_scaled=Ca_second*max_vp4;
  C3_second_scaled=C3_second*max_vp4;
  Cb_second_scaled=Cb_second*max_vp5;
  C2_second_scaled=C2_second*max_vp5;
  C4_second_scaled=C4_second*max_vp5;
  C1_third_scaled=C1_third*max_vp5;
  C1_pipi_third_scaled=C1_pipi_third*max_vp5;

% ---- Obtain minimum total capacitance in the second stage ---

% Among C1, C2, C4 and Ca(_second_scaled), divide all by the
smallest.
```

```
% Among C1_pipi, C3, and Cb(_second_scaled), divide by the smallest.

% ---- Scale the third stage after SWITCCAP simulations. -----

  Ca_third_scaled=Ca_third*max_vp6;
  C3_third_scaled=C3_third*max_vp6;
  Cb_third_scaled=Cb_third*max_vp7;
  C2_third_scaled=C2_third*max_vp7;
  C4_third_scaled=C4_third*max_vp7;

% ---- Obtain minimum total capacitance in the third stage ---

% Among C1, C2, and Ca(_third_scaled), divide by the smallest.

% Among C1_pipi, C3, C4, and Cb(_third_scaled), divide by the
smallest.
```

Part III: SWITCAP Program for Scaling and Filter Coefficients Simulation

```
TITLE: SIXTH-ORDER SC BIQUAD (IDEAL-OPAMPS)

TIMING;
PERIOD 3.3333E-8;
CLOCK PHI 1 (0 3/8);
END;

SUBCKT (1 5 9) biquadhighQ (K:P2 P:C1x P:C2x P:C3x P:C4x P:CAx P:CBx
P:CPIx);

S1 (1 2) P2;
S2 (2 0) #P2;
S3 (3 0) #P2;
S4 (3 4) P2;
S5 (5 6) #P2;
S6 (6 0) P2;
S7 (7 0) #P2;
S8 (7 8) P2;
S9 (9 10) P2;
S10 (10 0) #P2;
```

```
C1 (2 3) C1x;
C2 (3 10) C2x;
C3 (6 7) C3x;
C4 (4 9) C4x;
CA (4 5) CAx;
CB (8 9) CBx;
CPI (1 8) CPIx;
E1 (5 0 0 4) 1e5;
E2 (9 0 0 8) 1e5;
END;

SUBCKT (11 5 9) biquadlowQ (K:P2 P:C1x P:C2x P:C3x P:C4x P:CAx P:CBx
P:CPIx);

S1 (1 2) P2;
S2 (2 0) #P2;
S3 (3 0) #P2;
S4 (3 4) P2;
S5 (5 6) #P2;
S6 (6 0) P2;
S7 (7 0) #P2;
S8 (7 8) P2;
S9 (9 10) P2;
S10 (10 0) #P2;
S11 (11 1) P2;
C1 (2 3) C1x;
C2 (3 10) C2x;
C3 (6 7) C3x;
C4 (7 10) C4x;
CA (4 5) CAx;
CB (8 9) CBx;
CPI (1 8) CPIx;
E1 (5 0 0 4) 1e5;
E2 (9 0 0 8) 1e5;
END;

CIRCUIT;
```

```
V1 (1 0);
/*Before dynamic scaling*/
X1 (1 2 3) biquadlowQ (PHI 2.1262 0.2848 0.2848 0.2025 1 1 1);
X2 (3 4 5) biquadhighQ (PHI 1.0732 0.3248 0.3248 0.1765 1 1 0.9427);
X3 (5 6 7) biquadlowQ (PHI 11.6834 0.1900 0.1900 0.3370 1 1 1);
/*After scaling*/
/*X1 (1 2 3) biquadlowQ (PHI 1 1.4335 3.703 2.1669 6.1143 10.7004 1);*/
/*X2 (3 4 5) biquadhighQ (PHI 1 2.6441 2.8665 1.4372 7.7519 9.2682
1);*/
/*X3 (5 6 7) biquadlowQ (PHI 3.7497 1 8.2342 5.5259 13.9069 16.3965
1);*/
END;

ANALYZE SSS;
INFREQ 1 4e6 LOG 300;
SET V1 AC 1.0 0.0;
SAMPLE OUTPUT IMPULSE 1 3/8-;
PRINT VDB(2) VDB(3) VDB(4) VDB(5);
PRINT VDB(6) VDB(7);
END;
END;
```

Part IV: Simulate the Effects of Finite Op-Amp Gain/Bandwidth

```
TITLE: 6th-order LPF BIQUAD (NON IDEAL-OPAMP MODELS)

TIMING;
PERIOD 3.333E-8;
CLOCK PHI 1 (0 3/8);
CLOCK rq 1/100 (0 1/200)
END;

SUBCKT (1 2 3 4) opamp (P:a0);
  E1 (5 0 3 4) 1;
  E2 (1 2 8 0) a0;
  S1a (5 6) rq;
  S1b (8 6) #rq;
  S2a (8 7) rq;
  S2b (5 7) #rq;
```

```
  Ceq (6 7) 1.04167e-4;
  Cp (8 0) 8; /* Bandwidth controlling parameter*/
END;

SUBCKT (1 5 9) biquadhighQ (K:P2 P:C1x P:C2x P:C3x P:C4x P:CAx P:CBx
P:CPIx);

S1 (1 2) P2;
S2 (2 0) #P2;
S3 (3 0) #P2;
S4 (3 4) P2;
S5 (5 6) #P2;
S6 (6 0) P2;
S7 (7 0) #P2;
S8 (7 8) P2;
S9 (9 10) P2;
S10 (10 0) #P2;
C1 (2 3) C1x;
C2 (3 10) C2x;
C3 (6 7) C3x;
C4 (4 9) C4x;
CA (4 5) CAx;
CB (8 9) CBx;
CPI (1 8) CPIx;
xE1 (5 0 0 4) opamp(2000);
xE2 (9 0 0 8) opamp(2000);
END;

SUBCKT (1 5 11) biquadlowQ (K:P2 P:C1x P:C2x P:C3x P:C4x P:CAx P:CBx
P:CPIx);

S1 (1 2) P2;
S2 (2 0) #P2;
S3 (3 0) #P2;
S4 (3 4) P2;
S5 (5 6) #P2;
S6 (6 0) P2;
S7 (7 0) #P2;
```

```
S8 (7 8) P2;
S9 (9 10) P2;
S10 (10 0) #P2;
S11 (11 9) P2;
C1 (2 3) C1x;
C2 (3 10) C2x;
C3 (6 7) C3x;
C4 (7 10) C4x;
CA (4 5) CAx;
CB (8 9) CBx;
CPI (1 8) CPIx;
xE1 (5 0 0 4) opamp(2000);
xE2 (9 0 0 8) opamp(2000);
END;

CIRCUIT;

V1 (1 0);
/*without capacitance mismatches*/
X1 (1 2 3) biquadlowQ (PHI 0.5 0.71676 1.8515 1.0835 3.0571 5.3502
0.5);
X2 (3 4 5) biquadhighQ (PHI 0.5 1.322 1.4333 0.71859 3.8760 4.6341
0.5);
X3 (5 6 7) biquadlowQ (PHI 1.8749 0.5 4.1171 2.763 6.9535 8.1982 0.5);
/*with capacitance mismatches*/
/*X1 (1 2 3) biquadlowQ (PHI 0.5005807749102 0.7172081863109
1.8529235412651 1.0843828663141 3.0581122223560 5.3425772737829
0.4994215386226);*/
/*X2 (3 4 5) biquadhighQ (PHI 0.5000763755722 1.3217932424972
1.4304761979499 0.7172719968889 3.8714454914394 4.6310768558187
0.4997941648477);*/
/*X3 (5 6 7) biquadlowQ (PHI 1.8771135272584 0.5000207699676
4.1131684719799 2.7618695693500 6.9537708910291 8.1853492455430
0.5004532413102);*/
Cout (7 0) 8;
END;

ANALYZE SSS;
INFREQ 1 10e6 lin 3000;
```

```
SET V1 AC 1.0 0.0;
SAMPLE OUTPUT IMPULSE 1 3/8-;
PRINT VDB(3) VDB(5) VDB(7);

/*ANALYZE TRAN;
/*TIME 0+ 400 1/30;
/*SET V1 COSINE 0 1 1.5e6 0 0 0;
/*PRINT v(out) v(7) v(5) v(1);
/*PLOT v(out) v(Inp) v(opout)*/
END;
END;
```

Part V: HSPICE Program to Simulate the Op-Amp Macromodel

```
.options post
.OPTIONS RELTOL=1e-9 ABSTOL=1e-15 VNTOL=1e-6 * VATOL=1e-8
*.include models
.PARAM w1=10u L1=1u
* Subcircuit for a 741 op-amp
.subckt opamp 5 0 1 2
  * +in (=1) -in (=2) out+ (=3) out- (=4)
  rin 1 2 2meg
  Gmvgs 0 4 1 2 5m
  rgbw 4 0 800k
  cbw 4 0 5.305p
  eout 5 0 4 0 1
.ends opamp
*SWITCHES 1st section
mL2 L1 phi2 L2 g cmosn w=w1 L=L1
mL3 L2 phi1 g g cmosn w=w1 L=L1
mL4 L3 phi1 g g cmosn w=w1 L=L1
mL5 L3 phi2 L4 g cmosn w=w1 L=L1
mL6 L5 phi1 L6 g cmosn w=w1 L=L1
mL7 L6 phi2 g g cmosn w=w1 L=L1
mL8 L7 phi1 g g cmosn w=w1 L=L1
mL9 L7 phi2 L8 g cmosn w=w1 L=L1
mL10 L9 phi2 L10 g cmosn w=w1 L=L1
mL11 L10 phi1 g g cmosn w=w1 L=L1
mL12 out1 phi2 L9 g cmosn w=w1 L=L1
 *CAPS 1st stage
```

```
CL1 L2 L3 0.5p
cLa L4 L5 3.0571p
cL2 L3 L10 0.71676p
cL3 L6 L7 1.8515p
cL4 L10 L7 1.0835p
cLb L8 L9 5.3502p
cL1pp L1 L8 0.5p
*opamps 1st biquad
x11 L5 g g L4 opamp ;
x12 L9 g g L8 opamp ;
*SWITCHES 2nd section
mB1 b1 phi2 b2 g cmosn w=w1 L=L1
mB2 b2 phi1 g g cmosn w=w1 L=L1
mB3 b3 phi1 g g cmosn w=w1 L=L1
mB4 b3 phi2 b4 g cmosn w=w1 L=L1
mB5 b5 phi1 b6 g cmosn w=w1 L=L1
mB6 b6 phi2 g g cmosn w=w1 L=L1
mB7 b7 phi1 g g cmosn w=w1 L=L1
mB8 b7 phi2 b8 g cmosn w=w1 L=L1
mB9 b9 phi2 b10 g cmosn w=w1 L=L1
mB10 b10 phi1 g g cmosn w=w1 L=L1
mb11 b9 phi2 out2 g cmosn w=w1 L=L1
*CAPS 2nd stage
CB1 b2 b3 0.5p
cBa b4 b5 3.8760p
cB2 b3 b10 1.322p
cB3 b6 b7 1.4333p
cB4 b4 b9 0.71859p
cBb b8 b9 4.6341p
cB1pp b1 b8 0.5p
x21 b5 g g b4 opamp
x22 b9 g g b8 opamp
*SWITCHES 3rd section
*m1 a phi2 1 0 cmosn w=w1 L=L1
m2 1 phi2 2 g cmosn w=w1 L=L1
m3 2 phi1 g g cmosn w=w1 L=L1
m4 3 phi1 g g cmosn w=w1 L=L1
m5 3 phi2 4 g cmosn w=w1 L=L1
m6 5 phi1 6 g cmosn w=w1 L=L1
```

```
m7 6 phi2 g g cmosn w=w1 L=L1
m8 7 phi1 g g cmosn w=w1 L=L1
m9 7 phi2 8 g cmosn w=w1 L=L1
m10 9 phi2 10 g cmosn w=w1 L=L1
m11 10 phi1 g g cmosn w=w1 L=L1
m12 9 phi2 out g cmosn w=1u L=L1
*CAPS 3rd section
c1 2 3 1.8749p
ca 4 5 6.9535p
c2 3 10 0.5p
c3 6 7 4.1171p
c4 10 7 2.763p
cb 8 9 8.1982p
c1pp 1 8 0.5p
cLoad out 0 1p
*OPAMP
x31 5 g g 4 opamp
x32 9 g g 8 opamp
*connecting buffer
econ1 b1 0 out1 0 1
econ2 1 0 out2 0 1

vin L1 0 sin(1 1 1.5Meg 0 0 0);
Msample in phi2 L1 L1 cmosn w=w1 L=L1
vin L1 0 0
vphi1 phi1 0 pulse(0 5 0 0.01u 0.01u 0.030u 0.1u)
vphi2 phi2 0 pulse(0 5 0.05u 0.01u 0.01u 0.030u 0.1u)
Vg g 0 1

.tran 0.1u 30u 20u
.print v(out) v(L1) v(9)

*
include HP .5um Model files
*

.end
```

References

[1] D. Fried, "Analog sample-data filters," *IEEE Journal of Solid-State Circuits*, Vol. SC-7, No. 4, pp. 302–304, August 1972.

[2] R. W. Brodersen, P. R. Gray, and D. A. Hodges, "MOS Switched-capacitor filters," *Proceedings of IEEE*, Vol. 67, pp. 61–75, January 1979.

[3] P. E. Fleischer and K. R. Laker, "A family of active switched capacitor biquad building blocks," *The Bell Systems Technical Journal*, No. 58, pp. 2235–2269, 1979.

[4] K. Martin, "Improved circuits for the realization of switched-capacitor filters," *IEEE Trans. on Circuits and Systems*, Vol. CAS-27, No. 4, pp. 237–244, April 1980.

[5] K. Martin and A. S. Sedra, "Exact design of switched-capacitor bandpass filters using coupled-biquad structures," *IEEE Trans. on Circuits and Systems*, Vol. CAS-27, No. 6, pp. 469–478, June 1980.

[6] K. R. Laker and W. Sansen, *Design of analog integrated circuits and systems*, McGraw-Hill, New York, 1994.

[7] I. C. Jou, C. Y. Wu, and R. L. Liu, "The characteristic comparison of fully differential switched capacitor biquads," *Proceedings of IEEE International Symposium on Circuits and Systems*, Vol. III, pp. 1712–1715, Portland, OR, May 1989.

[8] T. Choi et al., "High-frequency CMOS switched-capacitor filters for communications application," *IEEE Journal of Solid-State Circuits*, Vol. SC-18, No. 6, pp. 652–664, December 1983.

[9] Q. Huang, "A novel technique for the reduction of capacitance spread in high Q SC circuits," *Proceedings of IEEE International Symposium on Circuits and Systems*, Vol. II, pp. 1249–1252, Helsinki, Finland, June 1988.

[10] K. Nagaraj et al., "Switched-capacitor circuits with reduced sensitivity to amplifier gain," *IEEE Trans. on Circuits and Systems*, Vol. CAS-34, No. 5, pp. 571–574, May 1987.

[11] G. M. Jacobs, D. J. Allstot, R. W. Brodersen, and P. R. Gray, "Design techniques for MOS switched-capacitor ladder filters," *IEEE Trans. on Circuits and Systems*, Vol. CAS-25, No. 12, pp. 1014–1021, December 1978.

[12] A. Fettweis et al. "MOS switched capacitor filters using voltage inverter switches," *IEEE Trans. on Circuits and Systems*, Vol. CAS-27, No. 6, pp. 527–538, June 1980.

[13] F. Montecchi, "On design of switched-capacitor filters with the voltage-inverter-switch approach," *Proceedings of IEEE International Symposium on Circuits and Systems*, Vol. II, pp. 1479–1482, Helsinki, Finland, June 1988.

[14] R. Gregorian and G. C. Temes, *Analog MOS integrated circuits for signal processing*, John Wiley & Sons, New York, 1986.

[15] G. C. Temes, "Switched-capacitor filters," *Class notes: Analog CMOS integrated circuit design*, Oregon State University, Corvallis, OR, 2002.

[16] W.-H. Ki and G. C. Temes, "Optimal capacitance assignment of switched-capacitor biquads," *IEEE Trans. on Circuits and Systems*, Vol. 42, No. 6, pp. 334–342, June 1995.

[17] G. C. Temes, "Finite amplifier gain and bandwidth effect in switched-capacitor filters," *IEEE Journal of Solid-State Circuits*, Vol. SC-15, No. 3, pp. 358–361, June 1980.

[18] K. Martin and A. S. Sedra, "Effects of the op-amp finite gain and bandwidth on the performance of switched-capacitor filters," *IEEE Trans. Circuits and Systems*, Vol. CAS-28, No. 8, pp. 822–829, August 1981.

[19] D. A. Johns and K. Martin, *Analog integrated circuits design*, John Wiley & Sons, New York, 1997.

[20] S. C. Fan et al., "Switched-capacitor filters using unit-gain buffers," *Proceedings of IEEE International Symposium on Circuits and Systems*, Houston, TX. pp. 334–337, May 1980.

[21] G. Fischer and G. Moschytz, "On the frequency limitations of SC filters," *IEEE Journal of Solid-State Circuits*, Vol. SC-19, No. 4, pp. 510–518, August 1984.

[22] L. E. Larson et al., "GaAs switched-capacitor circuits for high-speed signal processing," *IEEE Journal of Solid-State Circuits*, Vol. SC-22, No. 6, pp. 971–981, December 1987.

[23] F. Op't Eynde and W. C. Sansen, "Design and optimisation of CMOS wideband amplifiers," *Proceedings of IEEE Custom Integrated Circuits Conference*, pp. 25.7.1–25.7.4, San Diego, CA, May 1989.

[24] F. Op't Eynde and W. C. Sansen, "A CMOS wideband amplifier with 800 MHz gain-bandwidth," *Proceedings of IEEE Custom Integrated Circuits Conference*, pp. 9.1.1–9.1.4, San Diego, CA, May 1991.

[25] A. Baschirotto, F. Severi, and R. Castello, "A 200-Ms/s 10-mW switched-capacitor filter in 0.5-μm CMOS technology," *IEEE Journal of Solid-State Circuits*, Vol. 35, No. 8, pp. 1215–1219, August 2000.

[26] A. D. Plaza, "High-frequency switched-capacitor filter using unity-gain buffers," *IEEE Journal of Solid-State Circuits*, Vol. SC-21, pp. 470–477, June 1986.

[27] P. Wu, *Unit-gain buffer switched-capacitor filters—design techniques and circuit analysis*, Ph.D. dissertation, University of California, Los Angeles, CA, 1986.

[28] C. Y. Wu, P. H. Lu, and M. K. Tsai, "Design techniques for high-frequency CMOS switched-capacitor filters using non-op-amp-based unity-gain amplifiers," *IEEE Journal of Solid-State Circuits*, Vol. 26, pp. 1460–1466, October 1991.

CHAPTER 5

Switched-Capacitor Data Converters

5.1 Introduction

Data converters play a significant role in modern electronics systems because they are the essential tools interconnecting the analog world and the powerful digital signal processors (DSP). Since the 1970s, when complementary metal-oxide semiconductor (CMOS) fabrication technology was demonstrated as a competent candidate for cost-effective realizations of signal-processing functions in the digital domain, many novel integrated CMOS data converters have been invented, enabling the introduction of devices with prominent digital features such as digital video disc (DVD) and audio compact disc (CD) players/recorders, digital cameras, and asymmetric digital subscriber line (ADSL)/cable modems into our world.

Furthermore, with the fast advancement of wireless communications standards and technologies since the mid-1990s, the demands for low-cost CMOS data converters of high data rate, high linearity, and high dynamic range that consume very little power have increased dramatically [1][2]. A commercialized application of CMOS data converters to wireless communications can be found in today's wireless networking routers and relaying devices, which are built based on the wireless local area network (WLAN) standards such as IEEE 802.11/a/b/g.

Most recently, the common focus of many research endeavors in the industry has been on searching for the best CMOS-based method/standard to transmit broadband multimedia signals (e.g., digital audio, digital video, and digital photography) wirelessly. One of the noteworthy candidates is the ultra-wideband (UWB) short-range (10 meters or shorter) digital transmission system, which operates at a carrier frequency within the range of 3.1 GHz to 10.6 GHz, and has a minimum signal bandwidth of 500 MHz [3]. With such high carrier frequencies and data rates, the analog-to-digital (A/D) and digital-to-analog (D/A) interfaces of the UWB system must be capable of operating at a fast sampling clock rate for a conversion accuracy that can be readily achieved within the matching limitations of existing CMOS technologies. In addition, it is known that one of the biggest advantages as well as the major technical challenge of the UWB standard is that it is a wireless technology of low transmission power, enabling prominent features such as the seamless switching and the information security. However, in the meantime, the low-power transmission imposes stringent dynamic-range limits on the A/D converters. Specifically, it is extremely challenging to design a high-speed (gigahertz-level) ADC capable of distinguishing the low-power desired signal from strong blockers such as background noise and spectral images.

Before proceeding, it is instructive to categorize data converters into two main types: Nyquist-rate and oversampling. Despite the naming, in practice Nyquist-rate data converters seldom operate precisely at the input signal's Nyquist rate ($f_s = 2f_{in}$); otherwise impractically acute filtering would be required. Rather, Nyquist-rate converters are often clocked at 1.5 to 10 times the Nyquist rate.

On the other hand, as we will see in Section 5.5, oversampling data converters use clock frequencies that are much higher than the Nyquist rate (typically 16 to 512 times higher), thereby maintaining the input signal power while expanding the spectral bandwidth of the *white-noise-like* quantization noise power from $2f_{in}$ to ($2f_{in} \cdot$ OSR), where OSR stands for the oversampling ratio. In effect, the in-band ($f < 2f_{in}$) quantization noise power is reduced by OSR times in comparison with that of the configurations clocked at $2f_{in}$, thus the signal-to-quantization-noise-ratio (SQNR) is increased by $10lg$(OSR) in decibels.

More aggressive SQNR improvement can be achieved by using a delta-sigma modulator as the quantization noise shaper, which ideally *squeezes* the noise power into the high-frequency range of the spectrum without affecting the low-frequency input signal. The recent trend of designing high-performance oversampling data converters emphasizes the methodology of trading sophisticated delta-sigma modulator design (e.g., cascaded modulators equipped with multibit quantization and digital error calibration/correction) for a low OSR (e.g., 16, 12, 8, or even 4), and a high resolution ranging from 14 to 16 bits [4][5][6][7]. It is predicted that the design results will be particularly suitable for video processing and wireless communications applications.

This chapter deals with the fundamental properties and design principles of CMOS A/D and D/A converters and presents a variety of data converter architectures and circuit implementations. The chapter also emphasizes applications of switched-capacitor (SC) techniques to the design of integrated CMOS data converters.

Chapter Outline

Section 5.2 presents a number of important performance parameters useful for specifying an integrated ADC or DAC. Section 5.3 presents an overview of Nyquist-rate DAC architectures and circuits. The effects of capacitor mismatch on SC data converters are evaluated. An introduction to mismatch error cancellation techniques is provided as well. Section 5.4 details the architectures and the circuit implementations of Nyquist-rate ADCs. Various ADCs, including flash, two-step, pipelined, cyclic, and successive-approximation ADCs, are investigated. The critical design issues such as the capacitor mismatch, input offset, and finite op-amp gain are discussed. Finally, Section 5.5 includes the principles of oversampling data converters, with emphasis on the discussion of SC delta-sigma modulators.

5.2 Performance Parameters of Data Converters

DAC Specifications

The fundamental function of a D/A converter (DAC) is to transform a digital word M_{in} into an analog (or continuous) output signal N_{out} (strictly speaking, the output is not truely continuous but has a finite number of distinct values), and the input-output relationship is given by

$$N_{out} = K_{ref} \cdot \frac{M_{in}}{2^M} \tag{5.1}$$

K_{ref} is an analog reference whose dimension may be electric voltage, current, or charge. The digital input M_{in} may be represented by either a binary (also called *binary-weighted*) code or a thermometer code, depending on the application's requirements; however, note that these two coding formats are interchangeable. The magnitude of M_{in} can be expressed as

$$M_{in} = \sum_{i=1}^{M} A_i 2^{M-i} = 2^M \cdot \sum_{j=1}^{2^M-1} B_j \tag{5.2}$$

Here, A_i's are the binary-code bits while B_j's are the thermometer-code bits. In general, to represent 2^M different digital input values, a binary code only needs M

Table 5.1 Comparison chart: 3-bit binary versus 7-bit thermometer.

Binary Code			Thermometer Code						
A_1	A_2	A_3	B_1	B_2	B_3	B_4	B_5	B_6	B_7
0	0	0	0	0	0	0	0	0	0
0	0	1	0	0	0	0	0	0	1
0	1	0	0	0	0	0	0	1	1
0	1	1	0	0	0	0	1	1	1
1	0	0	0	0	0	1	1	1	1
1	0	1	0	0	1	1	1	1	1
1	1	0	0	1	1	1	1	1	1
1	1	1	1	1	1	1	1	1	1

bits, whereby a thermometer code needs up to $(2^M - 1)$ bits. Table 5.1 compares a 3-bit binary code and the equivalent 7-bit thermometer code.

Among the binary bits shown in the table, A_1 is the *most significant bit* (MSB), and A_3 is the *least significant bit* (LSB). Among the thermometer bits, B_1 and B_7 are the MSB and LSB, respectively. The LSB is defined to be the bit in M_{in} that controls the smallest possible change in the value of N_{out}, while the MSB controls the largest possible change.

Note that in the binary-weighted setting, the transition from $(A_1\ A_2\ A_3) = 011$ to 100 may cause the output N_{out} to temporarily fall to zero (if A_2 and A_3 go to 0 slightly before A_1 goes to 1) or jump to its maximum value (if A_2 and A_3 go to 0 slightly after A_1 goes to 1), thereby breaking the converter's *monotonicity* (i.e., the characteristic of the output always increasing with the input), and introducing nonlinearity errors. This phenomenon is called the *switching glitch* or the *major-carry effect.* In contrast, the switching glitch issue does not concern the thermometer-coded setting, in which only one bit is changed at a time. Thus, the thermometer-coded converter is guaranteed to be monotonic and the nonlinearity is minimized (assuming components such as capacitors match perfectly).

If the dimension of K_{ref} is electric voltage (i.e., $K_{ref} \equiv V_{ref}$), then N_{out} has the same dimension as that of K_{ref} (i.e., $N_{out} \equiv V_{out}$) since the digital word M_{in} in Equation (5.1) is dimensionless. From Equations (5.1) and (5.2) we can write the following:

$$V_{out} = V_{ref} \cdot \frac{M_{in}}{2^M} = \frac{V_{ref}}{2^M} \cdot \sum_{i=1}^{M} A_i 2^{-i} = \frac{V_{ref}}{2^M} \cdot \sum_{j=1}^{2^M - 1} B_j \tag{5.3}$$

As a convention, the ideal minimum change in V_{out} (i.e., the corresponding change in V_{out} when one LSB changes in $M_{in}/2^M$) is defined as

$$V_{LSB} = V_{ref} / 2^M \tag{5.4}$$

Also, if it is assumed that the digital word is binary coded, the maximum value of V_{out}, which is also referred to as the *full-scale output voltage* (V_{fs}) and obtained when $A_i \equiv 1$, $i = 1, 2, \ldots, M$, can then be expressed as

$$V_{fs} = V_{ref} \sum_{i=1}^{M} 2^{-i} = V_{ref}\left(1 - 2^{-M}\right) = V_{ref} - V_{LSB} \tag{5.5}$$

The term *resolution* is often used to specify the number of digital input bits (in the case of a DAC) or output bits (in the case of an ADC). For instance, the resolution of the DAC discussed earlier is M bits. The term *accuracy* is used to specify the precision of the actual transfer curve (i.e., the curve that connects each of the output values) in comparison with the ideal one. From Equation (5.4), we see that the resolution of a DAC can be determined once the ratio (V_{ref}/V_{LSB}) is known. However, the proper representation of the accuracy is a little more involved because one has to take *offset errors*, *gain errors*, and *nonlinearity errors* into account [8][9][10][11]. In practice, a circuit's accuracy is usually presented in *effective number of bits* (ENOB).

The term *offset error* is defined to be the nonzero output voltage when the digital input equals 0 . . . 0—that is, the deviation of the output voltage (when $M_{in} = 0 \ldots 0$) from its ideal value (zero). It can be expressed in units of LSBs as

$$\varepsilon_{offset} = \frac{V_{out}}{V_{LSB}}\Big|_{0\ldots0} \tag{5.6}$$

The term *gain error* is defined as the difference between the slopes of the actual and the ideal transfer curves in units of LSBs, which is given by

$$\varepsilon_{gain} = \underbrace{\left(\frac{V_{out}}{V_{LSB}}\Big|_{1\ldots1} - \frac{V_{out}}{V_{LSB}}\Big|_{0\ldots0}\right)}_{\text{actual slope}} - \underbrace{\left(2^M - 1\right)}_{\text{ideal slope}} \tag{5.7}$$

Note that ideally when $M_{in} = 1 \ldots 1$, $V_{out} = V_{fs}$. Hence, the ideal slope of the DAC transfer curve is given by dividing (V_{fs}/V_{ref}) by 2^M, and the result is equal to ($2^M - 1$) based on Equation (5.5).

The term *nonlinearity error* refers to the deviation from the ideal transfer curve after both the offset and gain errors have been removed. The nonlinearities of a data converter are usually expressed as a combination of two items: *integral nonlinearity error* (INL) and *differential nonlinearity error* (DNL).

INL is often defined as the maximum deviation of the DAC transfer curve from a straight line connecting its end points [8][9]. However, in the literature, an alternative measurement called the *best-fit test* is widely used. In such a test, the difference between a certain output and its *best-fit* value (a point on the best-fit straight line) is recorded as the INL error at that point in units of LSBs [11]; as a result, an illustrative INL curve can be plotted based on these recorded errors. The latter way of defining INL is employed in this book.

DNL is normally defined, also in units of LSBs, as the deviation of the *vertical* step between two adjacent outputs from its ideal value (i.e., 1 LSB). In a manner similar to that presented earlier, a DNL curve is often used to illustrate the linearity performance of the converter. In addition, it is of practical interest to note that the DNL error between levels N and $(N + 1)$ can be found by simply subtracting the INL error of level N from that of level $(N + 1)$.

EXAMPLE 5.1

For a 3-bit *unipolar* (i.e., single-sign) DAC with $V_{ref} = 4V$, what is the maximum value of the output voltage (V_{fs}) in an ideal situation? If the following measured output voltages (in volts) are recorded: {−0.005, 0.497, 0.999, 1.498, 2.004, 2.503, 3.002, 3.504}, find the offset error, gain error, maximum INL error, and maximum DNL error (all in units of LSBs). Lastly, calculate the ENOB of the DAC.

Solution: Based on Equations (5.4) and (5.5), we can immediately obtain the following:

$$V_{fs} = V_{ref} - V_{LSB} = 4 - (4/8) = 3.5\,V \tag{5.8}$$

Utilizing Equation (5.6), we can obtain the offset error:

$$\varepsilon_{offset} = \frac{-0.005}{V_{LSB}} = -0.01\,LSB \tag{5.9}$$

Utilizing Equation (5.7), we can obtain the gain error:

$$\varepsilon_{gain} = \frac{3.504 + 0.005}{V_{LSB}} - (2^3 - 1) = 0.018\,LSB \tag{5.10}$$

To determine the maximum INL and DNL errors, we need to derive all the INL and DNL errors first. The offset and gain errors must be removed from the measured output values before we start calculating the INL or DNL errors. Assuming the ith item in the measured output sequence is V_{out_i}, we can write its scaled value (free of offset and gain errors; in units of LSBs) as follows [9][11]:

$$V_{nout\_i} = \left(\frac{V_{out\_i}}{V_{LSB}} - \varepsilon_{offset} \right) - \left(\frac{i-1}{2^3 - 1} \right) \cdot \varepsilon_{gain} \tag{5.11}$$

After some calculations, we obtain all the eight scaled output values (in units of LSBs):

{0.000, 1.001, 2.003, 2.998, 4.008, 5.003, 5.999, 7.000}

Next, the INL errors can be obtained by calculating the difference between the ideal values: {0, 1, 2, . . . , 7} and the preceding. The results are as follows:

{0, +0.001, +0.003, −0.002, +0.008, +0.003, −0.001, 0}

Thus, the maximum INL error is equal to 0.008 LSB (<0.5 LSB). The DNL errors can be found by calculating the difference between adjacent INL errors and the results are

{+0.001, +0.002, −0.005, +0.010, −0.005, −0.004, +0.001}

So the maximum DNL error is equal to 0.01 LSB (<0.5 LSB).

The ENOB of the DAC can be obtained by comparing the largest deviation of the measured voltages from the ideal values (in this case, 0.005 V) with the full-scale output voltage V_{fs}. In other words, the ENOB is specified in such a way that the largest deviation is less than the full-scale value divided by 2^{ENOB}. As the result, the ENOB is given by

$$ENOB = \log_2 \left(\frac{V_{fs}}{\varepsilon_{\max}} \right) = \log_2 \left(\frac{3.5}{0.005} \right) \approx 9.45 \text{ bits} \tag{5.12}$$

Interestingly, the ENOB is larger than the specified resolution of the DAC (i.e., 3 bits). This implies the circuit is so well designed that its precision is comparable to that of a DAC built to a much higher resolution (i.e., about 9.45 bits).

The foregoing specification of the DAC's ENOB is also referred to as the *absolute accuracy* since it has taken all errors (offset, gain, and nonlinearity) into account. In comparison, the *relative accuracy* accounts for the nonlinearity errors only (i.e., free of offset and gain errors), and it can be found by substituting ε_{max} (i.e., 0.005 V) in the foregoing equation with the magnitude of the maximum INL error (i.e., $0.008 \times 0.5 = 0.004\,\text{V}$). Consequently, the relative accuracy reads a larger ENOB than the absolute accuracy does.

Note that in the foregoing example, ε_{max} *happens to* have the same magnitude as that of the offset error, which is the smallest possible output signal. Thus, Equation (5.12) in a sense interprets the definition of *dynamic range* (DR), which is the ratio of the largest possible to the smallest possible input/output signal values (here, it is the ratio with respect to the output signal). Often, DR is expressed in decibels:

$$DR = 20\lg\left(\frac{V_{fs}}{V_{\min}}\right)\text{dB} \tag{5.13}$$

where V_{min} stands for the smallest possible output signal. So the DAC in Example 5.1 has a dynamic range of 56.9 dB. It is also of interest to derive the following:

$$\frac{DR}{ENOB} \cong \frac{56.9}{9.45} \approx 6.02\ \text{dB/bit} \tag{5.14}$$

As seen earlier, a thermometer-coded DAC is guaranteed to be *monotonic*. This is because its maximum DNL error rarely exceeds 1 LSB. Monotonicity is often regarded as a qualitative feature rather than a quantitative parameter; hence the detailed quantitative analysis of monotonicity is waived here for brevity.

There are many other important performance parameters useful for specifying a DAC. For instance, the term *signal-to-noise-ratio* (SNR) stands for the ratio of the largest possible signal power to the noise power (both expressed in RMS values). For a DAC, SNR and dynamic range (DR) are usually interchangeable. Additionally, data converters suffer from the *sampling time uncertainty* or *aperture jitter*, in a way similar to that we saw in relation to testing sample-and-hold (S&H) circuits in Chapter 3. We will discuss both SNR and aperture jitter in more detail when we study ADC specifications shortly.

ADC Specifications

With a fundamental function complementary to that of a DAC, an M-bit A/D converter (ADC) is used to transform an analog input C_{in} (e.g., a sinusoidal waveform) into an M-bit digital output words D_{out} (strictly speaking, the output has a staircase waveform rather than a sequence of digital bits). The input-output relationship of an ideal M-bit ADC is given by

$$K_{ref} \cdot \frac{D_{out}}{2^M} = C_{in} \pm \varepsilon_q, \quad -\frac{K_{ref}}{2^{M+1}} \le \varepsilon_q \le \frac{K_{ref}}{2^{M+1}} \tag{5.15}$$

Here, K_{ref} is an analog reference whose dimension may be electric voltage, current, or charge. The digital output D_{out} may be represented by either a binary code or a thermometer code. Yet the most important item in the foregoing equation is ε_q, known as the *quantization noise*; it is defined as the difference between the actual analog input C_{in} and the value of the staircase output ($K_{ref} \cdot D_{out}/2^M$). Assuming both C_{in} and K_{ref} are electric voltages (i.e., V_{in} and V_{ref}) and rearranging Equation (5.15), we can obtain the following:

$$D_{out} = 2^M \cdot \frac{V_{in}}{V_{ref}} + E_q \tag{5.16}$$

This indicates that theoretically the quantization noise E_q (in units of LSBs) can be modeled as an additive noise source to the ADC. Ideally, the magnitude of E_q shall be limited to within −0.5 LSB, and +0.5 LSB, otherwise a code transition point may be missing (i.e., a missing code).

Figure 5.1 illustrates a circuit using a pair of *complementary* A/D and D/A converters, which are specified to the same resolution and using the same reference voltage V_{ref}. From Equation (5.16), it can be found that this pair of ADC and DAC is used to extract the quantization noise V_q.

Additionally, note that the errors induced by the quantization noise normally concern ADCs only as the quantization noise in essence originates from the ambigu-

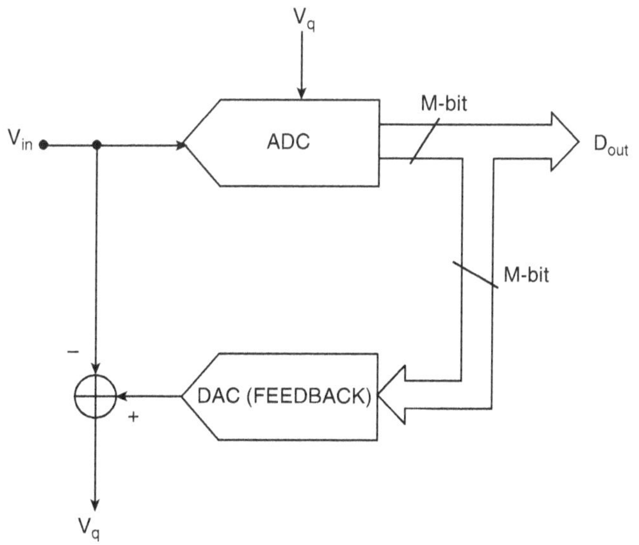

Figure 5.1 Extracting quantization noise through a pair of complementary A/D and D/A.

ity that multiple analog input levels are converted into the same digital code [9][11]. This makes physical sense since for an ADC, the analog input signal comes with an infinite number of levels while the output signal has only a few discrete values; in comparison, for a DAC, both input and output have a finite number of levels, which are corresponded in a "one-to-one" fashion. For example, if the DAC's input has 2^M discrete values, then 2^M discrete output values will be produced correspondingly. That is, ideally the DAC output is *well defined* by the input, and as a result, no signal ambiguity will occur.

The term *resolution* defines the number of digital output bits of an ADC. The *accuracy* of an ADC may also be expressed as the *effective number of bits* (ENOB), and is closely related to the specification of *dynamic range.*

Most ADC parameters can be related to the corresponding DAC parameters in a complementary manner. The *offset error* of an ADC is defined as the deviation of the first digital code transition point (i.e., when D_{out} jumps from 0 . . . 0 to 0 . . . 1) from its ideal position (0.5 LSB). It is given by (in units of LSBs)

$$\varepsilon_{offset} = \left.\frac{V_{in}}{V_{LSB}}\right|_{0\ldots1} - \frac{1}{2} \tag{5.17}$$

Similarly, the term *gain error* of an ADC is given by (in units of LSBs)

$$\varepsilon_{gain} = \underbrace{\left(\left.\frac{V_{in}}{V_{LSB}}\right|_{1\ldots1} - \left.\frac{V_{in}}{V_{LSB}}\right|_{0\ldots1}\right)}_{actual\ slope} - \underbrace{(2^M - 2)}_{ideal\ slope} \tag{5.18}$$

The *integral nonlinearity error* (INL) of an ADC may be defined as the maximum deviation of the code transition points on the ADC transfer curve from a straight line, which connects the origin and the final transition point (when D_{out} becomes 1 . . . 1), after both offset and gain errors are removed [8][9]. Alternatively, the difference between each actual code transition point and its *best-fit* position is recorded in units of LSBs and used for plotting an INL curve [11].

The *differential nonlinearity error* (DNL) of an ADC is defined, also in units of LSBs, as the deviation of the *horizontal* step (or code width) between two adjacent code transition points from its ideal value (i.e., 1 LSB). Similarly, a DNL curve is often used to illustrate the linearity performance of the converter. Also, the DNL error between code transition points N and $(N + 1)$ can be found by simply subtracting the INL error of point N from that of point $(N + 1)$.

The complement to the DAC's monotonicity in the ADC case is called *missing code*. Similar to the DAC, an ADC is guaranteed *not* to miss any digital codes if the maximum DNL error is less than 1 LSB. However, if the magnitude of the quantization noise (E_q) exceeds the range of −0.5 LSB to +0.5 LSB, a missing code will happen.

One of the most important ADC parameters is the *signal-to-noise ratio* (SNR). Note that when the input signal is a sinusoid, harmonic distortions may occur in the ADC's output signal. The parameter that takes both noise and harmonic distortion into account is called the *signal-to-noise-plus-distortion ratio* (SNDR). For an ADC, SNR is normally greater than SNDR. For the maximum sinusoidal input with a peak-to-peak value equal to V_{ref}, the expression of SNR is given by (in decibels)

$$SNR = 20\lg\left(\frac{v_{in\_max}}{v_n}\right) \tag{5.19}$$

Here, V_{in_max} is the RMS value of the maximum input signal, which is usually equal to that of V_{ref}, and V_n is the RMS value of the noise power. If it is assumed that the input signal frequency is high, then the quantization noise power can be approximated to be uniformly distributed between $-0.5V_{LSB}$ and $+0.5\ V_{LSB}$. After some mathematical manipulations, we can then obtain the following:

$$SNR = 20\lg\left(\frac{v_{in\_max}}{v_n}\right) = 20\lg\left(\frac{2^{M-1}V_{LSB}/\sqrt{2}}{V_{LSB}/\sqrt{12}}\right) = 6.02M + 1.76\ \text{dB} \tag{5.20}$$

The preceding equation specifies the SNR limit for an M-bit ADC (without the helps from oversampling or noise shaping). However, SNR usually decreases with the magnitude of the input signal, and the degradation in SNR is given by

$$\Delta SNR = 20\lg\left(\frac{v_{in}}{v_{in\_max}}\right) \tag{5.21}$$

For instance, if it is assumed that an ADC has a reference voltage V_{ref} of 4 V, and that the peak-to-peak value of V_{in} is equal to 80 mV, then SNR will be reduced from its largest possible value by about 34 dB! Furthermore, if the peak-to-peak value of V_{in} is reduced to 1 V_{LSB}, then the degradation in SNR will be as much as 6.02M dB, and the resultant SNR is only 1.76 dB.

In practice, the resolution of a Nyquist-rate ADC (without oversampling or noise shaping) is usually specified to equal or approximate the ENOB. This is different from the DAC case. For instance, recall that in Example 5.1 the DAC is specified to a 3-bit resolution, but its ENOB indeed reaches 9.45 bits.

To get a feel for the aforementioned difference, consider the fact that the definition of SNR (or ENOB) for an ADC must deal with the quantization noise, which is dictated by how many actual bits of resolution can be physically built to, whereas in the scenario of a DAC, the quantization noise is not a major concern, and the specification of its SNR (or ENOB) emphasizes other circuit imperfections such as gain and nonlinearity errors, which rely on component matching accuracy rather than on the number of input digits.

The last noteworthy source of error is called the *sampling-time uncertainty* or *aperture jitter*. Both DAC and ADC are subject to this type of error. But it raises a more serious concern in the design of ADCs than of DACs, especially when it comes to high-speed (on the order of 50 MHz) and medium-to-high-resolution (8–14 bits) ADCs.

Furthermore, note that the essential characteristic of an ADC is sampling continuous-time signals, which is the same as that of an S&H. As a result, Equation (3.68) can be used to quantify the error due to aperture jitter in an ADC setting. Assuming that the input V_{in} applied to an M-bit ADC is a sinusoid with a peak-to-peak voltage of V_{ref}, and that the input signal frequency is f, we can write the following for the ADC based on Equation (3.68):

$$\varepsilon_{aperture\_jitter}(t) = \sigma \cdot \left| \frac{dV_{in}(t)}{dt} \right| = \frac{V_{ref}}{2} \cdot 2\pi \cdot f \cdot \sigma \cdot \left| \cos(2\pi \cdot f \cdot t) \right| \tag{5.22}$$

where σ stands for the aperture jitter at time t. As explained in Chapter 3, the maximum error voltage occurs at the zero crossing ($t = 0$), which is given by

$$\varepsilon_{aperture\_jitter\_\max} = V_{ref} \cdot \pi \cdot f \cdot \sigma = 2^{M} \cdot V_{LSB} \cdot \pi \cdot f \cdot \sigma \tag{5.23}$$

In practice, it is desirable that this maximum error voltage be kept less than 0.5 V_{LSB}; otherwise the ADC's accuracy will be corrupted. Rearranging (5.23), we obtain the following:

$$\sigma \le \frac{1}{2^{M+1} \cdot \pi \cdot f} \tag{5.24}$$

EXAMPLE 5.2

Consider a 9-bit Nyquist-rate ADC that is used to convert a 250-MHz sinusoidal input signal with a peak-to-peak voltage value of V_{ref}. What is the maximum allowable value of the sampling-time uncertainty according to Equation (5.24)? What is the maximum allowable value of the sampling-time uncertainty if the output needs to maintain a 9-bit ENOB?

Solution: Based on Equation (5.24), we can obtain the following:

$$\sigma_{ap\_max} = \frac{1}{2^{10} \cdot \pi \cdot 250 \times 10^6} \approx 1.24 \text{ ps} \tag{5.25}$$

From Equation (5.20) we obtain the equivalent SNR for a 9-bit ENOB:

$$SNR = 6.02 \times 9 + 1.76 = 55.94 \text{ dB} \tag{5.26}$$

Utilizing Equation (3.69), we write the theoretical SNR limit due to aperture jitter:

$$SNR_l = -20\lg\left(2\pi \cdot f \cdot \frac{\sigma_l}{\sqrt{2}}\right) = -20\lg\left(2\pi \cdot 250 \times 10^6 \cdot \frac{\sigma_l}{\sqrt{2}}\right) \geq SNR \tag{5.27}$$

Hence,

$$\sigma_l \leq 1.44 \text{ ps} \tag{5.28}$$

The foregoing development indicates that Equation (5.24) provides a tight sufficient condition for maintaining a 9-bit accuracy (ENOB) when aperture jitter is the dominant source of error in the ADC. However, in practice many other errors such as the quantization noise or component mismatch must be taken into account.

INL, DNL, and Quantization Noise

It is of practical interest to clarify the relationships among INL, DNL, and quantization noise. First, note that a data converter (either a DAC or an ADC) with an *M*-bit resolution must keep the maximum absolute values of INL and DNL less than (or equal to) 1/2 LSB; otherwise the true resolution (i.e., accuracy) that it can provide will be less than *M* bits. For example, a 3-bit DAC with an absolute INL or DNL that is greater than 1/2 LSB but smaller than 1 LSB actually has an ENOB equivalent to that of a *perfectly* accurate 2-bit DAC. If the absolute INL or DNL exceeds 1 LSB, then the resultant ENOB will be equal to only 1 bit, and also, the converter may become nonmonotonic!

This problem may be better understood through the following example. If we assume that the INL errors (starting from the LSB and in units of LSBs) of the 3-bit DAC are given by {0, 0.5, –0.5, 0.5, –0.5, 0.5, –0.5, 0}, then the output level that corresponds to the input binary code 001 will be indistinguishable from that corresponding to 010, and it can be found that similar signal ambiguities will also occur between 011 and 100 as well as between 101 and 110. Thus, although the DAC is physically built to a 3-bit resolution, it fails to resolve eight discrete levels; instead it can only resolve five levels. In other words, the converter is equivalent to an ideal 2-bit DAC (rounded from 2.324-bit).

Second, after the offset and gain errors of a data converter are removed, which nonlinearity error shall be considered the limiting factor in determining the ENOB of a data converter, INL or DNL? In other words, whose maximum absolute value shall be set to 1/2 LSB in the worst-case analysis?

The answer greatly depends on the type of codes that the data converter employs. Although there are many other digital coding schemes such as Gray coding, in this book we only discuss the two most commonly seen coding schemes in data conversion applications: binary and thermometer. A general rule of thumb is noted as follows: If the data converter is built to facilitate a binary-weighted digital signal (i.e., a binary-weighted converter), then the DNL requirement is more stringent than the INL requirement, meaning DNL should be the limiting factor. In contrast, for a thermometer-coded converter, the INL requirement is more difficult to meet, thus INL should be the limiting factor.

Third, note that the quantization noise exists even in an ideal ADC, whereas the gain error, DNL, or INL occurs only when there are nonidealities in the circuit. As mentioned earlier, the inevitable existence of the quantization noise in an ADC makes physical sense since the analog input signal has an infinite number of levels, while the output signal is a discrete-time signal with a finite number of levels. In an ideal ADC, the quantization noise error does exist but is no greater than 1/2 LSB (an absolute value), and it manifests itself as a saw-tooth waveform in the time-domain, which is symmetrically alternating between –1/2 and +1/2 LSB. In a real ADC with circuit imperfections, the resulting DNL and INL errors distort the saw-tooth waveform from its ideal geometrical shape, and as the result, it becomes asymmetric.

A few other performance parameters, such as *glitch impulse area* and *settling time* for DACs and *bit-error-rate* (BER) and *conversion time* for ADCs, can be seen in the literature. Their detailed descriptions are waived here for brevity. As a final note, the two most noteworthy industry standards dealing with the performance measurements of data converters are IEEE Standard 764-1984 [12] and IEEE Standard 1241-2000 [13].

In summary, this section presented the key performance parameters of A/D and D/A data converters. The mathematical formulae useful for quantifying these parameters were provided.

5.3 Nyquist-Rate DACs

Integrated Nyquist-Rate DACs

Recall that for a DAC, the digital input word M_{in} in Equation (5.1) is dimensionless, and the analog output signal N_{out} has the same dimension as that of K_{ref}.

When the dimension of K_{ref} is electric voltage, the converter is called the *voltage-mode* (or *ladder*) DAC. A voltage-mode DAC is essentially a voltage meter (that is, *potentiometer*), and it is often realized based on the resistor-string approach [14][15].

A 2-bit thermometer-coded resistor-string DAC is shown in Figure 5.2(a). As the schematic shows, the on/off operations of the switches, which are built of MOS transistors, are controlled by the read-out bits of the digital binary-to-thermometer decoder. Note that a 4-bit instead of a 3-bit thermometer code is applied such that the maximum value of V_{out} is limited to $V_{ref} - V_{LSB}$, or, equivalently, 0.75 V_{ref}.

Generally speaking, to realize an M-bit thermometer-coded resistor-string DAC, 2^M resistors (all equal sizes) are needed. Thus, the number of resistors, the amount of power dissipation/silicon area, and the RC time constant increase exponentially with M, making this configuration unsuitable for low-power, high-speed, or small form-factor applications.

An improved resistor-string DAC that requires only $2^{(M/2+1)}$ resistors for an M-bit resolution is reported by Holloway [14]. The reduction in the number of resistors is achieved by using two resistor strings, each of which contains $2^{(M/2)}$ equal resistors. In the meanwhile, the M-bit accuracy is maintained by using the second string to *interpolate* between two adjacent nodal voltages on the first string, which is equivalent to substituting each resistor in the first string with a string of $2^{(M/2)}$ resistors. Therefore, the effective number of resistors is given by $2^{(M/2)} \cdot 2^{(M/2)} = 2^M$, although only $2^{(M/2+1)}$ resistors are actually used. To quantify, consider an 8-bit resistor-string DAC; the double-string configuration saves up to 224 resistors as compared to the single-string configuration.

Nevertheless, the large time constant due to the switched-RC (C represents a combination of the MOS transistors' gate-to-source and gate-to-substrate capacitances as well as the parasitic capacitances) network still restricts the achievable speed of a resistor-string DAC.

When the dimension of K_{ref} in Equation (5.1) is electric current, the converter is called the *current-mode* (or *current-steering*) DAC, in which multiple current sources are used to distinguish between different digital inputs [16][17].

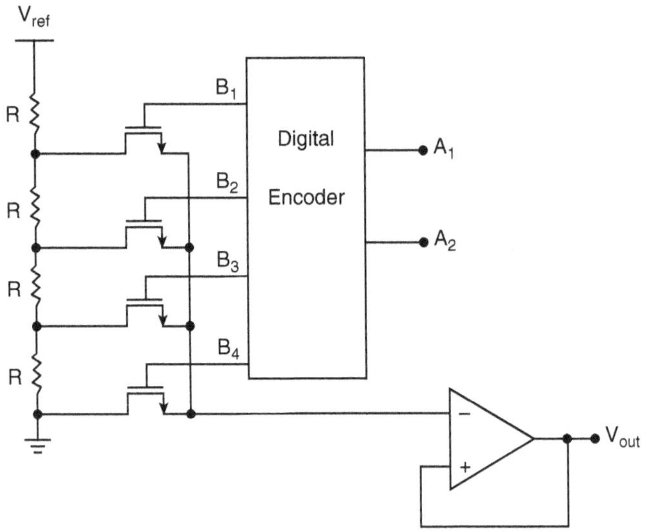

(a)

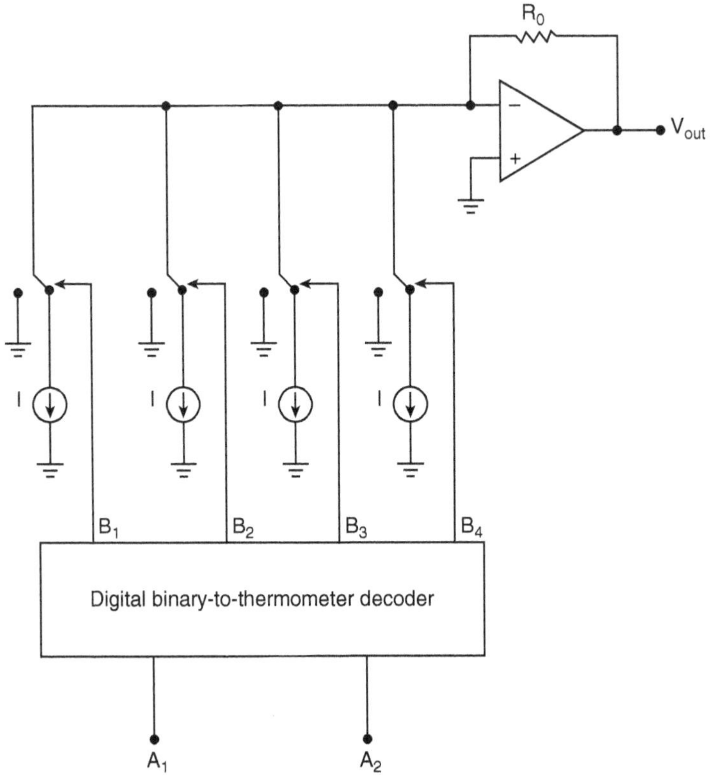

(b)

Figure 5.2 Non-SC thermometer-code DACs. (a) Resistor-string DAC. (b) Current-steering DAC.

A 2-bit thermometer-coded current-steering DAC is shown in Figure 5.2(b). In general, this type of DAC architecture employs 2^M equal current sources for transforming an M-bit binary input to avoid overloading. As shown, a 2-bit input binary code is decoded into the equivalent 4-bit thermometer code, with each bit directing the output of a current source, either to the op-amp or to ground. And the currents sent to the inverting terminal of the op-amp are superposed and converted into an analog voltage by the feedback resistor R_0.

In general, the conversion accuracy of a current-steering DAC greatly depends on the matching accuracy of the current sources as they are not exactly the same in reality. To resolve the nonlinearity problems due to mismatch errors between current sources, Schouwenaars et al. [17] reported a dynamic current matching technique, and a detailed description of the technique can be found in [18]. The basic idea of this technique is to make each current source *copy* the output of a single reference current source I_{ref} in a periodic manner. In effect, the mismatch errors between current sources are minimized for they are calibrated by the same current reference.

The *current-copying* operation (or *calibration*) of each current source may be initiated when the corresponding thermometer bit goes to either zero or one, and it usually takes less than one-tenth of a conversion period (i.e., the output settling time).

In most cases, it is desired that the calibration be made transparent to the user (i.e., a *background calibration*), so an additional current source is needed to substitute for the one that is undergoing calibration—the previously *unused* 2^Mth current source is now up to the job. Examination of the literature will show that there are various background calibration techniques for current-mode Nyquist-rate DACs. Besides [18], the noteworthy references include [19][20][21].

Finally, the number of current sources and the amount of power dissipation in the thermometer-code configuration increases exponentially with M. As a result, when $M > 6$, *segmented* current-steering DACs that contain thermometer-coded MSBs and binary-weighted LSBs are normally used in practice. The interested reader is referred to the literature for examples.

Nyquist-Rate SC DACs

When the dimension of K_{ref} in Equation (5.1) is electric charge, the converter is called the *switched-capacitor* (SC) (or *charge-mode*) DAC as it is often constructed based on SC techniques. A 2-bit thermometer-coded SC DAC is shown in Figure 5.3(a). This type of SC DAC is often referred to as the *parallel* SC DAC because the input digital bits are read in a parallel fashion.

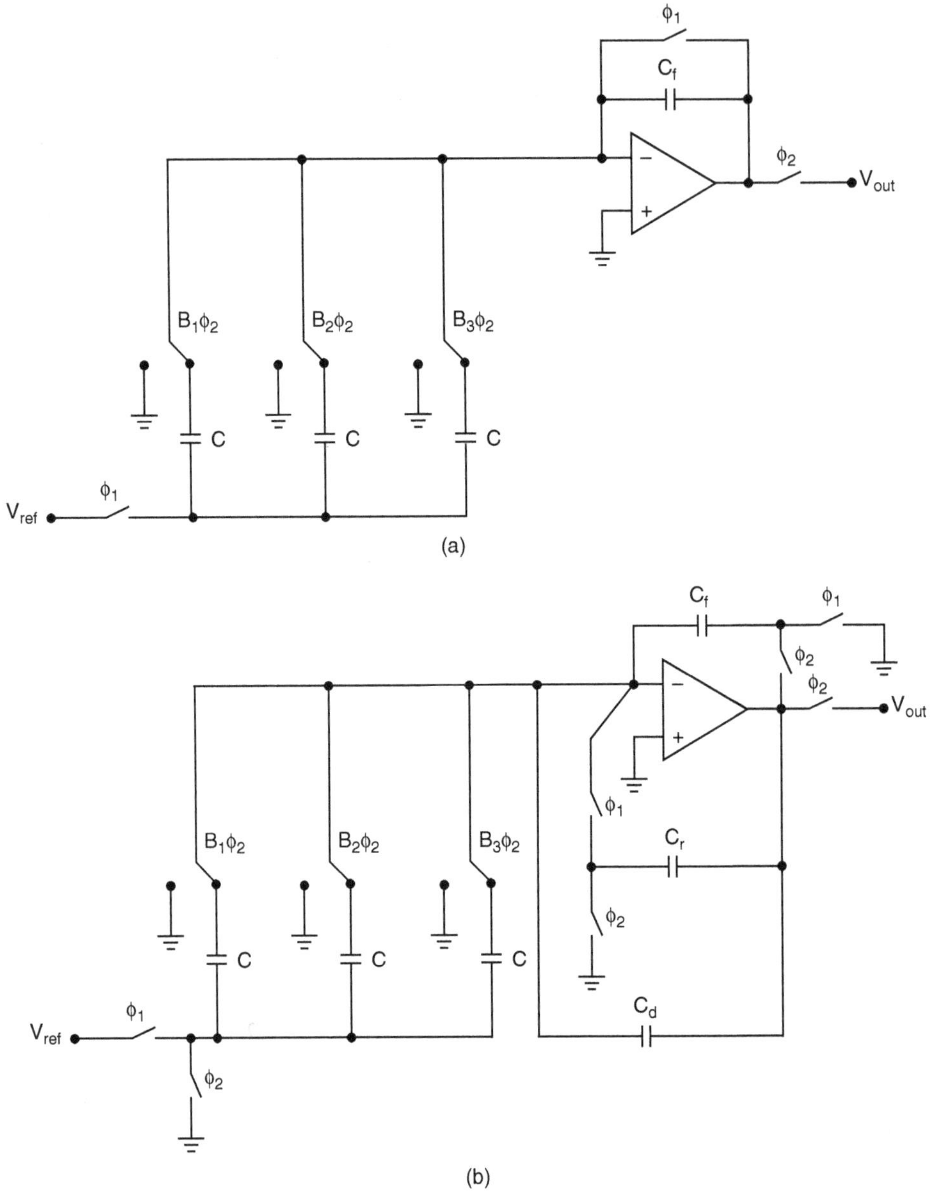

Figure 5.3 (a) Thermometer-coded SC DAC. (b) Improved SC DAC.

The circuit operates as follows. The input digital word is a thermometer code with bits B_1, B_2, and B_3. When $\Phi_1 = 1$, the DAC is being reset, the feedback capacitor C_f (C_f is equal to $4C$ in this case) is discharged, and the three input capacitors (all equal sizes, C) are charged by the reference voltage V_{ref}. Next, when $\Phi_2 = 1$, the DAC is in the conversion mode, and the charges on the first k activated ($k \leq 3$) input capacitors are coupled onto C_f, generating an output voltage $V_{out} = (k/4)\ V_{ref}$.

Apparently, this SC DAC is sensitive to parasitic capacitances, which normally contribute to a gain error. And when the input Φ_1 switch is turned off, a charge dependent on V_{ref} is injected onto each input capacitor, which is then sent to the output and causes harmonic distortions. The DAC also suffers from some other circuit imperfections such as input offset error, finite op-amp gain, and finite op-amp bandwidth (see Chapter 7). Moreover, the op-amp output must slew between $(k/4)\ V_{ref}$ and an input offset voltage, which is very close to 0 V, every time the DAC's operation mode changes (i.e., from the reset mode to the conversion mode or vice versa).

To alleviate these problems, an improved SC DAC shown in Figure 5.3(b) can be used. This SC DAC is constructed from the SC amplifier reported by Martin et al. [22].

In accounting for the many useful features that this ingenious design is able to offer, we can immediately find that the arrangement of input switches not only guarantees the circuit to be parasitic insensitive, but also makes it possible to change the sign of the DAC output. Also, the charge injection error can be canceled by simply adopting a fully differential configuration.

Here, the errors due to the input offset voltage V_{off} (and the flicker noise) of the op-amp are also removed by charging the activated input capacitors and the feedback capacitor C_f to the input offset voltage during $\Phi_2 = 1$, and canceling them out during the following $\Phi_1 = 1$ [22]. In such an arrangement, the finite op-amp gain effect is alleviated as well.

Furthermore, the slew rate requirement for the op-amp is greatly relaxed through the use of a reset capacitor C_r. The basic idea here is that instead of abruptly shorting the input and output terminals of the op-amp and making the output voltage drop near 0 V once $\Phi_1 = 1$, we use the capacitor C_r, which has *memorized* the output voltage at the end of the previous conversion mode ($\Phi_2 = 1$), to insert a voltage step between the input and output terminals as Φ_1 jumps to 1. In effect, between modes, the op-amp's output voltage needs to alternate between V_{out} and $(V_{out} - V_{off})$, thereby reducing the slew magnitude to only V_{off}. Although theoretically the capacitance of C_r would not affect the transfer function of the SC DAC [22], it needs to be assigned properly such that the tradeoff between capacitance spread and silicon area is balanced.

Interestingly, an additional feedback capacitor C_d is connected around the op-amp. It is an optional *deglitching* capacitor according to Matsumoto and Watanabe [23], that forms a feedback loop in case all the switches around the op-amp are open. The capacitance of C_d is normally quite small to avoid a high-frequency charge leakage, which could become a more pronounced problem when the input digital codes are changing fast.

Another important type of charge-mode DAC is called the *serial* or *cyclic* (also referred to as the *algorithmic*) DAC, which uses only a small number of components such as capacitor and voltage buffer to perform the D/A conversion. The term *serial* suggests that the digital input must be read in a serial or one-by-one manner, while the other term *cyclic* indicates that the same circuit is used for transforming all the digital bits in a recycling manner. A simple M-bit cyclic SC DAC is shown in Figure 5.4.

The circuit operates as follows. Prior to the conversion, the two equal-valued capacitors (C_1 and C_2) are discharged to ground via the reset switches, which are controlled by the clock pulse Φ_r; they are then turned off and will not be on until all the digital bits are converted. The input digital word is converted from the parallel into the serial format, and it is assumed that the least significant bit (LSB) B_M is the first bit to be converted. When $\Phi_1 = 1$, C_1 is either charged to the reference voltage V_{ref} or discharged to ground, depending on the value of B_M (1 or 0). Next, when $\Phi_2 = 1$, C_1 and C_2 are connected in parallel and the charge on C_1 acquired during the previous "$\Phi_1 = 1$" phase is thus shared equally between C_1 and C_2, resulting in an output voltage at the end of the first bit-conversion cycle, which is given by

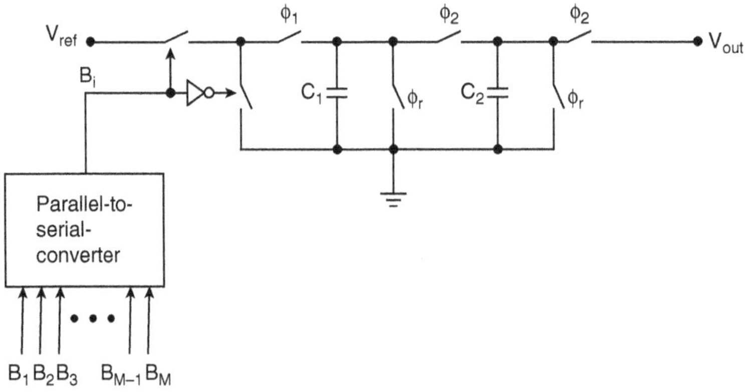

Figure 5.4 Cyclic SC DAC.

$$V_{out}(1) = B_M\left(\frac{V_{ref}}{2}\right) \tag{5.29}$$

We realize from the preceding that the SC configuration around C_1 and C_2 is equivalent to a sample-and-hold (S&H) with a one-half voltage gain.

After Φ_2 is turned off, the next least significant bit B_{M-1} enters. Applying the principle of charge conservation to the system, we can find the output voltage at the end of the second bit-conversion cycle, which is given by

$$V_{out}(2) = \frac{(B_{M-1}V_{ref}C_1 + B_M V_{ref}C_1/2 + B_M V_{ref}C_2/2)}{(C_1 + C_2)} = \left(B_{M-1} + \frac{B_M}{2}\right)\left(\frac{V_{ref}}{2}\right) \tag{5.30}$$

This process continues till the end of the Mth bit-conversion cycle, and it can be found that the desired analog output voltage corresponding to the complete input word is given by

$$V_{out}(N) = \left(\sum_{i=1}^{M}\frac{B_i}{2^i}\right)V_{ref} \tag{5.31}$$

Thus, if the system clock cycle is equal to T, then it will take a cyclic DAC MT to convert an M-bit digital word. As a result, cyclic DACs are normally used for slow and medium-speed D/A conversions in practice.

A well-known derivative of cyclic DAC is called the *pipelined* DAC, which trades multiple conversion stages (normally M stages for an M-bit conversion) for a readout rate M-time faster than that of the cyclic configuration. In other words, the analog voltage given by Equation (5.31) will show up at the DAC output every clock cycle T (after the first MT for *warming up*; see Section 5.4). The pipelined methodology is also widely adopted in the design of ADCs; hence, we will explore its key merits in Section 5.4.

Lastly, cyclic DACs with more sophisticated structures have been developed to resolve circuit imperfections such as parasitic capacitances or clock feedthrough errors—for example, the DAC introduced by Matsumoto and Watanabe [24]. The interested reader is referred to the literature for more details.

Matching Accuracy of Data Converters

As discussed earlier, for the thermometer-coded current-mode DAC, the mismatches betweens its current sources will introduce errors to the output; thus, calibration will be needed to correct these errors. Consider an M-bit thermometer-coded current-mode DAC; if we assume that each current source has an equal *absolute* mismatch error $|\Delta I|$ and the sum of all the mismatch errors are equal to zero [18], then it can be shown that the worst-case absolute INL (in units of LSBs) is given by

$$|INL|_{\max} = \frac{2^{M-1} \cdot |\Delta I|}{I} \tag{5.32}$$

where I is the ideal current, and it represents 1 LSB in a thermometer-coded situation ($I > 0$). As shown in Section 5.2, the worst-case INL presented earlier must be less than 1/2 LSB to prevent the DAC's accuracy from being corrupted. Thus, the maximum allowable current source mismatch is given by

$$|\Delta I|_{\max} = \frac{I}{2^M} \tag{5.33}$$

To quantify, if it is assumed that I is set to $4\,\mu$A, and that M is set to 10 bits, then the maximum allowable mismatch error will be as small as 3.9 nA!

As mentioned in Section 5.2, DNL is a less critical limiting factor compared to INL in a generic thermometer-coded configuration. Here, we can find that the worst-case absolute DNL (in units of LSBs) is given by

$$|DNL|_{\max} = \frac{|\Delta I| + I}{I} - 1 = \frac{|\Delta I|}{I} \tag{5.34}$$

Equating the preceding to 1/2 LSB, we realize that maximum allowable current source mismatch is as large as $0.5I$. Again, this shows that the DNL requirement is much easier to fulfill than the INL requirement in the thermometer-coded DAC.

In a manner similar to that of the current-mode DAC, we can obtain the maximum allowable resistor mismatch for an M-bit thermometer-coded resistor-string DAC. For the sake of brevity, only the result is provided as follows:

$$|\Delta R|_{\max} = \frac{R}{2^M} \tag{5.35}$$

As for an SC data converter, the achievable accuracy (ENOB) of the converter is normally dictated by errors due to mismatches between capacitors. The best capacitor matching accuracy (i.e., $\Delta C/C$) of existing CMOS technologies is about 0.02%. Theoretically speaking, the maximum ENOB of a Nyquist-rate thermometer-coded SC CMOS data converter is about 12-bit, if no error cancellation is applied [9]. In fact, when other noise sources are taken into account, it is hard to obtain an ENOB larger than 9 bits (i.e., $\Delta C/C \approx 0.2\%$).

EXAMPLE 5.3

The circuit shown in Figure 5.5 can be used to measure the mismatch error between two nominally equal valued capacitors used in SC DACs or ADCs. As the schematic shows, the switches are controlled by two nonoverlapping clock phases, Φ_1 and Φ_2. Describe the operation of the circuit, and examine the relationship between the voltage at node A (i.e., V_a) and the reference voltage V_{ref} (assume $V_{ref} = 1\,\text{V}$). If the measured V_a is a square wave whose magnitude is toggling between 0 V (when $\Phi_1 = 1$) and 1 mV (when $\Phi_2 = 1$), and it is assumed that capacitor mismatch is the dominant source of error in the SC converter, estimate ENOB of the converter. (Consider both thermometer-coded and binary-weighted coding schemes).

Solution: The circuit operates as follows. When $\Phi_1 = 1$, the voltage across C_1 is charged to V_{ref} while C_2 is discharged to ground. Next when $\Phi_2 = 1$, node A is floating, and C_1 and C_2 are in series. Applying the principle of charge conservation, we can find that when Φ_2 jumps to 1, the charge flows to the left side of node A should cancel the charge flows to the right—that is

$$C_2(V_a - V_{ref}) + C_1(V_a + V_{ref}) = 0 \tag{5.36}$$

Thus, the relationship between V_a and V_{ref} is given by

$$V_a = \left(\frac{C_2 - C_1}{C_2 + C_1}\right) \cdot V_{ref} = \frac{\Delta C}{(C_2 + C_1)/2} \cdot \left(\frac{V_{ref}}{2}\right) \tag{5.37}$$

When $\Phi_2 = 1$, $V_a = 1.0\,\text{mV} > 0$, hence $C_2 > C_1$. From the foregoing equation, we notice that $(2V_a / V_{ref})$ determines the capacitor matching accuracy ($\Delta C/C$), where C is the nominal capacitance. Thus, if it is assumed that these capacitors are used in a thermometer-coded SC data converter, and that the capacitor mismatch is the dominant source of noise, then the resultant ENOB can be given by

$$ENOB_{ther\_max} = -\log_2\left(\frac{\Delta C}{C}\right) = -\log_2\left(\frac{2V_a}{V_{ref}}\right) \approx 9\ bits \tag{5.38}$$

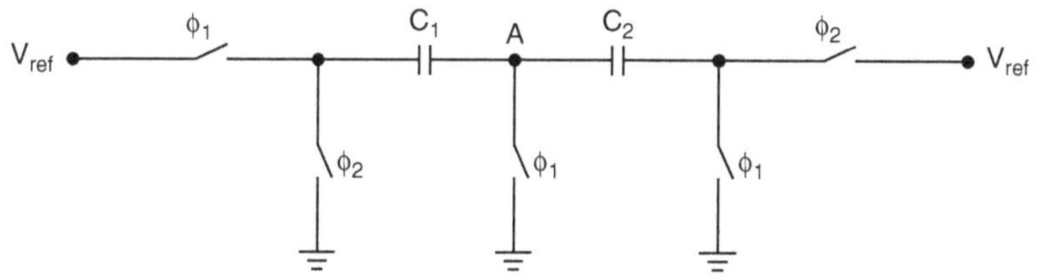

Figure 5.5 Circuit used to measure the mismatch between C_1 and C_2.

If these capacitors are used in a binary-weighted SC data converter, then we need to *refresh* the relationship between ENOB and ($\Delta C/C$) first. As mentioned in Section 5.2, for a binary-weighted converter, DNL is the limiting factor of ENOB, thus, we shall look into the expression of DNL in terms of ($\Delta C/C$). It can be found that the worst-case DNL of a binary-weighted converter tends to occur in the *major-carry* area when the digital input changes from 011 . . . 11 to 100 . . . 00.

Consider an M-bit binary-weighted SC DAC. If we assume that the actual input capacitors are $\{C - \Delta C, 2C - 2\Delta C, \ldots 2^{(M-2)}C - 2^{(M-2)}\Delta C, 2^{(M-1)}C + 2^{(M-1)}\Delta C\}$[8][9], then the worst-case absolute DNL (in units of LSBs) is given by

$$|DNL|_{\max} = \frac{1}{C}\cdot\left[2^{M-1}(C+|\Delta C|) - \sum_{i=1}^{M-1} 2^{i-1}(C-|\Delta C|)\right] - 1 = (2^M - 1)\cdot\frac{|\Delta C|}{C} \qquad (5.39)$$

Keeping the maximum DNL less than 1/2 LSB, we get

$$\frac{|\Delta C|}{C} \le \frac{1}{2^{M+1} - 2} \qquad (5.40)$$

Thus, ENOB (assuming in this case it is equivalent to M) can be found,

$$ENOB_{bi\_\max} = \log_2\left(\frac{C}{\Delta C} + 2\right) - 1 \approx 8 \text{ bits} \qquad (5.41)$$

In comparison with the thermometer-coded configuration, the preceding binary-weighted configuration loses by 1 bit of ENOB. Interestingly, if the actual input capacitors are precisely controlled so that their capacitances are expressed as: $\{C - \Delta C, 2C - 2\Delta C, \ldots 2^{(M-2)}C - 2^{(M-2)}\Delta C, 2^{(M-1)}C + 2^{(M-1)}\Delta C - \Delta C\}$, then it can be shown that the sum of the individual capacitor mismatch errors will be equal to zero; what's more, in both cases (i.e., thermometer-coded and binary-weighted), the resulting ENOB will be equal to 9 bits. However, the last item in the preceding sequence of capacitances (i.e., $2^{(M-1)}C + 2^{(M-1)}\Delta C - \Delta C$) is difficult to implement in practice and may require expensive post-fabrication calibrations such as laser trimming.

To alleviate the problems caused by mismatch errors between circuit components (capacitors in particular), various *mismatch error cancellation* techniques have been reported. *Reference refreshing* [25] and *ratio-independent* [26][27] techniques were developed to make the circuit's accuracy (e.g., gain and nonlinearity errors) independent of capacitor mismatch. However, they usually require high-performance op-amps with large gain-bandwidth products [27]. More recently, it was reported that the ratio-independent technique can be used in combination with a gain- and offset-insensitive design [28] or a correlated double sampling scheme [29] to relax the accuracy requirements for the op-amps.

In addition, the *capacitor error averaging* [30] approach can be used to cancel the capacitor mismatch; however, it does so at the price of reducing the conversion rate by one-third and doubling the power budget.

The capacitor trimming [31][32] technique can be used to tune a capacitor in an SC circuit till it matches the other capacitor(s), by shunting the tuned capacitor with a small trimming capacitor. However, in practice, this trimming capacitor is usually realized by an array of capacitors controlled by digital logic, thereby increasing the cost and area. Moreover, since the capacitor trimming cannot be done without interrupting the normal A/D conversion (otherwise the output digital codes will be corrupted), it is therefore a *foreground calibration* and not appropriate for high-speed applications in general.

As an alternative to the previous discussion, the idea of calibrating mismatch errors in the background (*background calibration*) has gained increasing popularity because the background calibration can be performed without interrupting the converter's normal operation. As mentioned earlier, in current-mode data converters [18], a background calibration technique is employed to correct matching errors between current sources, whereas in SC data converters, a similar technique can be used to correct capacitor mismatch errors.

The background calibration of an SC data converter can be done in either the analog or the digital domain. As the name suggests, digital calibration utilizes digital circuitry to remove errors due to component mismatching. Many state-of-the-art digital calibration techniques utilize oversampled delta-sigma modulators to shape the power density of mismatch errors into the higher frequency range, and then use low-pass filters to filter them out, hence the name *mismatch-shaping.* But the speed limit due to the oversampling is a major concern. We will explore a few digital mismatch-shaping techniques in Section 5.5 when we study oversampling SC data converters.

As for the analog background calibration, the quantity of practical techniques is relatively small, compared to what we can find in the literature today with respect to its digital counterpart. This is partially due to the fact that analog calibration requires extra analog circuitry such as calibration DACs and comparators, which apparently must be more accurate or linear themselves than the components undergoing calibration. In addition, the relatively low fabrication cost and technology scalability/portability of digital circuitry has made the digital calibration a more attractive choice than the analog calibration.

An interesting analog calibration technique was reported by Moon et al. [33]. The technique uses a thermometer-coded SC DAC similar to that in Figure 5.3(a) as the demonstration platform. The basic idea is to adaptively calibrate the charge delivered via each input branch by using a *variable* reference voltage for each branch (similar to the current copying and the reference refreshing techniques reported in [18] and [25], respectively), until all branch charges accord with a single reference charge value, which is rectified by an accurate capacitor (i.e., the nominal C). In effect, the deviation of a branch charge from its ideal value due to capacitor mismatch is corrected.

Further investigation will show that in the SC DAC described in [33], the reference voltage of each input branch is refreshed every 2^M clock cycles (for an M-bit thermometer-coded conversion). Also, 2^M rather than $(2^M - 1)$ input branches are required, with the extra input branch substituting for the one undergoing calibration, which is similar to the substitution of the calibrated current source described in [18].

To conclude, this section presented the circuit implementations of non-SC and SC-type integrated D/A converters. The discussion investigated the problematic capacitor mismatch and quantified its impact on the data converter's accuracy. Finally, various mismatch error cancellation techniques were introduced.

5.4 Nyquist-Rate ADCs

Flash ADCs

The process of a standard analog-to-digital (A/D) conversion is normally classified into two distinct operations: sampling and quantization. Either sampling (often accompanied by holding) or quantization can be used to transform a signal with a continuous amplitude into a set of discrete levels.

Sampling is a time-based operation that is typically controlled by a system clock, and it emphasizes capturing the input data at certain discrete moments for the convenience of subsequent operations (e.g., quantization), whereas quantization utilizes a reference (voltage or current) to categorize the input data into discrete groups, regardless of whether the data have been sampled. In other words, sampling acquires and preconditions the input signal, while quantization generates an expression of the input signal using a series of digital bits, based on the result of comparing the input signal's amplitude with the reference. Thus, sampling is often realized by sample-and-hold (S&H) circuits, while quantization is normally done by comparators.

Nevertheless, in some A/D converter (ADC) architectures, the dividing line between sampling and quantization fades out. As a result, sampling and quantization can be done simultaneously and no dedicated S&H circuits would be required. A well-known example is called the *flash* (or *parallel*) ADC architecture. As the name suggests, these ADCs operate at a very high speed since they *flash* out the digital bits in a *parallel* fashion.

Figures 5.6(a) to Figure 5.6(c) illustrate three different ways to implement a 2-bit flash ADC. All three circuits operate in a similar manner as follows. In each circuit, the positive input of a comparator is connected to a certain node on the resistor string (V_{ri}), while the negative input is connected to the input analog signal V_{in}. If V_{in} is larger than V_{ri}, then the corresponding comparator sends a logic low (0) to the following digital logic circuitry. Otherwise, a logic high (1) will be sent.

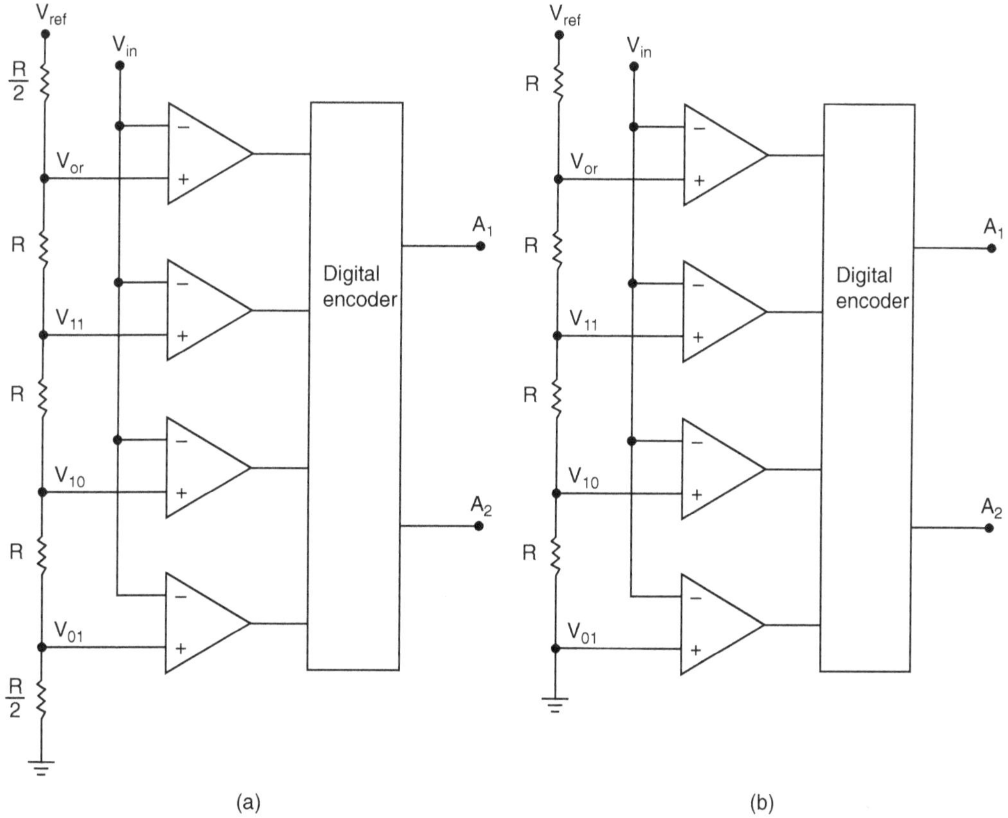

Figure 5.6 Three different realizations of a 2-bit flash ADC.

Figure 5.6 ***Continued***

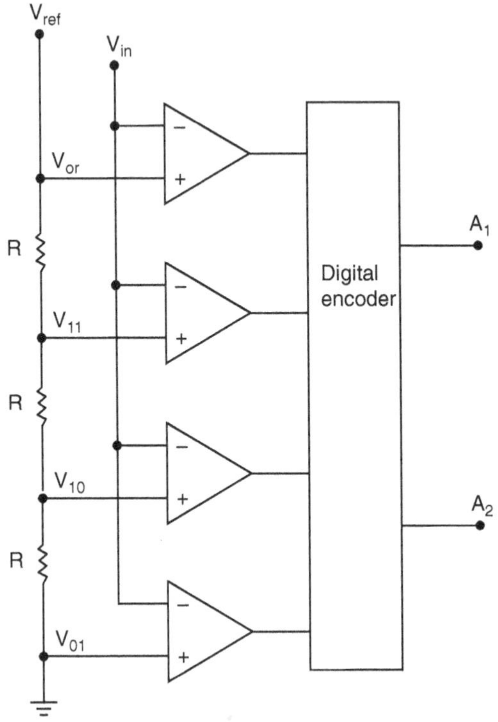

(c)

After the comparison, a parallel 4-bit thermometer code is sent to the digital logic circuitry, where the 4-bit input is transformed into a 3-bit *true* thermometer word (with the overloading bit removed) by three two-input NAND gates. The result is then passed through a 3-to-2 encoder, and the digital output is a 2-bit binary code $\{A_1A_2\}$.

Note that if the difference between V_{in} and V_{ri} is small, then the comparator needs a sufficient voltage gain to amplify this difference so that its logic output is guaranteed to be well defined. In practice, each comparator of a flash ADC is typically composed of an analog preamplifier and a latch, with the former *tracking* and amplifying $(V_{in} - V_{ri})$ during the first half clock cycle and the latter detecting and *holding* the instantaneous polarity of a preamplified $(V_{in} - V_{ri})$ during the second half clock cycle (i.e., in a nonoverlapping fashion [34]).

Qualitatively speaking, the preceding comparator constitutes a *pseudo*-S&H, which samples a voltage difference and then holds its *polarity*. However, it is faster than a normal S&H because the preamplifier does not need to spend time settling its output to a specified voltage level, since only the polarity of the output signal matters. Therefore, in a flash ADC, sampling and quantization are typically carried out simultaneously, and no dedicated front-end S&H circuits are needed.

EXAMPLE 5.4

For the flash ADCs illustrated in Figures 5.6(a–c), find out for what values of V_{in} the digital output of each ADC will change without overloading, and discuss the results.

Solution: Let us start with the converter in Figure 5.6(a). The current that flows from the reference voltage V_{ref} to ground is given by $I = V_{ref}/4R$. Starting from the bottom, we denote the voltages appearing at the positive inputs of the four comparators to be V_{01}, V_{10}, V_{11}, and V_{or}, respectively. The denotation V_{or} stands for *over-range voltage*, which is equivalent to the maximum allowable input voltage without overloading the ADC. By inspecting the circuit diagram, we can write the following threshold values of V_{in} for which the digital output code will change:

$$V_{01} = \frac{V_{ref}}{8},\ V_{10} = \frac{3V_{ref}}{8},\ V_{11} = \frac{5V_{ref}}{8},\ V_{or} = \frac{7V_{ref}}{8} \tag{5.42}$$

We now can find that the first code transition happens when V_{in} equals $V_{ref}/8$. After that, the distance between two adjacent code transition points is equal to $V_{ref}/4$, and the over-range voltage is equal to $7V_{ref}/8$.

Similarly, for the converter in Figure 5.6(b), we can write the following:

$$V_{01} = 0,\ V_{10} = \frac{V_{ref}}{4},\ V_{11} = \frac{V_{ref}}{2},\ V_{or} = \frac{3V_{ref}}{4} \tag{5.43}$$

Here, the first code transition happens when V_{in} equals 0 V instead of $V_{ref}/8$. After that, the distance between two adjacent code transition points is equal to $V_{ref}/4$, and the over-range voltage is equal to $3V_{ref}/4$, which is smaller than $7V_{ref}/8$.

And for the converter in Figure 5.6(c), we have

$$V_{01} = 0,\ V_{10} = \frac{V_{ref}}{3},\ V_{11} = \frac{2V_{ref}}{3},\ V_{or} = V_{ref} \tag{5.44}$$

The first code transition also happens when V_{in} equals 0 V. After that, the distance between two adjacent code transition points is equal to $V_{ref}/3$, which is larger than $V_{ref}/4$, and the over-range voltage is equal to V_{ref}, which is larger than both $7V_{ref}/8$ and $3V_{ref}/4$.

Comparing one converter with another, we realize that the first flash ADC realization shown in Figure 5.6(a) is advantageous over the remainders, mainly because it provides four discrete output levels between 0 V and V_{or}, whereas the other two provide only three. As a result, in an ideal situation, the absolute quantization noise of the first ADC remains less than or equal to 0.5 LSB, while the others have an absolute quantization noise that is equal to 1 LSB. In addition, the first configuration can be used to realize a *bipolar* or *midtread* flash ADC (i.e., a converter that operates with both positive and negative analog signals), while the other two cannot.

Despite their differences, all three converters shown in Figure 5.6 share a common characteristic—that is, each of them includes four comparators. Generally speaking, to realize an M-bit flash ADC, 2^M comparators are required.

Alternatively, we can save one comparator by reversing the polarity of each comparator. That is, the negative input of each comparator is connected to a node on the resistor string, and the positive input is connected to V_{in}. In effect, to realize an M-bit flash ADC, $(2^M - 1)$ comparators are needed. Note that once the polarity of the comparators is reversed, the thermometer code will be inverted as well; hence, the digital logic circuitry needs to be changed correspondingly.

However, the number of comparators is still too large. For instance, an 8-bit flash ADC requires at least 255 comparators. Such a large number of comparators dissipate high power and occupy a large silicon area. Also, they result in a large parasitic capacitance at the input terminal (V_{in}), which limits the overall speed of the flash ADC and complicates the design of the circuit preceding the ADC.

Furthermore, besides the resistive matching accuracy (i.e., $\Delta R/R$), the conversion accuracy of the flash ADC greatly depends on the precision of the comparators (e.g., offset and metastability errors [34]), which becomes more difficult to maintain as M increases.

Adopting the approach discussed in Section 5.3, we can find that the worst-case INL of an M-bit flash ADC (in units of LSBs) is given by

$$|INL|_{\max} = 2^{M-1} \cdot \left|\frac{\Delta R}{R}\right| + 2^M \cdot \frac{|V_{off}|}{V_{ref}} \tag{5.45}$$

Here, we assume that each resistor has an absolute mismatch error of $|\Delta R|$, that the sum of all the resistor mismatch errors is equal to zero, and that V_{off} is the maximum input offset voltage of the comparator. Thus, even if the resistors were all perfectly matched, it would still be impractical to use flash ADCs for obtaining high resolutions due to the comparator offset error. For instance, assume that there is no resistor mismatch, and that a 12-bit flash ADC is desired ($V_{ref} = 5$ V), then the maximum input offset voltage must be kept less than (or equal to) 0.61 mV to keep the INL below 1/2 LSB, which is nearly impossible to achieve given the existing CMOS technologies.

In comparison with a traditional flash ADC, an *interpolating* flash ADC has a reduced number of preamplifiers inside the comparators; thus, the total input capacitance is reduced by an *interpolating factor* (K), while the parallel characteristic of a flash ADC is maintained. For instance, a traditional 4-bit flash ADC requires sixteen preamplifiers, whereas an interpolating ADC with $K = 4$ only needs four.

Interpolation may be implemented using one of the following approaches: resistor interpolating [35], capacitor interpolating [36], and current-mirror interpolating [37]. All three approaches, although named differently, share the same methodology, which is to slice the (voltage or current) difference between two adjacent preamplifiers' outputs into K equal pieces so that the number of preamplifiers is reduced by K. Interestingly, examination of the resistor interpolating structure will show that it is somewhat similar to the double resistor-string technique shown earlier [14].

Also note that an M-bit interpolating flash ADC still needs 2^M latches after the interpolating stage for carrying out the polarity detection. The number of the latches can be reduced by using the *folding* technique. Specifically, given a *folding factor* F, the number of the latches needed is reduced to $2^{(M-F)}$. Both interpolating and folding are incorporated into one flash ADC sometimes—for instance, in [38]—but they can also be used individually. The reader is referred to the literature for more details on folding flash ADCs.

The design of state-of-the-art flash CMOS ADCs emphasizes the specifications that combine very high sampling rates, from a few hundred megahertz [39][40] up to several gigahertz [41][42][43][44], with ENOBs in the range between 5 and 8 bits. Fast 6-bit flash CMOS ADCs [39][40][41][42] find use primarily in applications such as the read-write channels of a magnetic data storage system, which is ever pushing the speed limit. Fast 8-bit flash CMOS ADCs such as [43][44] can be used for instrumentations and wireless/wireline communications. The ongoing commercialization of ultra-wideband (UWB) technologies also stirs the development of CMOS ADCs that will be capable of providing a moderate accuracy (4 to 6 bits) across a wide signal bandwidth (at least 500 MHz). At the time of this writing, flash CMOS ADC is the best architecture of choice for this potential application.

Due to the inherent characteristics of flash ADCs such as high clock rate and low resolution, they are seldom built based on SC configurations. However, in some capacitor-interpolating or folding flash ADCs, multistage SC preamplifiers or comparators [11] have been used to reduce the clock feedthrough as well as input-referred offset errors. The use of multistage SC comparators, although providing a better accuracy performance, may limit the overall speed of the flash ADC. Specifically, a cascade of M comparators has a time constant approximately given by (assuming all comparators are identical [11])

$$\tau \cong \frac{4ML^2 A_0}{3\mu_n V_{eff}} \tag{5.46}$$

where L is the length of the input transistor(s), A_0 is the dc gain of each comparator, and V_{eff} is the effective gate-source voltage of the input transistor(s). For example, if $M = 3$, $A_0 = 10$, $\mu_n = 0.05\,\text{m}^2/\text{Vs}$, $V_{eff} = 0.7\,\text{V}$, and the flash ADC is implemented in a 0.35-μm CMOS technology (here the nominal value of L is equal to $0.35 \times \sqrt{2} \approx 0.495\mu\text{m}$), then from the preceding formula we can find that the smallest possible time constant due to the multistage comparators is about 0.28 nS. If we assume that it would take 7τ for the outputs of all comparators to settle within 0.1% of their final values, then the maximum achievable clock rate is approximately 510 MHz.

Furthermore, the value of V_{eff} will be decreased as the power supply voltage continues shrinking from 3.3 V down to 2.5 V, 1.8 V, or even below 1.0 V, thereby further reducing the speed. On the other hand, aggressive submicron CMOS scaling—for instance, 0.18- or 0.12-μm CMOS technology—can help significantly increase the speed. Also, it is a common practice that the tradeoff between the comparator's dc gain (A_0) and the total number of stages (M) be optimized for the best performance.

As a final note, there are many other practical design issues regarding flash ADCs than those presented here, such as bobble/sparkle error, clock jitter, and flashback error. It is not the intention of this book to list them exhaustively. The interested reader is referred to the literature for more details. Good references on these issues include [9][11].

Two-Step ADCs

As mentioned in the previous subsection, for implementing an ADC that operates at the fastest possible clock rates with a resolution in the range of 4 to 8 bits, the flash architecture is clearly the best choice. However, it normally requires a large number of small-offset comparators and dissipates a lot of power.

The *two-step* ADC (sometimes it is also called the *subranging* ADC when the subtraction is not adopted) architecture was originally developed for providing 8 to

10 bits resolution at sampling rates of tens of megahertz [45]. However, in the past few years, thanks to the fast advancements of *n*-step (n ≥ 3) pipelined ADCs, the two-step ADCs' formerly exclusive *speedway* has been occupied by the new *n*-step ADCs operating at high sampling rates (from 10 MHz to 80 MHz) with a high resolution ranging from 10 to 15 bits. That is, the lower-speed portion of the two-step ADC's performance range has been gradually merged with the higher-speed portion of the *n*-step ADC's performance range.

As a new technological trend, most state-of-the-art CMOS two-step ADCs [46][47] are targeting at the higher-speed performance range, namely, with higher sampling rates (from 80 MHz to 200 MHz) and a medium resolution ranging from 7 to 9 bits. These specifications are appropriate for applications such as wireless LAN, Ethernet communications, instrumentation, and interfaces for processing uncompressed digital video signal, for instance digital visual interface (DVI).

The conceptual block diagram of a 10-bit two-step ADC is shown in Figure 5.7. As we can see, there are three subconverters embedded in the two-step ADC. The 4-

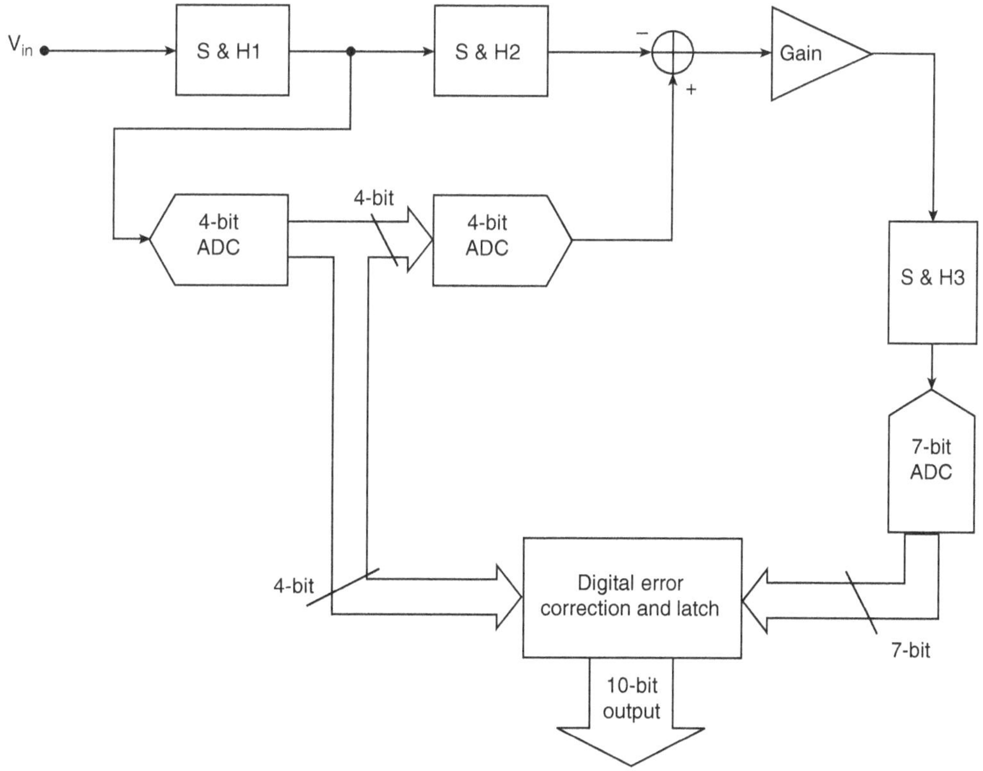

Figure 5.7 Block diagram of a 10-bit two-step ADC.

bit ADC (also called sub-ADC) is normally implemented based on the flash ADC architecture (often with interpolating or folding). The 4-bit DAC or a sub-DAC, which may be implemented as a thermometer-coded SC DAC similar to that shown in Figure 5.3(b) or a hybrid (i.e., with reference resistors and capacitors) DAC, recovers the 4-bit digital word to an analog signal. Recalling the model of quantization noise extraction shown in Figure 5.1, we can find that the quantization noise due to the 4-bit sub-ADC (or *residue*) is obtained by using a subtraction unit. The quantization noise is then sent to the following *residue amplifier*. The third data converter is a 7-bit sub-ADC in flash structure.

The 10-bit two-step ADC operates as follows. The 4-bit sub-ADC carries out a *coarse* conversion of the input signal, and the resulting estimate of the four MSBs is sent to the error correction module. Meanwhile, the estimate of the four MSBs is also transformed back into an analog signal by the 4-bit sub-DAC, which is then subtracted from the sampled-and-held input V_{in}. Next, the output of the subtraction unit is amplified by a gain stage before entering into the 7-bit sub-ADC. The 7-bit sub-ADC then carries out a *fine* conversion. The 7-bit digital output is the estimate of the 6 LSBs, and it is combined with the 4-bit output of the first sub-ADC in the digital error correction module. The final encoded output is a 10-bit digital signal.

As the schematic shows, there is a front-end S&H circuit denoted as S&H1. This S&H circuit is an essential building block of the two-step ADC. Without this S&H, the analog input signal cannot change more than 1/2 V_{LSB} during the time when the coarse conversion and subtraction are in process (we denote this time interval as t_c). If it is assumed that a sinusoidal input with a peak-to-peak voltage magnitude of V_{ref} and a frequency of f_{in} is applied to the input of an M-bit two-step ADC, we can find the worse-case limit of t_c, which is given by

$$t_c \leq \frac{V_{ref}/2^{M+1}}{(V_{ref}/2)\cdot(2\pi\cdot f_{in})} = \frac{1}{2^{M+1}\cdot\pi\cdot f_{in}} \tag{5.47}$$

If $M = 10$ and $f_{in} = 20\,\text{MHz}$, the maximum allowable value of t_c will be only about 8 ps, which makes it impossible for the coarse conversion or subtraction to finish successfully. Therefore S&H1 is necessary; additionally, it must have an accuracy of no less than M bits or the output of following stages will deviate significantly from the desired value.

In comparison, S&H2 is not as critical; it is an optional component. It simply creates an extra delay so that the residue amplifier will have sufficient time to settle before the next input comes in. This, however, indicates that a two-step ADC tends to suffer from latency induced by the time it takes to settle the residue amplifier and complete the fine conversion.

In comparison with a 10-bit single-stage flash ADC, the number of comparators in the two-step ADC shown in Figure 5.7 is considerably smaller, which is equal to 142 when the 4–7 combination (i.e., 4-bit and 7-bit sub-ADCs) is adopted. In addition, if a 5–5 combination is used, then the number of comparators will drop to only 62. As the result, the power dissipation, die area, and input capacitance loading are reduced significantly. However, the even-splitting configuration (e.g., 5–5) is not popular in practice for the following reason. Assume that the desired resolution of the two-step ADC is M bits. If the true resolutions (i.e., accuracies) of the first and second sub-ADCs are expressed as M_1 bits and M_2 bits, respectively, then the relationship between the first and second stages in terms of accuracy can be expressed as

$$\frac{2^{M1+M2-1}}{2^M} \leq \frac{1}{2^G} \cdot \frac{\frac{V_{ref}}{2^{M2}}}{V_{ref}/2^M} \tag{5.48}$$

The left part of the preceding represents the maximum residue error at the subtraction unit's output in units of LSBs, assuming that 1 LSB corresponds to $(V_{ref}/2^M)$. The right part, also in units of LSBs, represents the minimum input step that the second sub-ADC is able to resolve. In addition, here the closed-loop gain of the residue amplifier is expressed as 2^G (in units of V/Vs) for convenience. Note that the preceding is a tight sufficient condition to guarantee that the residue from the first stage is successfully resolved by the second sub-ADC.

After a few rearrangements, we obtain the following relaxed but still sufficient condition.

$$G = 2(M - M_2) - M_1 + 1 \tag{5.49}$$

For the even-splitting configuration (i.e., $M = 10$ and $M_1 = M_2 = 5$), from the preceding equation we get $G = 6$. As a result, the residue amplifier needs to provide a closed-loop gain of 64 V/V at clock rates on the order of 10 megahertz. Moreover, it can be shown that the open-loop gain of the residue amplifier that keeps the absolute gain error less than or equal to 1/2 LSB [9][10] is given by

$$A_{opl} = A_{cld}\left(2^{M2+1} + 1\right) \tag{5.50}$$

where A_{opl} and A_{cld} represent the open-loop and closed-loop gain of the residue amplifier, respectively. Thus, the 5–5 combination requires an open-loop gain of

4160 V/V, or, equivalently, 72.4 dB, which is not easy to achieve under a sampling clock of 10 MHz or higher. Alternatively, if it is assumed that $M_1 = 5$ and $M_2 = 6$, then the amplifier's closed-loop and open-loop gains will be reduced to 16 V/V and 66.3 dB, respectively. Figure 5.7 shows that $M_1 = 4$ and $M_2 = 7$. In this case the residue amplifier's closed-loop gain will be equal to only 8 V/V, and the open-loop gain is still about 66.3 dB.

Interestingly, a quick survey of all the possible $\{(M_1, M_2)\}$-combinations will show that the foregoing 4–7 and 5–6 combinations are the two most appropriate choices for a 10-bit implementation, since both greatly ease the design of a high-speed residue amplifier in terms of stability and the gain-bandwidth product. Furthermore, it is instructive to know that the number of resolution (M_1) for the first sub-ADC tends to affect the ADCs' *spurious-free-dynamic-range* (SFDR) performance [48].

As new-generation two-step ADCs find primary use in wireless communications and other high-speed (50 MHz and above) conversion applications, the SFDR specifications of these ADCs have been developed recently [48][49]. The SFDR of an ADC is defined to be the difference in decibels between its full-scale (FS) fundamental and the maximum spurious tone in the output spectrum.

For a specific ADC, SFDR is typically higher than *signal-to-noise-plus-distortion-ratio* (SNDR), since SFDR primarily deals with spectral spurs, which are normally dictated by the inherent nonlinearity of the quantization process (i.e., the input and the quantization noise are somewhat *correlated*) and by the interstage gain error. It can be proved that the SFDR of a two-step ADC is dependent on the bit number of its first sub-ADC (M_1), such that

$$SFDR \approx 9M_1 - 20\lg\varepsilon - c \tag{5.51}$$

where ε is the relative interstage gain error, and c is an offset value that ranges from 0 for low resolutions to 6 for high resolutions [48]. Based on this equation, for every bit that M_1 gains, theoretically SFDR is increased by 9 dB. Thus, when SFDR is more critical than SNDR—for instance, in an integrated wireless receiver with a front-end 10-bit two-step ADC—the 5–6 combination will be a better choice than the 4–7 combination. The rigorous proof of Equation (5.51) can be found in [49].

In accounting for all the components shown in Figure 5.7, we find that each of S&H1, S&H2, the subtraction unit, and the sub-DAC requires a 10-bit accuracy, whereas each of the residue amplifier, S&H3, and the second sub-ADC requires a 7-bit accuracy. The sub-ADC in the first stage only needs to be a true 4-bit flash ADC—that is, with 4-bit resolution and 4-bit accuracy (ENOB), thanks to S&H1 and the digital error correction module [11]. Thus, the most critical analog building block in a two-step ADC is the front-end S&H (S&H1), which must be capable of sampling with a high precision at a high clock rate. In [48], a simple and yet high-performance integrated SC S&H appropriate for CMOS two-step ADCs was reported.

Pipelined ADCs

The *pipelined* (or *n-step*) ADC may be considered the generalization of the two-step ADC. Figure 5.8 illustrates a conceptual block diagram of a 10-bit pipelined ADC. As the schematic shows, the 10-bit pipelined ADC consists of an input S&H and nine cascaded (or pipelined) ADC stages. In each stage (except for the last stage, whose bit number is often greater than that of the remainders), the input signal is converted into a digital code by a sub-ADC (usually a flash ADC with a resolution of no more than 4 bits), and then the digital code is read into the shift register array.

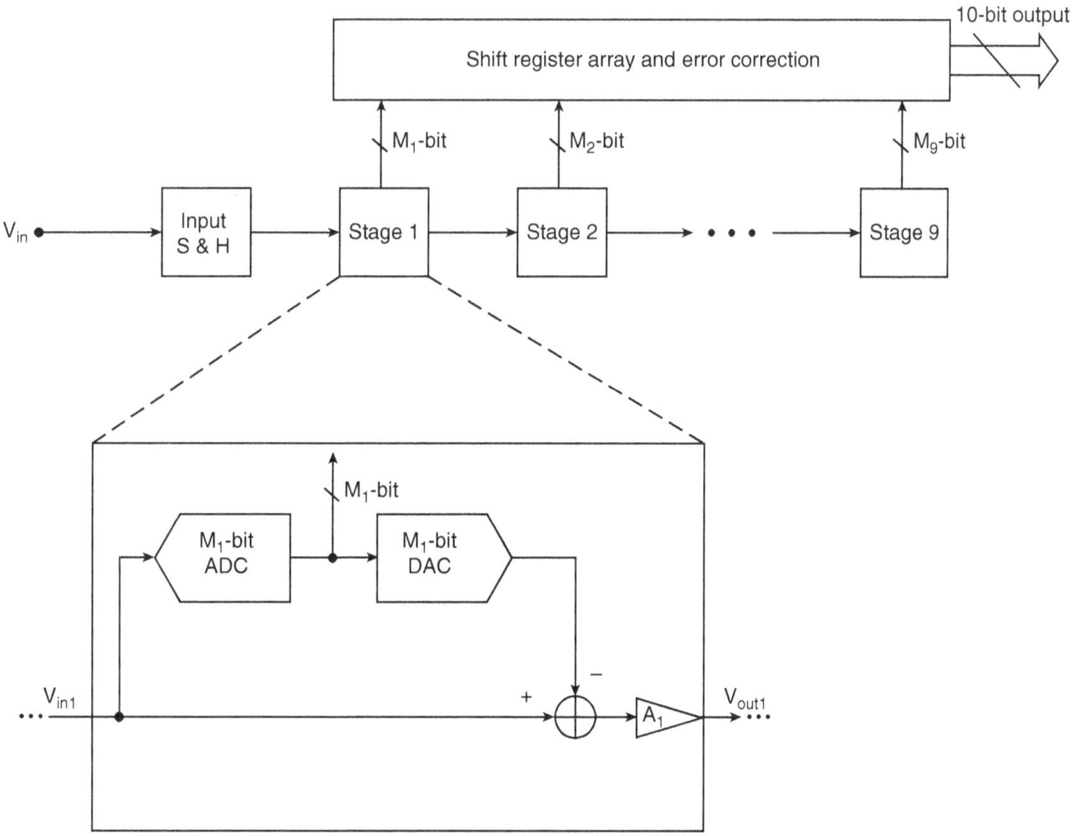

Figure 5.8 Block diagram of a 10-bit pipelined ADC.

In the meantime, the output digital code is converted back into an analog signal by a sub-DAC (usually an SC or hybrid DAC). Similar to a two-step ADC, the quantized analog signal is then subtracted from the input signal, and the residue signal is amplified by a gain factor and passed on to the next stage.

All sub-ADCs (or DACs) of stages 1 to 8 are normally designated to have the same bit resolution (i.e., $M_1 = M_2 = \ldots = M_8$). In practice, the bit number of each subconverter is usually chosen to be either 1 or 1.5; the latter is more popular for it dramatically desensitizes the pipelined ADC's performance to the circuit imperfections such as comparator offsets [50][51]. The principle of 1.5-bit/stage configuration will be explained shortly.

Before proceeding, note that an M-bit pipelined ADC introduces a latency of M clock cycles. That is, the output of the first stage (i.e., the MSB) needs to wait ($M - 1$) clock cycles till the output of the final stage (i.e., the LSB) is determined in the Mth clock cycle.

However, all nine stages never stop processing new input samples, and the ideal processing rate of each stage is equal to the clock rate. This indicates that the overall conversion rate of the ADC is determined by the operation of each individual stage and independent of the total number of stages. Thus, ideally the pipelined ADC could operate at a very high sampling rate (comparable to that of the flash ADC). Nevertheless, in reality the pipelined ADC is seldom used in place of the flash ADC because the yield is uneconomically low (consider the extra design and fabrication cost due to the S&Hs and residue amplifiers).

A prominent feature of the pipelined ADC architecture is that the circuit's complexity increases *linearly*, as opposed to *exponentially* (e.g., flash and two-step), with the number of resolution. Use the 10-bit pipelined ADC shown in Figure 5.8 as an example: If the 1-bit/stage configuration is used for building the converters in stages 1 to 8 and a 2-bit flash ADC is employed by stage 9, then only 11 comparators are needed, which is considerably less in comparison with 1023 (flash) and 142 (two-step).

Nowadays, the pipelined ADC architecture is a strong contender in low-cost (i.e., small silicon area, low power, and high technology scalability/portability) and high-performance (i.e., high resolution and high speed) A/D conversion applications.

Next, we investigate the implementation of the pipelined ADC. In most cases, the subconverters, subtraction unit, and residue amplifier of each stage can be combined and implemented as a single SC circuit block, which is called the *multiplying digital-to-analog converter* (MDAC).

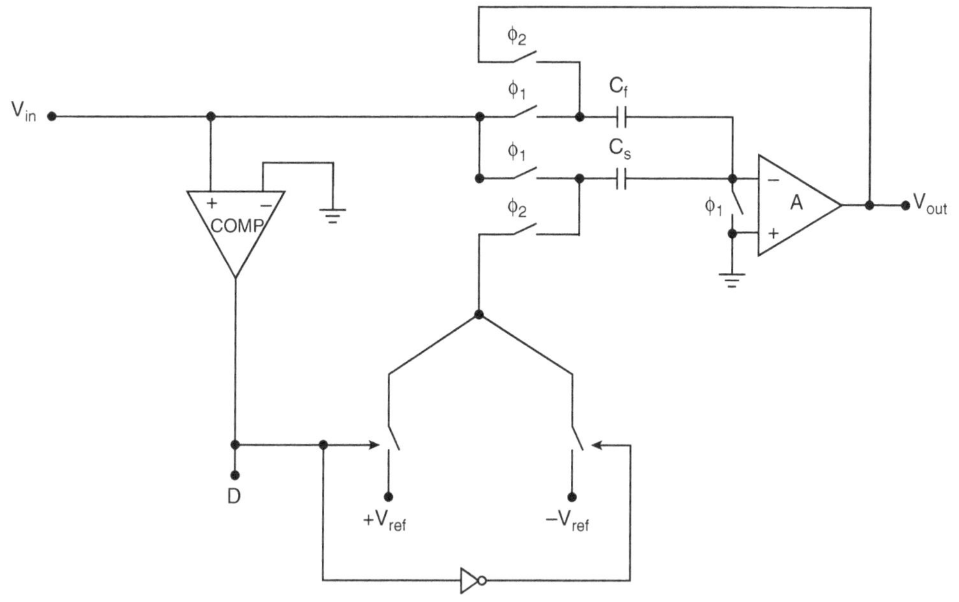

Figure 5.9 1-bit SC pipelined stage.

First, consider the 1-bit pipelined stage shown in Figure 5.9. Although a single-ended configuration is illustrated here for simplicity, in practice the fully differential version is normally used to maximize the circuit's common-mode performance. In the schematic, the two-level comparator ("COMP") realizes a 1-bit sub-ADC, and the digital output D connects either $+V_{ref}$ or $-V_{ref}$ to the sampling capacitor C_s, realizing a 1-bit sub-DAC.

The circuit operates as follows. When $\Phi_1 = 1$, the op-amp is reset (the reset switch is connected from the op-amp's negative input to ground instead of its output, so that the op-amp's intermode slewing requirement is relaxed), and both C_s and the feedback capacitor C_f are charged by V_{in}. Next, when $\Phi_2 = 1$, C_f and C_s are coupled together, and C_s is connected to either $+V_{ref}$ or $-V_{ref}$, depending on the comparator's output D. If it is assumed that the op-amp has an infinite gain, and that the comparator does not have an input offset error, then the input/output relationship of the 1-bit pipelined stage can be expressed as

$$V_{out} = \begin{cases} \dfrac{(C_s + C_f)}{C_f} \cdot V_{in} - \dfrac{C_s}{C_f} \cdot V_{ref},\ if\ V_{in} > 0 \\ \dfrac{(C_s + C_f)}{C_f} \cdot V_{in} + \dfrac{C_s}{C_f} \cdot V_{ref},\ if\ V_{in} < 0 \end{cases} \tag{5.52}$$

If C_s is equivalent to C_f, then the function is rewritten as

$$V_{out} = \begin{cases} 2V_{in} - V_{ref}, \text{ if } V_{in} > 0 \\ 2V_{in} + V_{ref}, \text{ if } V_{in} < 0 \end{cases} \tag{5.53}$$

This indicates that the quantization noise of the 1-bit sub-ADC (also known as the residue), which may be expressed as either ($V_{in} + 0.5V_{ref}$) or ($V_{in} - 0.5V_{ref}$), depending on the polarity of V_{in}, is multiplied by 2. Also, note that the SC configuration around the op-amp actually realizes an S&H function.

Although an interstage gain of 2 is used here (as in most textbooks and papers) to simplify the analysis, note that this gain factor is not optimal. It has been reported that an optimization of the interstage gain would be necessary in many technical challenging scenarios [51][52], for instance, implementing a low-voltage ($V_{dd} \leq 2.5$ V) and high-resolution ($M \geq 12$ bits) CMOS pipelined ADC with a stringent power budget (less than 120-mW; analog portion only). It is not the author's intention to cite the many mathematically interesting but intense derivations from the excellent work by Cline [52], which is an important part of ongoing research at the time of this writing. The interested reader is encouraged to explore this topic in the reference.

Next, based on the transfer function of Equation (5.53), we can draw the transfer curve of an ideal 1-bit pipelined stage, which is illustrated in Figure 5.10(a).

However, in practice nonidealities in the circuit such as the comparator offset, capacitor mismatch, finite op-amp gain, or charge injection cause the actual transfer curve to deviate from the ideal. First, as indicated in Figure 5.10(b), the comparator

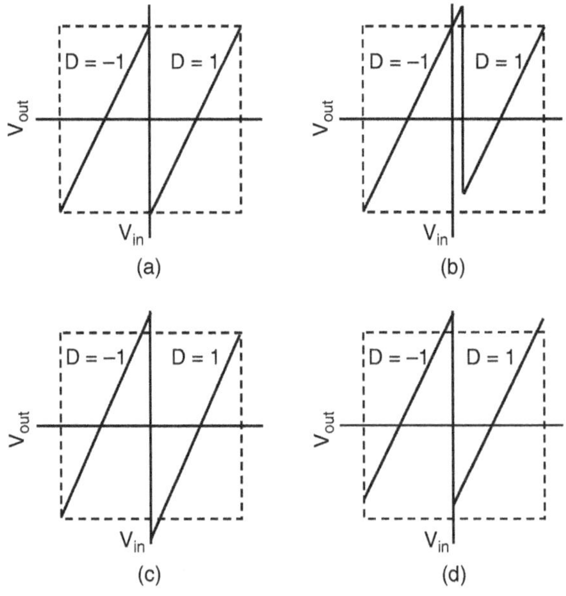

Figure 5.10 Transfer curves of a 1-bit pipelined stage. (a) Ideal. (b) With comparator offsets. (c) With capacitor mismatch and/or finite op-amp gain. (d) With charge injection.

offset shifts the transition point and causes the residue to exceed the *resolvable range* of the following stages [50], which is confined within the rectangle shown in Figure 5.10(b). In such a case, if a wrong decision of D is made, then the resultant wrong V_{out} is likely to cause the next stage to make a wrong decision. The same pattern goes on till the last stage (consider an analogy to the domino theory/effect).

Second, the capacitor mismatch and the finite op-amp gain can introduce an interstage gain error (i.e., the gain deviates from 2). This interstage gain error changes the slope of the curve, as shown in Fig. 5.10(c). Third, the charge injection error causes the transfer curve to shift vertically, as shown in Figure 5.10(d).

As Figures 5.10(b) through 5.10(d) illustrate, a common effect of the preceding nonidealities is that they all cause the stage residue to go beyond the resolvable range of the following stages. This *out-of-range* problem significantly increases the possibility of missing codes, which is also an indication that the linearity of the pipelined ADC has been corrupted.

To alleviate the foregoing problems, we can use the 1.5-bit pipelined stage in Figure 5.11. As the schematic shows, in comparison with the 1-bit stage in Figure 5.9,

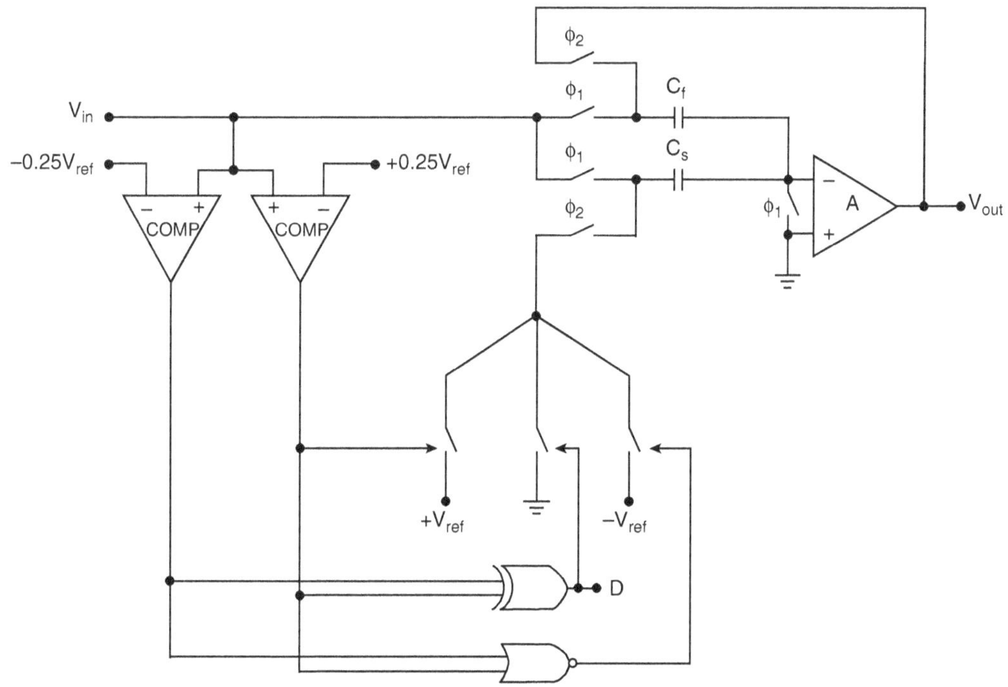

Figure 5.11 1.5-bit SC pipelined stage.

this circuit makes use of two comparators in addition to two logic gates (XOR and NOR) to realize the subconverters. In this configuration, when $\Phi_2 = 1$, C_s may be connected to one of three potentials: $+V_{ref}$, $-V_{ref}$, and ground, depending on the comparator's output. Similar to the preceding analysis of the 1-bit pipelined stage, we can find the ideal input/output relationship of the 1.5-bit pipelined stage as follows:

$$V_{out} = \begin{cases} 2V_{in} - V_{ref},\ if\ V_{in} > 0.25V_{ref} \\ 2V_{in},\ if\ -0.25V_{ref} < V_{in} < 0.25V_{ref} \\ 2V_{in} + V_{ref},\ if\ V_{in} < -0.25V_{ref} \end{cases} \tag{5.54}$$

Here, we also assume that the capacitance of C_s is equal to that of C_f. As Equation (5.54) indicates, in the 1.5-bit/stage configuration, the input threshold voltage (i.e., the input signal voltage level that causes the transfer function to change) is no longer located at the origin; instead, in this case we have two input threshold voltages: one is located at $+0.25V_{ref}$ and the other is located at $-0.25V_{ref}$. Additionally, the preceding function implies that a 1.5-bit stage provides three decision levels as compared to the two levels provides by a 1-bit stage. Next, we discuss how the 1.5-bit stage works. Consider a 3-bit pipelined ADC that is composed of two cascade 1.5-bit stages. We can find the mapping from the input signal level to the corresponding 3-bit binary output code in Table 5.2.

The decision levels listed in Table 5.2 are specified based on the assumption that a 1.5-bit stage generates digital words of 00, 01, and 10, in response to input signals that satisfy $V_{in} < -0.25V_{ref}$, $-0.25V_{ref} < V_{in} < +0.25V_{ref}$, and $V_{in} > +0.25V_{ref}$, respectively. For instance, if the input signal voltage level is between $-0.625V_{ref}$ and $-0.375V_{ref}$, then the first 1.5-bit stage will generate a digital word of 00, and an analog residue voltage (V_{out}) that ranges between $-0.25V_{ref}$ and $+0.25V_{ref}$ according to Equation (5.54). Consequently, the second 1.5-bit stage will generate a digital word

Table 5.2 Mapping of V_{in} to the 3-bit binary output.

V_{in}	Decision levels	3-bit Binary codes
Between $-V_{ref}$ and $-0.625V_{ref}$	0000	000
Between $-0.625V_{ref}$ and $-0.375V_{ref}$	0001	001
Between $-0.375V_{ref}$ and $-0.125V_{ref}$	0010 and 0100	010
Between $-0.125V_{ref}$ and $+0.125V_{ref}$	0101	011
Between $+0.125V_{ref}$ and $+0.375V_{ref}$	0110 and 1000	100
Between $+0.375V_{ref}$ and $+0.625V_{ref}$	1001	101
Between $+0.625V_{ref}$ and $+V_{ref}$	1010	110

of 01, which is lined up with the previous digital word (i.e., 01) to form the decision level: 0001, as presented in the foregoing table. Next, by using simple *overlap-and-add* [50][51] digital logic—that is, use a shift register to overlap the LSB of the first stage's output with the MSB of the second stage's output, and use an adder to add them up—we can obtain a 3-bit binary word: 001, which is the final digital output for this particular input signal.

However, when the input signal satisfies $-0.375V_{ref} < V_{in} < -0.125V_{ref}$ (i.e., within the range that encompasses one of the input thresholds, $-0.25V_{ref}$), we may obtain either one of two decision levels (i.e., 0010 and 0100), depending on which side of the threshold the input signal belongs to. Specifically, if the input signal satisfies $-0.375V_{ref} < V_{in} < -0.25V_{ref}$, then the resulting decision level is marked at 0010; on the other side, if it satisfies $-0.25V_{ref} < V_{in} < -0.125V_{ref}$, then the decision is 0100. Nevertheless, it is easy to find that both decision levels lead to the same 3-bit binary code (i.e., 010). In other words, for a V_{in} near the threshold voltage (i.e., $-0.375V_{ref} < V_{in} < -0.125V_{ref}$), the decision made by the comparators in each 1.5-bit stage may go either way, and it does not affect the final digital output or the overall accuracy of the ADC. A similar argument also applies to the case when $+0.125V_{ref} < V_{in} < +0.375V_{ref}$. This is a desirable feature for the design of pipelined ADCs, because it allows us to build ADCs that are insensitive to the wrong decision level that mainly results from comparator dc offsets, thereby greatly easing the design of the comparators.

Unfortunately, although a pipelined ADC that is built on the 1.5-bit/stage configuration is typically insensitive to comparator offset errors, it is still subject to the nonlinearity errors induced by the finite op-amp gain, capacitor mismatch, and charge injection. For instance, if it is assumed that the op-amp used in the MDAC has a finite gain of A, then the transfer function derived earlier in Equation (5.54) can be rewritten as

$$V_{out} = \begin{cases} (2V_{in} - V_{ref}) \cdot \left(1 - \dfrac{2}{A+2}\right), & \text{if } V_{in} > 0.25V_{ref} \\ 2V_{in} \cdot \left(1 - \dfrac{2}{A+2}\right), & \text{if } -0.25V_{ref} < V_{in} < 0.25V_{ref} \\ (2V_{in} + V_{ref}) \cdot \left(1 - \dfrac{2}{A+2}\right), & \text{if } V_{in} < -0.25V_{ref} \end{cases} \quad (5.55)$$

Thus, the interstage gain deviates from its ideal value, which in this case is equal to 2, by $4/(A + 2)$. Notice that the mismatch between C_s and C_f is not taken into account in the foregoing function, while in practice it cannot be neglected since the interstage gain greatly depends on the capacitance ratio (assuming the ratio-independent technique [26] is not used here).

Many techniques have been proposed to remove the interstage gain error due to the finite op-amp gain and capacitor mismatch errors. Among them are the analog calibration/correction techniques (see Section 5.3). However, they are primarily used to cancel capacitor mismatch errors, and they require extra analog components and a more complex circuit design.

One of the predominant accuracy enhancement techniques for pipelined ADCs nowadays is called the *digital self-calibration* [51][52]. The key point of this technique is to have the ADC measure and correct its own gain error as well as other nonlinearity errors, all in the digital domain. In comparison with the conventional capacitor trimming technique, which measures the errors in the analog domain by using a trimming capacitor, the digital self-calibration technique uses the 1.5-bit/stage configuration to record the digital output of each stage, estimates the average interstage gain error, and removes the error using simple digital logic [51]. Thanks to the powerful digital signal processor, the digital self-calibration is usually technology scalable and insensitive to process or other environment/temperature variations.

However, similar to the capacitor trimming technique, the digital self-calibration technique is also a foreground calibration [51][52], due to the serial timing arrangement of the measurement and the calibration. In practice, the digital self-calibration is normally done during the power-up or standby mode to avoid interrupting the normal A/D conversion.

Since the late 1990s, various background calibration techniques have been reported, enabling the calibration and the standard A/D conversion to be performed separately and simultaneously. A straightforward solution is to replace the pipelined stage undergoing calibration with a dedicated redundant stage, which is similar to the methodology of calibrating the current-mode DAC [18] and SC DAC [33]. However, the price is more silicon area, power, and, most important, a speed penalty.

Many state-of-the-art pipelined ADCs incorporate the correlation algorithm into their background calibration schemes [53][54][55]. The basic idea is that the analog input signal is first modulated with a signed binary pseudo-random sequence such as {+1, 0, –1, 0, +1 . . .}, which can be generated by using multiple D-flip-flops and a feedback XOR gate. Next, the modulated input signal travels through the pipelined stages. Then the ADC's output digital signal is multiplied by the same signed binary pseudo-random sequence in the digital domain, thereby extracting or demodulating the nonlinearity errors including the gain error out of the output signal. Lastly, the errors can be canceled by simply subtracting them from the ADC's output. As the result, no extra analog circuit blocks are needed.

Thanks to the effectiveness and simplicity, the correlated-based calibration scheme has become a prevalent topic in the field of high-performance data converters (both in industry and academia). In a few works, for instance [7], the feasibility of adopting correlation-based calibration to cancel the nonlinearity DAC errors due to the use of a multibit quantizer in the delta-sigma ($\Delta\Sigma$) modulator has been explored.

As a final note with respect to the accuracy enhancement for a pipelined ADC, we should know that there is an alternative technique to the aforementioned 1.5-bit/stage design. That is, we can design a multibit first stage and build the remaining stages either on the 1-bit/stage or 1.5-bit/stage configuration. Intuitively speaking, the higher is the first stage's bit number (i.e., resolution), the higher the achievable accuracy will be and the less the ADC's performance will rely on the digital calibration. In addition, the use of a multibit first stage can help reduce the total number of pipelined stages; hence, the total power consumption of the ADC may be reduced [56].

However, it can be found that a multibit sub-ADC in the first stage (typically a flash ADC) tends to have a larger input capacitive load since more input comparators are used. In such a case, the input S&H circuit of the ADC must be able to drive this large capacitive load; hence, the total power consumption will be increased. Moreover, a multibit first stage requires a residue amplifier with a high gain and a large bandwidth to reduce the accuracy requirements for the succeeding stages. In other words, the resolution of the first stage needs to be optimized to balance the tradeoff between the resolution, circuit's complexity, and power consumption. An interesting calibration-free pipelined ADC with a 4-bit first stage can be found in [57].

In summary, there are many solutions to implementing a high-performance pipelined ADC. The choice of the best ADC is seldom absolute and greatly depends on the specific applications and the technologies available. In practice, it is convenient to adopt a figure of merit (FOM) in evaluating the ADC's performance. To evaluate pipelined ADCs useful for high-speed and high-resolution applications, the following FOM is often adopted [56]:

$$FOM = \frac{Power \cdot V_{dd}}{2^{ENOB} \cdot f_{clk}} \tag{5.56}$$

Here, V_{dd} is the power supply voltage, and f_{clk} is the system clock frequency. The smaller is the preceding FOM, the more appropriate the ADC will be for applications that require a high speed and a high resolution.

Cyclic ADCs

A *cyclic* (or *algorithmic*) A/D converter, like its D/A counterpart introduced in Section 5.3, uses the same analog circuitry to perform the conversion in a recycling fashion. Thus, the cyclic ADC is typically chosen for applications that require a low power consumption and a small chip area.

A conceptual schematic of an M-bit cyclic ADC is shown in Figure 5.12. Note that the cyclic ADC looks almost identical to the 1-bit pipelined stage shown in Figure 5.9, except for the feedback from the subtraction unit to the input sampling switch.

The cyclic ADC operates as follows. During the first cycle after reset, the input V_{in} is sampled by the front-end S&H, and is compared with 0 V. The result of the comparison is the sign bit, which is read into the shift register. During the next cycle, the held V_{in}, or, equivalently, $V(1)$ is doubled by the multiply-by-two residue amplifier, and the result is subtracted by either $+V_{ref}$ or $-V_{ref}$, depending on the output of the

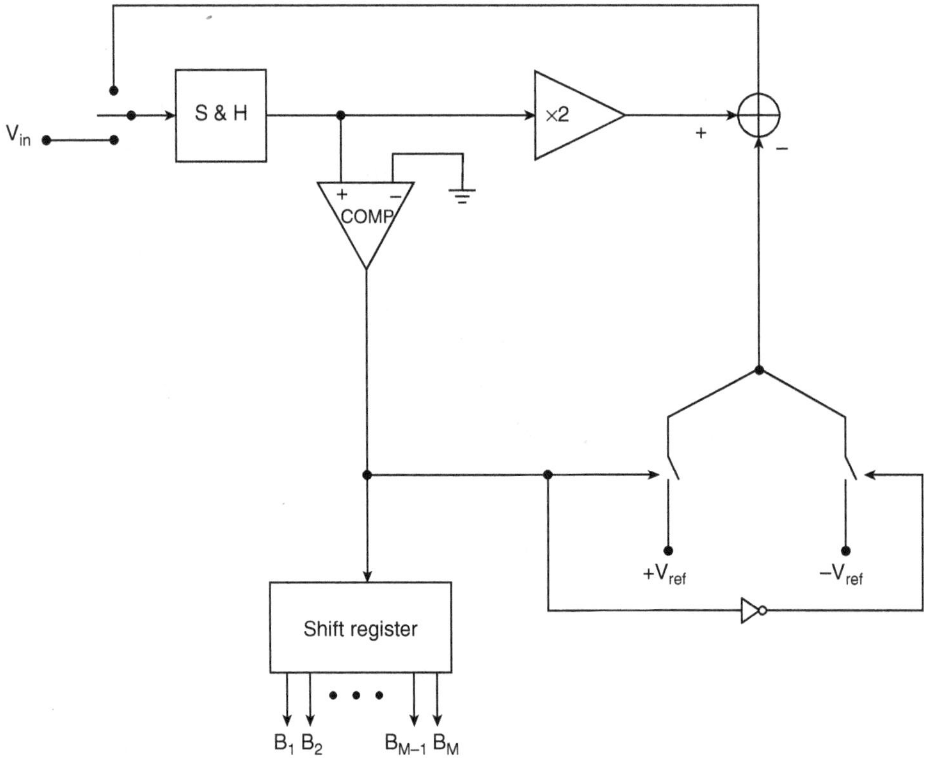

Figure 5.12 Cyclic ADC.

comparator $B(1)$. For instance, if it is assumed that the comparator outputs a 1 (i.e., $V(1) > 0\,\text{V}$), then the residue voltage will be given by $V(2) = 2V(1) - V_{ref}$. Next, $V(2)$ is sampled by the front-end S&H, and the foregoing procedure is repeated to decide the MSB and the new residue voltage $V(3)$, which may be given by either $[2V(2) - V_{ref}]$ or $[2V(2) + V_{ref}]$, depending on the output of the comparator $B(2)$. This recursive operation continues till all M bits (from the MSB to the LSB) are decided. Thus, the operation of the ADC follows an algorithm, which is described by [24][26][27]

$$V(i+1) = 2V(i) + (-1)^{bi}\, V_{ref},\ for\ i = 1, 2, \ldots, M \text{ and } b_i = \begin{cases} 1 & if \quad V(i) \geq 0 \\ 0 & if \quad V(i) < 0 \end{cases} \tag{5.57}$$

From the preceding function, we realize that the cyclic ADC works in a serial manner; hence, it is a *serial* ADC, as opposed to the *parallel* flash ADC and the *serial-in-parallel-out* two-step and pipelined ADCs. As a result, standard cyclic ADCs normally operate at low sampling rates ranging from 5 to 500 kHz.

As for the achievable resolution of a cyclic ADC, the major limiting factor is the gain error of the multiply-by-two (×2) amplifier, which is mainly due to capacitor mismatch. Ideally, the MDAC shown in Figure 5.9 could be used to realize the multiply-by-two amplifier of the cyclic ADC. However, the effect of mismatch between the sampling capacitor C_s and feedback capacitor C_f (shown in Figure 5.9) is more pronounced in the cyclic ADC than in the pipelined ADC, because the cyclic ADC uses the same stage repeatedly. Intuitively speaking, a pseudo-positive feedback loop of the gain error exists in the cyclic ADC; hence, the error's RMS value tends to aggregate over time and eventually corrupt the system.

One of the best solutions to the problem regarding the multiply-by-two amplifier's gain error is called the *ratio-independent technique* [24][26]. The essence of this technique is to transfer electric charge from the input to output *virtually* through only one capacitor (C_s), thereby allowing the circuit's voltage gain to be independent of capacitance ratios. Although the actual circuit implementation requires an extra capacitor (C_f), it is merely a depositary of charge; thus, its capacitance and relation to C_s are not important.

A ratio-independent multiply-by-two SC amplifier is shown in Figure 5.13 [11][26]. The input V_{in} comes from the output terminal of the preceding front-end S&H, and the output V_{out} goes to the subtraction circuit. As the schematic shows, four nonoverlapping clock cycles are needed to decide the output, and some switches are on during more than one cycle. For example, the denotation $\Phi_1 + \Phi_3$ means that the switch is turned on during Φ_1 *and* Φ_3 cycles. Table 5.3 lists the values of the voltages

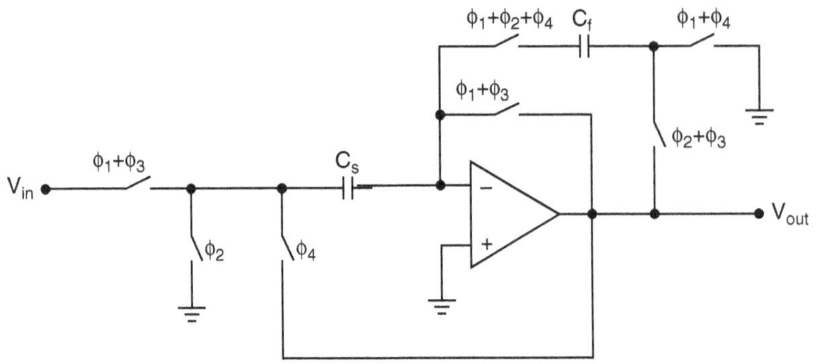

Figure 5.13 Ratio-independent multiply-by-two SC amplifier.

Table 5.3 Operation of the circuit shown in Figure 5.13.

Voltages	Phase Φ_1	Phase Φ_2	Phase Φ_3	Phase Φ_4
V_{Cs}	$V_{in} - V_{off}$	$-V_{off}$	$V_{in} - V_{off}$	$2V_{in} - V_{off}$
V_{Cf}	V_{off}	$V_{off} - V_{in}(C_s/C_f)$	$V_{off} - V_{in}(C_s/C_f)$	V_{off}
V_{out}	V_{off}	$V_{in}(C_s/C_f)$	V_{off}	$2V_{in}$

in the circuit during different cycles. (It is assumed that the op-amp has an input offset voltage V_{off}. V_{Cs} and V_{Cf} represent the voltages across C_s and C_f, respectively.)

Comparing Figure 5.13 with Table 5.3 and applying the charge reservation principle, we can find that the circuit operates in the following four steps.

Step 1. Sample the input by C_s.

Step 2. Deposit the charge on C_s to C_f.

Step 3. Sample the input by C_s for the second time.

Step 4. Withdraw the saved charge from C_f and transfer it back to C_s to obtain an output voltage equal to $2V_{in}$.

Additionally, it can be shown that the circuit is insensitive to the op-amp's input offset.

As we can see, four clock cycles are needed to complete one multiplication. Thus, an M-bit cyclic ADC needs at least $4M$ clock cycles to complete an M-bit conversion (the nonideal settling of the op-amp may cause additional time). For example, theoretically a 14-bit cyclic ADC will need 56 clock cycles to decide a 14-bit word.

In [27], an offset-insensitive multiply-by-two SC amplifier that can complete one multiplication within three clock cycles is reported. However, the reduction of conversion time is made possible by including three instead of two capacitors in the SC amplifier.

Moreover, both of the previous multiply-by-two amplifiers suffer from the effects of the insufficient op-amp gain, or, equivalently, the finite op-amp gain. If the op-amp gain is not sufficiently large, the negative input terminal of the op-amp can no longer be considered a virtual ground, and thus an additional gain error will occur. Chin and Wu [28] reported a gain-insensitive ratio-independent SC amplifier to resolve this gain error. It was claimed that in their design the minimum linear gain requirement for the op-amp would be reduced from $(6M + 16)$ dB to $(3M + 6)$ dB for an M-bit cyclic ADC—at the cost of minimum $7M$ clock phases to complete a full conversion. For instance, a 14-bit cyclic ADC will require a minimum op-amp dc gain equal to only 48 dB but at least 98 clock cycles to complete a 14-bit conversion.

Implementing high-resolution (10 to 14 bits) CMOS cyclic ADCs that can operate at higher sampling rates (500 kHz to 5 MHz) is an actively ongoing research. One of the noteworthy results is a 10-bit CMOS cyclic ADC clocked at 3 MHz, which was reported by Kitagawa et al. [58].

Successive-Approximation ADCs

The successive-approximation ADC is often considered the *dual* to the cyclic ADC since it also uses only one conversion stage. However, in contrast to the cyclic ADC's operation of doubling the residue voltage $V(i)$, the successive-approximation ADC halves the effective search space to decide each bit. A conceptual block diagram of a successive-approximation ADC is shown in Figure 5.14.

The ADC operates as follows. During the first cycle after reset, the input V_{in} is sampled by the front-end S&H and is compared with 0 V. The result of the comparison is the sign bit, and it is read into the successive-approximation register (SAR). Next, the SAR sets the DAC's output to $0.5V_{ref}$, which is then compared with V_{in}. If it is assumed that the comparator outputs a 1 (i.e., $V_{in} > 0.5V_{ref}$), then SAR will decide the MSB to be 1, and the DAC's output will become $0.75V_{ref}$. Otherwise, the MSB will be 0 and the DAC will output $0.25V_{ref}$. If we assume that $0.75V_{ref}$ is compared with V_{in} to decide the next bit, then SAR may set DAC's new output to either $0.875V_{ref}$ or $0.625V_{ref}$, depending on the current output of the comparator. The operation continues in this way till all M bits (from the MSB to the LSB) are determined by the SAR.

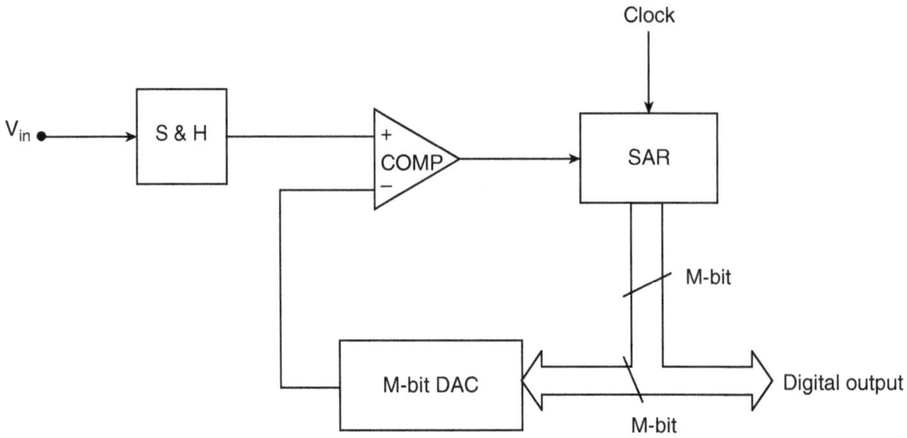

Figure 5.14 Successive-approximation ADC.

In a word, in each clock cycle the SAR divides the search space in two for locating one bit. As a result, M clock cycles will be required to complete an M-bit conversion. In comparison with its cyclic dual, the successive-approximation ADC typically has a faster conversion rate.

In the successive-approximation ADC shown in Figure 5.14, the most critical analog circuit block is the M-bit DAC, because its performance directly affects the accuracy and the speed of the ADC. One of the most widely adopted approaches to implement the DAC is using a binary-weighted charge-redistribution DAC, which was reported by McCreary et al. [59].

A major drawback of the binary-weighted charge-redistribution DAC is the large silicon area needed to layout the capacitors. In general, an M-bit charge-redistribution DAC requires $(M + 1)$ capacitors, and the total capacitance is given by

$$C_{total} = \sum_{i=0}^{M} 2^i C \tag{5.58}$$

where C is the unit-element capacitor. Also, the large capacitance tends to result in a high power consumption if the ADC is operating at a high sampling rate.

As an area-efficient alternative to the binary-weighted charge-redistribution DAC, an SC DAC that requires only three unit-element capacitors (regardless of the required resolution) is shown in Figure 5.15 [60]. The nominal capacitances of all three capacitors (C_1, C_2, and C_3) are equal to C. The switches denoted by Φ_r are turned on only during the reset mode. The on/off operations of Φ_x and Φ_y are con-

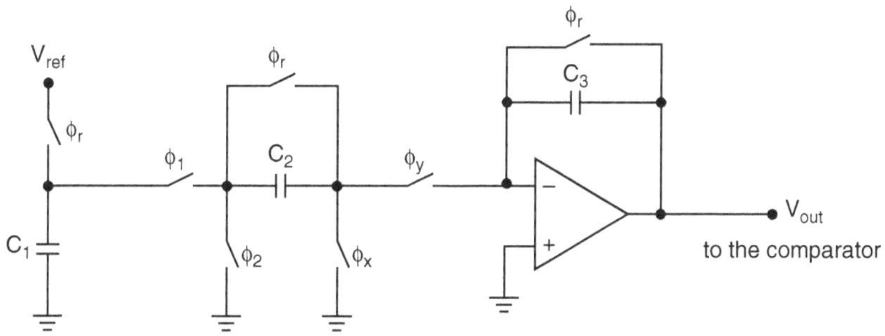

Figure 5.15 Area-efficient SC DAC.

trolled by the digital output of the SAR, or, equivalently, D in the following respective ways:

$$\begin{cases} \Phi_x = D \cdot \Phi_1 + \overline{D} \cdot \Phi_2 \\ \Phi_y = \overline{D} \cdot \Phi_1 + D \cdot \Phi_2 \end{cases} \tag{5.59}$$

In other words, a switch may be turned on during either Φ_1 or Φ_2 cycles, depending on the value of D. For instance, if $D = 1$, then the Φ_x switch will be turned on only when $\Phi_1 = 1$, while the Φ_y switch will be turned on only when $\Phi_2 = 1$.

The SC DAC operates as follows. During the reset mode, the capacitor C_1 is charged by the reference voltage V_{ref}, while both C_2 and C_3 are discharged. Next, $\Phi_1 = 1$, and the first digit from the SAR is forced to be equal to 1 (for the sign bit). As a result, the Φ_x switch is turned on, and now C_1 shares its charge with C_2. Prior to $\Phi_2 = 1$, the nominal voltages across C_1 and C_2 are both settled to $0.5V_{ref}$. Next, $\Phi_2 = 1$, and D is still kept at 1; thus, the Φ_y switch is turned on, the charge on C_2 is now deposited to the feedback capacitor C_3, and a *positive* output voltage equal to $0.5V_{ref}$ is obtained.

During the following Φ_1 cycle, either the Φ_x or Φ_y switch may be turned on, depending on the next digital output from the SAR. For example, if the next D is equal to 0, which means V_{in} is smaller than $0.5V_{ref}$, then the Φ_y switch is turned on while the Φ_x switch is not. As a result, the stored charge on C_1 (i.e., $0.5V_{ref}C_1$) is shared with C_2 and, meanwhile, an inverted copy of the new charge on C_2 is transferred to C_3. Therefore, the resultant output voltage is given by $(0.5V_{ref} - 0.25V_{ref}) = 0.25V_{ref}$. On the other hand, if the next D is equal to 1, then the resultant output voltage will be equal to $0.75V_{ref}$. The operation continues in the same way till the Mth output voltage is determined. In total, M clock cycles are required to complete an M-bit conversion.

However, like many other SC Nyquist-rate data converters, the preceding SC DAC suffers from circuit imperfections such as the capacitor mismatch, the nonideal op-amp (e.g., input offset and finite gain), parasitic capacitance, and charge injection. Although the latter two can be alleviated by adopting a fully differential structure and a more careful layout, the cancellation of the remainders is nontrivial. We learned in Section 5.3 that the errors due to capacitor mismatch can be removed using analog or digital calibrations. In addition, the problems due to the op-amp's imperfections can be effectively resolved by employing advanced techniques such as *autozeroing* and *correlated double sampling* [61]. We will look into these accuracy-enhancement techniques in Chapter 7.

Before leaving this section, note that there are other Nyquist-rate ADCs that are realizable using the SC techniques, such as the *integrating* or *incremental* ADCs and the *time-interleaved* ADCs, with the former being the slowest type of ADC and the latter being the second fastest type of ADC (between the flash and two-step ADCs, although the speed boundaries are becoming vague [47]). For the sake of brevity, their characteristics are not studied in this book. For classic examples of SC implementations, the reader is referred to references such as [62] for the incremental type, and [63] for the time-interleaved type.

5.5 Oversampling Data Converters

Nyquist Rate versus Oversampling

As we learned earlier in this chapter, the achievable accuracy of a Nyquist-rate data converter is limited by the matching accuracy of the analog components (e.g., capacitors or current sources) in the circuit, which normally has a minimum relative error of about 0.02%. As a result, most Nyquist-rate converters are not capable of achieving ENOBs over 12-bit calibrations. However, applications such as digital audio and instrumentations usually demand data converters with ENOBs as large as 20 or even 24 bits. Even with effective calibrations, the integrating data converter, which is the most accurate (and the slowest) Nyquist-rate data converter, can hardly achieve an accuracy as high as 24 bits.

In contrast to the Nyquist-rate data converters, oversampling data converters are capable of achieving ENOBs over 20 bits without any special calibration circuitries; however, this occurs at the cost of applying sampling rates much higher than the Nyquist rate (typically by a factor between 16 and 512). In other words, oversampling data converters trade speed for accuracy.

In addition, for a Nyquist-rate ADC, undesired out-of-band signals tend to reside near the desired signal cutoff frequency. As the result, a highly selective antialiasing filter (AAF) is often required to precede the ADC. In comparison, an oversampling ADC uses a sampling frequency that is much higher than the signal cutoff frequency, thus the possibility of aliasing is trivial, and the use of a high-order AAF is not mandatory. Furthermore, the decimation filter, which can be easily implemented in the digital domain, provides additional low-pass filtering at the oversampling ADC's output. In a word, oversampling helps relax the selectivity requirement for the antialiasing filter.

Another important difference between the Nyquist-rate and the oversampling data converters is that the Nyquist-rate converter is memoryless while the oversampling converter has memory. Specifically speaking, the Nyquist-rate converter generates one output for each instantaneous input, regardless of the earlier inputs, whereas each output of the oversampling converter depends on all previous inputs.

As discussed in Section 5.2, in an ADC, the origin of the quantization noise is the signal ambiguity due to the incidence of converting multiple analog input levels that the ADC cannot distinguish into the same digital code. Intuitively speaking, the more input samples are taken into account by the ADC to decide the digital output, the less severe the effect of the ambiguity on the ADC's accuracy, or its quantitative equivalence—the quantization noise—will be. This qualitatively explains why the signal-to-quantization-noise-ratio (SQNR) can be increased by increasing the oversampling ratio (OSR), which is conventionally defined as follows (f_s is the sampling frequency and f_0 is the signal bandwidth):

$$OSR = \frac{f_s}{2f_0} \tag{5.60}$$

Similar to the derivation of Equation (5.20), it can be found that the quantization noise is approximated to a random sequence uniformly distributed across [$-0.5V_{LSB}$, $+0.5V_{LSB}$], and its power is given by

$$\sigma_p^2 = \frac{1}{V_{LSB}} \int_{-0.5VLSB}^{+0.5VLSB} q^2 dq = \frac{V_{LSB}^2}{12} \tag{5.61}$$

Based on the well-known *white-noise assumption* of the quantization noise [64], the power spectral density of the quantization noise is assumed to be evenly distributed across the frequency range between $-0.5f_s$ and $+0.5f_s$. What oversampling essentially does from a frequency-domain prospective is to define a much smaller signal bandwidth ranging between $-(0.5/\text{OSR})f_s$ and $+(0.5/\text{OSR})f_s$; as a result, only a small portion of the noise power is included within the signal bandwidth and counted in the calculation of SNR (for the moment we assume that the quantization noise is the dominant source of noise in the system; hence, we may use the term SNR in place of SQNR). Or, oversampling can be considered equivalent to stretching the quantization noise power across the frequency spectrum (given a fixed signal bandwidth f_0), thereby reducing the noise power in the signal band.

Since the noise power is assumed to be constant across the frequency spectrum (i.e., like a white noise), we can write the aforementioned small portion of the noise power as the following:

$$\sigma_p^2\Big|_{oversampled} = \frac{V_{LSB}^2}{12} \cdot \frac{1}{OSR} \tag{5.62}$$

Thus, from Equation (5.20) we obtain

$$SNR\Big|_{oversampled} = 6.02M + 1.76 + 10\lg(OSR) \tag{5.63}$$

The last term in this equation represents the SNR improvement thanks to the oversampling operation. Note that every time the OSR is doubled, the maximum SNR will be increased by 3 dB, or, equivalently, it has an ENOB gain of 0.5 bits/octave.

Nevertheless, in most cases, it is impractical to obtain a high SNR through the straight oversampling. For instance, if it is assumed that a 1-bit quantizer (i.e., $M = 1$) is used to achieve an 86-dB SNR (i.e., 14-bit), then from the foregoing equation we can find that an OSR of about 66,374,307 will be required. If the desired Nyquist rate is equal to 10 kHz, then the required clock sampling frequency will be as high as 663.74 GHz! As one can imagine, the sampling frequency will be amazingly high if a 20-bit resolution is needed.

Therefore, to the practical implementation of a high-resolution data converter, pure oversampling is not adequate, and an important architectural modification is required to facilitate a more aggressive SNR improvement. This modification is called *noise shaping* (or *loop filtering*), which is the topic of the next subsection.

Noise Shaping and Stability

In an oversampling data converter that is capable of shaping noise, the noise-shaping operation is normally performed by the well-known *delta-sigma* (or $\Delta\Sigma$) *modulator*, which makes use of negative feedback(s) to suppress the in-band quantization noise power, thereby effectively improving the SNR performance.

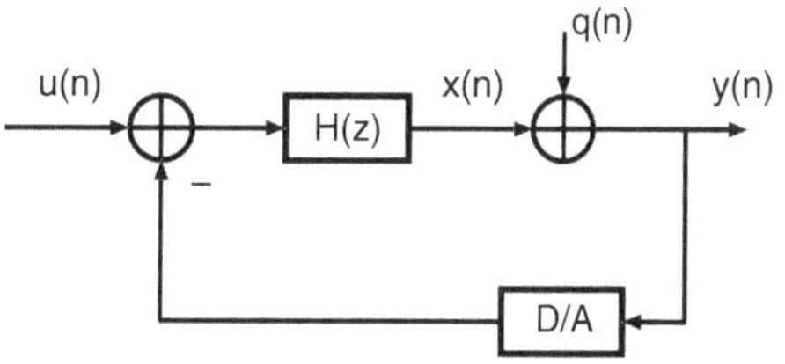

Figure 5.16 Block diagram of a ΔΣ modulator.

The conceptual block diagram of a ΔΣ modulator is shown in Figure 5.16. It should be mentioned that the additive white noise assumption [64] is adopted here as before—that is, the quantization noise q is assumed to be independent of the *main* input signal u and uniformly distributed across $[-0.5V_{LSB}, +0.5\ V_{LSB}]$.

Note that in Figure 5.16, the quantizer is approximated to an adder, that is, the inherent nonlinear characteristics of the quantization process, which were mentioned in the discussion of the two-step ADC, are neglected here. Thus, the entity shown in Figure 5.16 is often called the *linearized* model of the ΔΣ modulator.

Based on the linearized model of the general ΔΣ modulator, we can explore the principle of noise shaping using the z-domain analysis. First, let us write the z-domain expression of the modulator's output $Y(z)$ as follows:

$$Y(z) = STF(z)\cdot U(z) + NTF(z)\cdot Q(z) \tag{5.64}$$

Here, $U(z)$ and $Q(z)$ are the z-domain expressions of the input signal and the quantization noise, respectively. $STF(z)$ is the *signal transfer function*, and $NTF(z)$ is the *quantization noise transfer function*. Solving the above equation for $STF(z)$ and $NTF(z)$, we can write their corresponding expressions in terms of $H(z)$, which are given by

$$\begin{cases} STF(z) = \dfrac{Y(z)}{U(z)}\Big|_{Q(z)\equiv 0} = \dfrac{H(z)}{1+H(z)} \\ NTF(z) = \dfrac{Y(z)}{Q(z)}\Big|_{U(z)\equiv 0} = \dfrac{1}{1+H(z)} \end{cases} \tag{5.65}$$

If it is assumed that the transfer function of the loop-filter, $H(z)$, is realized by a delayed noninverting SC integrator—that is, $H(z) = 1/(z-1)$—then we can obtain

$$\begin{cases} STF(z) = \dfrac{H(z)}{1+H(z)} = z^{-1} \\ NTF(z) = \dfrac{1}{1+H(z)} = 1 - z^{-1} \end{cases} \tag{5.66}$$

The foregoing transfer functions indicate that the main input signal is simply delayed by one clock cycle, whereas the quantization noise power is suppressed or shaped in the low-frequency range (i.e., $|z| \to 1$). A better way to understand shaping of the quantization noise is to find the magnitude responses of the preceding functions, which are given by

$$\begin{cases} |STF(z)| = |z^{-1}| = |\cos\Omega - j\sin\Omega| = 1 \\ |NTF(z)| = |1 - z^{-1}| = |1 - \cos\Omega + j\sin\Omega| = 2\left|\sin\dfrac{\Omega}{2}\right| \end{cases} \tag{5.67}$$

As the preceding equations indicate, the magnitude of $NTF(z)$ increases with the normalized frequency Ω, which is defined as $\Omega = 2\pi f/f_s$. Thus, the quantization noise power is reduced in the lower frequency range (i.e., where Ω is small) and pushed to the higher frequency domain (i.e., where Ω is large). This phenomenon may be analogous to squeezing toothpaste. However, the magnitude of $STF(z)$ is always equal to 1; hence, the input signal is intact.

Figure 5.17 illustrates the transformation from the conceptual ADC model shown in Figure 5.1 into a first-order noise-shaping ΔΣ ADC (the decimation filter is not shown), and the *before-and-after* quantization noise power distributions.

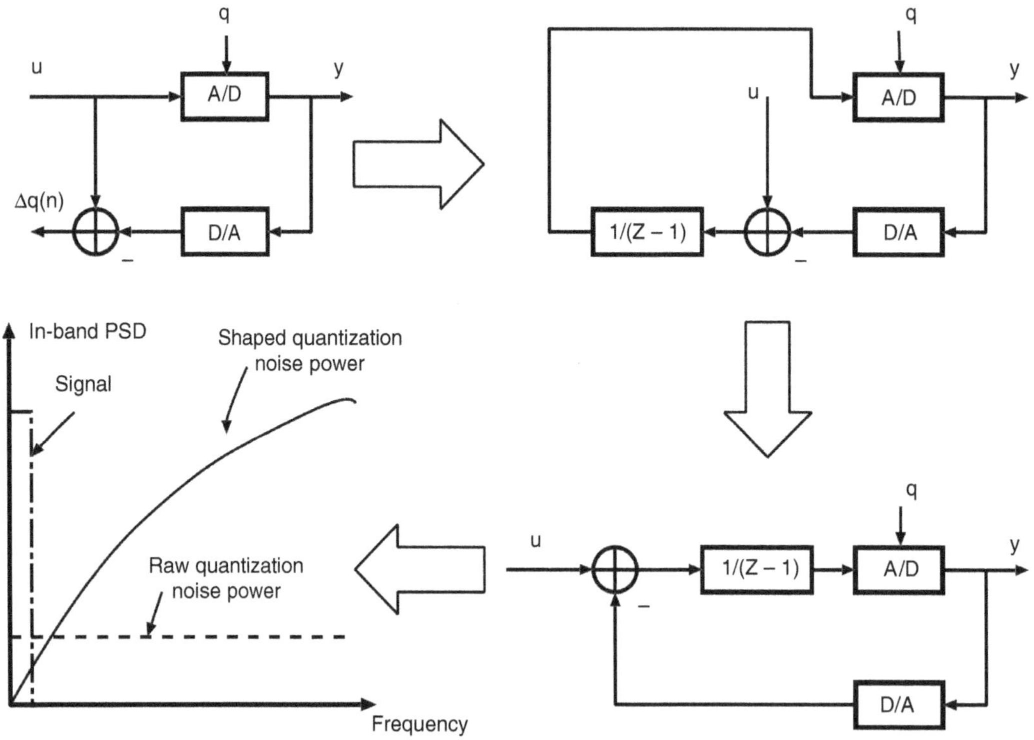

Figure 5.17 First-order noise shaping.

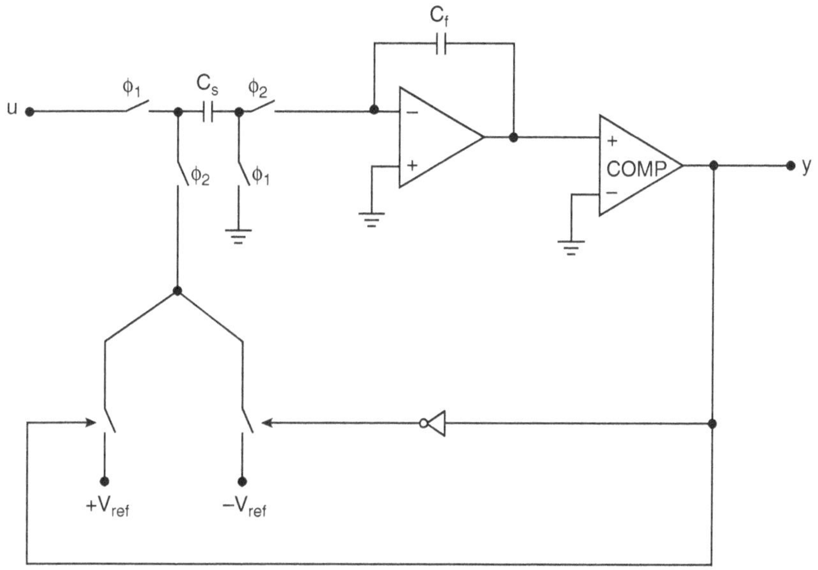

Figure 5.18 First-order noise-shaping ΔΣ modulator using a 1-bit quantizer.

The SC realization of a first-order ΔΣ modulator is shown in Figure 5.18. Note that for simplicity the decimation filter is not shown. Although only the single-ended circuit is shown, in practice a fully differential configuration is required.

As shown in the schematic, the loop filter is realized by a noninverting parasitic-insensitive SC integrator, and the quantization is performed by a two-level comparator. As introduced earlier, the 1-bit feedback DAC is realized by connecting either $+V_{ref}$ or $-V_{ref}$ to the negative input terminal, depending on the comparator's output.

Thanks to the first-order noise shaping, the SNR of the converter is improved. Taking Equations (5.64) and (5.67) into consideration, we can derive the maximum SNR of the first-order noise-shaping ΔΣ ADC (in decibels), which is given by

$$SNR\Big|_{1st-order} = 6.02M + 1.76 + 30\lg(OSR) - 10\lg\left(\frac{\pi^2}{3}\right) \tag{5.68}$$

Therefore, doubling the OSR of a first-order noise-shaping modulator provides a 9-dB increase in the maximum SNR, or equivalently, an ENOB gain of 1.5 bits/octave. However, this improvement is still not adequate to the practical realization of, say, a 96-dB SNR (i.e., 16-bit). That is, if we assume that a 1-bit quantizer (i.e., $M = 1$) is used, then from the preceding equation we can find that an OSR of about 1297 will be required to achieve the 96-dB SNR. If the desired Nyquist rate is equal to 10 kHz, then the resultant clock sampling frequency will reach 12.97 GHz, which is still too high for most applications that require low power consumption.

Thus, a more aggressive noise-shaping topology is needed to achieve a higher resolution without significantly increasing the sampling rate (or OSR). If we assume that the ΔΣ modulator always uses a 1-bit quantizer (we will study modulators that employ multibit quantizers later), then the only remaining option will be to increase the effective order (L) of the noise-shaping loop filter.

The next logical step is to build a ΔΣ modulator with the second-order noise shaping capability. A second-order ΔΣ loop filter provides the following quantization noise transfer function:

$$NTF(z) = (1 - z^{-1})^2 \tag{5.69}$$

A well-known implementation of a second-order ΔΣ modulator was reported by Boser and Wooley [65]. The loop filter in the ADC is usually realized by two cascaded delayed SC integrators, with a weighted DAC feedback going into each integrator's negative input. This type of structure is often referred to as the *distributed-feedback topology.*

An alternative second-order topology that consists of two forward paths and one DAC feedback was reported by Silva et al. [66]. The basic idea is to distribute the input signal u directly to the quantizer's input terminal through a dedicated forward path, so that u will be canceled out at the input of the first integrator; hence, no input signal is passed through the two delayed SC integrators. In other words, the integrators process the quantization noise only, which is assumed to be independent of u. Therefore, the nonlinearity errors due to the signal-dependent interferences such as the charge injection are greatly reduced.

Despite their structural differences, both of the preceding second-order topologies share the same expression for the maximum SNR, which is given by [11][64]

$$SNR\big|_{2nd-order} = 6.02M + 1.76 + 50\lg(OSR) - 10\lg\left(\frac{\pi^4}{5}\right) \tag{5.70}$$

Doubling the OSR of a second-order noise-shaping ΔΣ modulator provides a 15-dB increase in the maximum SNR, or equivalently, an ENOB gain of 2.5 bits/octave. In comparison with the first-order noise shaping, the OSR required for realizing a 16-bit resolution has dropped from 1297 to about 106.

To generalize, it can be found that the maximum SNR of an Lth-order noise-shaping ΔΣ modulator is given by

$$SNR\Big|_{Lth-order} = 6.02M + 1.76 + (20L+10)\lg(OSR) - 10\lg\left(\frac{\pi^{2L}}{2L+1}\right) \tag{5.71}$$

This indicates that the maximum SNR will increase with the OSR at the rate of $(6L + 3)$ dB/octave, or equivalently, an ENOB gain of $(L + 0.5)$ bits/octave. To quantify, consider a third-order noise-shaping ΔΣ modulator with a 1-bit quantizer. It can be calculated that an OSR of only 36 is adequate to achieve a 96-dB SNR (i.e., a 16-bit resolution). Note that M is still equal to 1.

As suggested by the foregoing development, given a noise-shaping ΔΣ modulator with a fixed SNR specification, the higher is its loop order, the lower the required OSR would be. However, once the loop order exceeds 2 (i.e., $L > 2$), the design of a ΔΣ modulator using a 1-bit quantizer faces another challenge—that is, the modulator may become unstable [67].

The root of the instability problem is a combination of the nonlinear characteristics with respect to the 1-bit quantizer, and the magnitude response of $NTF(z)$. It is well known that a 1-bit quantizer outputs either +1 or −1 depending on the polarity of the input signal, rather than on its absolute value. In other words, the quantizer is nonlinear and provides only two output options, although the input values may vary actively.

Recalling the magnitude of a first-order $NTF(z)$ given by Equation (5.67) and the linearized model shown in Figure 5.16, we can find that in a first-order modulator the magnitude of the quantization noise q is limited between −2 and +2; hence, the quantizer's input is limited between −3 and +3 (assuming the quantizer has a dc gain equal to 1). In comparison, it can be shown that the magnitude of the $NTF(z)$ in a third-order ΔΣ modulator is given by

$$|NTF(z)| = \left|2\sin\frac{\Omega}{2}\right|^3 = 8\times\left|\sin\frac{\Omega}{2}\right|^3 \tag{5.72}$$

Now the magnitude of the quantization noise q can be of any value that resides within the range between −8 and +8, and the range of the quantizer's input values is thus given by [−9, +9]. Since the third-order modulator has a wider range of input values (positive and negative) compared to that of the first-order modulator, the former has a higher probability of generating a long string of 1's or (−1)'s (i.e., of being unstable) than the latter. For example, if we assume that the quantizer's input x are constantly changing but are somehow restricted within the positive domain (i.e., $x > 0$), then the quantizer will keep on outputting 1's, regardless of the value of each input. As a result, the average quantization noise power is increased continuously, and the SNR is degraded accordingly. Eventually, the SNR may drop to or even below zero.

To test whether a ΔΣ modulator is stable or not, we can examine its output sequence in the time-domain (i.e., looking for long 1's or −1's). Alternatively, we can use a spectrum analyzer or a computer with a data acquisition kit to record its output sequence and then plot the "SNR *vs.* input frequency" curve using the fast Fourier transform (FFT). If we find an abrupt drop of SNR in the curve, then the modulator is likely to be unstable.

The design of a high-order ($L > 2$) and 1-bit-quantization ΔΣ modulator with a guaranteed stability is nontrivial, and in most cases, computer-aided design (CAD) techniques are employed to simulate the modulator's behavior. The simulation results are often compared with some criteria for decision making. A widely adopted stability criterion is often referred to as *Lee's rule* [67], which states that the out-of-band (i.e., $f > 2f_0$) magnitude of $NTF(z)$ should be no more than 2 to ensure stability. However, examination of the existing practical high-order modulators (with 1-bit quantizers) will show that this rule is only a rule of thumb—that is, sometimes a modulator following the rule still becomes unstable when certain inputs are applied to it. On the other hand, a number of high-order modulators disobey the rule but indeed turn out to be stable at all times.

One point to emphasize here is that the stability of a high-order and 1-bit-quantization ΔΣ modulator is difficult to judge based on pure mathematical calculations. At the time of this writing, the most reliable approach is to run computer-based behavioral simulations using as many input options as possible. Two most convenient (in the author's opinion) computer programs for simulating SC ΔΣ modulators are as follows: the MATLAB simulation toolbox written by Schreier [68] and the SWITCAP program developed at Columbia University [69]. These two programs can be used in combination, with the former determining the appropriate modulator topology and system coefficients based on the specifications and the latter performing the switch-level simulation. Appendix 5.1 presents exemplary MATLAB and SWITCAP source codes for simulating a fourth-order audio ΔΣ ADC.

Types of ΔΣ Modulators

Delta-sigma (ΔΣ) modulators can be roughly categorized into the following types: low-order ($L < 3$) 1-bit single-stage, high-order ($L \geq 3$) 1-bit single-stage, 1-bit multistage (each stage uses a 1-bit quantizer), multibit single-stage, multibit multistage (each stage uses a multibit quantizer), 1-bit/multibit multistage (all stages use 1-bit

Table 5.4 Summary of different types of ΔΣ modulators: a report card.

Modulator type	SNR *vs.* OSR	SNR *vs.* Power	Stability	DAC linearity	Circuit simplicity
Low-order 1-bit single-stage	C–	C	A	A	A
High-order 1-bit single-stage	B+	A–	C	A	A–
1-bit multistage	B+	A–	A	A	B+
Multibit single-stage	A–	B+	A	C	C+
Multibit multistage	A	B	A	C	C–
1-bit/multibit multistage	A–	A	A	B+	C
Multibit/1-bit multistage	A	B+	A	B	C

quantizers except for the *last* stage, which uses a multibit quantizer), and multibit/1-bit multistage modulators (all stages use 1-bit quantizers except for the *first* stage, which uses a multibit quantizer). Each type of ΔΣ modulator has its own share of advantages and disadvantages. Table 5.4 summarizes the advantages and disadvantages for each type. Note that the term *power* in the table means the total power consumption, which accounts for both analog and digital portions of the circuit.

ΔΣ Modulators with Single-Bit Quantization

Since the late 1980s, ΔΣ modulators have been primarily used for the high-resolution conversion of narrow-band (or low-frequency) signals, such as telephony voice recognition ($f_0 < 3\,\text{kHz}$) and digital audio processing ($f_0 < 20\,\text{kHz}$). For these applications, a high-order 1-bit single-stage topology is often preferred because the circuitry is relatively straightforward and the resultant 1-bit feedback DAC is inherently linear (we will discuss the nonlinearity of a multibit DAC and its impact on the modulator's performance later).

However, due to the aforementioned potential stability problem, the search for a high-order, 1-bit quantization, and single-stage ΔΣ topology with a guaranteed stability is not an easy task. Research endeavors since the early 1990s have offered a variety of topologies. In this book, only two practically proven topologies are chosen as the subject of the following discussion for the sake of brevity.

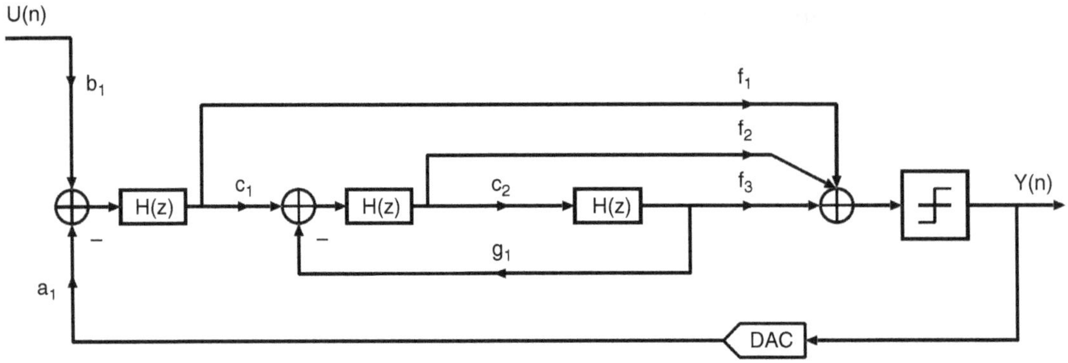

Figure 5.19 Third-order DQIR ΔΣ topology.

The first topology is shown in Figure 5.19. Note that a third-order configuration is used here as a demonstration platform, and each $H(z)$ block is realized by a delayed and noninverting SC integrator. In practice, the delayed integrator is often preferred because it provides sufficient time for the op-amp to settle, thereby reducing the op-amp's transient requirements. The structure is called the *distributed-quantizer-input-plus-internal-resonator* (DQIR) ΔΣ topology for it has three distributed forward paths that lead toward the quantizer, and one internal resonator, which is formed by a loop consisting of two $H(z)$ blocks and a subtraction unit.

The internal resonator spreads two of the three $NTF(z)$'s zeros from dc (i.e., $z = 1$) to certain frequencies within the signal bandwidth, for reducing the in-band noise. The system coefficients such as c_1, c_2, g_1, and f_2 can be adjusted to optimize the zero placements and minimize the in-band noise power, which is often called the zero optimization [64][70]. After the adjustment, the resultant new $NTF(z)$ response looks similar to that of an inverse-Chebyshev high-pass filter. Note that the modulator needs an analog summer in front of the quantizer, which is often realized by an SC gain stage.

A fifth-order version of this DQIR modulator was reported by Sooch et al. [71], which has five delayed SC integrators and two separate internal resonators. In [71], the modulator is implemented as part of an 18-bit DAC. It accepts an 18-bit digital input signal sequentially and transforms it into a 1-bit sampled-data output. A fourth-order SC low-pass filter is placed after the modulator to attenuate the out-of-band noise. Since the SC filter's output is still a sampled-data signal (i.e., staircase), a voltage buffer or *smoother* (shown in Figure 5.20) is needed to convert the switched-capacitor output to a true continuous-time signal.

The voltage buffer operates as follows. When $\Phi_1 = 1$, the input capacitor C_1 is charged to the input voltage V_{in}. Next $\Phi_2 = 1$, and the previous saved charge on C_1 is shared with the feedback capacitor C_2. Following the charge reservation principle, we can obtain the z-domain transfer function of the voltage buffer, which is given by

$$H(z) = \frac{V_{out}(z)}{V_{in}(z)} = \frac{C_1 z^{-1}}{(C_1 + C_2) - C_2 z^{-1}} \tag{5.73}$$

It can be shown that the preceding transfer function has a dc gain equal to 1, and a zero at infinity.

The second topology is shown in Figure 5.21. It was first proposed in [67] as an *interpolative* modulator, and it can also be referred to as the *distributed-forward-feedback-plus-internal-resonator* (DFFIR) ΔΣ topology for it has four distributed forward paths, three distributed feedback paths, and one internal resonator. The inclusion of the distributed forward and feedback paths brings a certain degree of freedom in realizing the *NTF*(z) and *STF*(z) [64][70]. Similar to the previous DQIR

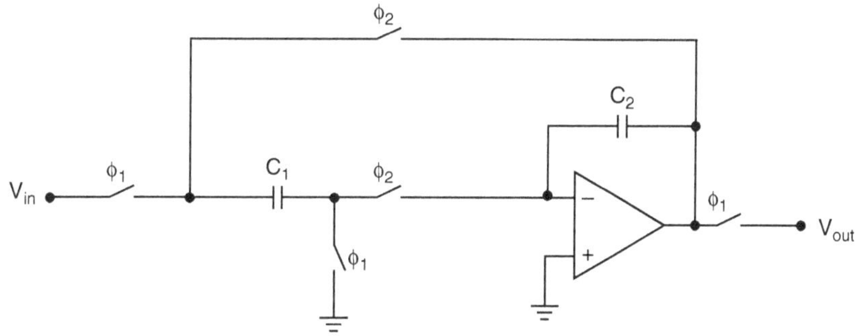

Figure 5.20 Sampled-data to continuous-time voltage buffer.

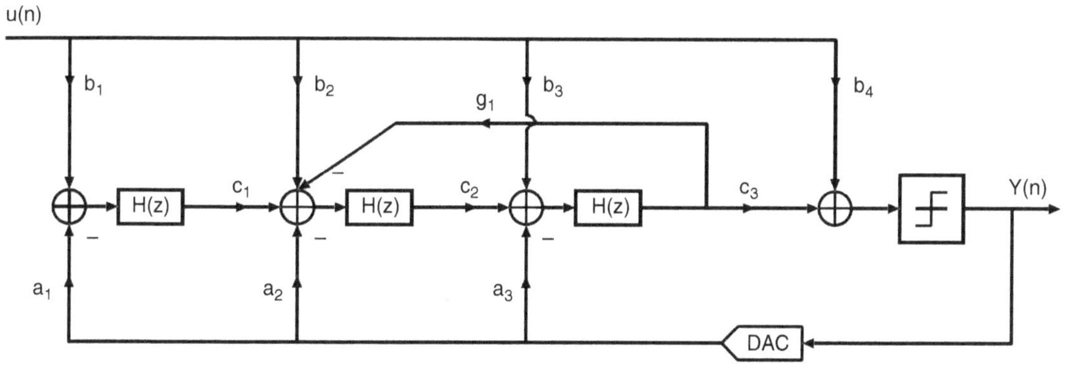

Figure 5.21 Third-order DFFIR ΔΣ topology.

topology, the internal resonator can be used to spread the zeros and reduce the in-band noise.

Note the second integrator in the schematic may have a high dc output voltage level for four signals are attached to its input node. As a result, the op-amps in the second and third integrators may become saturated, thereby limiting the dynamic range of the overall system. Dynamic range scaling (see Chapter 4) technique is often employed to find the optimal system coefficients for the best dynamic range performance.

A fifth-order realization of the preceding DFFIR topology was reported by Ferguson et al. [72], which has five delayed SC integrators and two separate internal resonators. The reported SNR performance of the modulator is comparable to that of the fifth-order DQIR modulator in [71].

As Table 5.4 indicates, the advantages of the single-bit multistage (or cascaded) $\Delta\Sigma$ modulator include the robust stability, guaranteed DAC linearity, and good SNR versus power consumption figure. In practice, the two often seen topologies are the 2–2 cascaded and the 2–1–1 cascaded $\Delta\Sigma$ topologies.

Figure 5.22 shows a 2–2 cascaded $\Delta\Sigma$ topology. The left part of the diagram illustrates the analog circuitry, which is composed of two cascaded second-order 1-bit modulators. Note that the quantization noise of the first stage (multiplied by an interstage gain factor f_1) is fed into the second stage. The right part of the diagram shows the digital error cancellation logic. It can be proved by using the z-domain analysis that ideally the quantization noise generated by the first stage would be canceled, as long as the following conditions are met simultaneously [73]:

$$b_1 = b_3,\ b_2 = b_4,\ a_2 = a_4,\ \text{and}\ a_1 = a_3 + g_1 \tag{5.74}$$

The specific numerical values of these system coefficients can be determined by applying dynamic range scaling on the basis of the interstage gain factor f_1 (i.e., set f_1 to 1 and scale the others). As mentioned in Chapter 4, the task is accomplished by scaling the system coefficients so that the maximum levels at the op-amps' outputs are equalized.

EXAMPLE 5.5

Analyze the 2–2 cascaded $\Delta\Sigma$ modulator shown in Figure 5.22 using the z-domain analysis. Find the necessary conditions for the system to cancel the quantization noise from the first stage, and modulate the quantization noise from the second stage

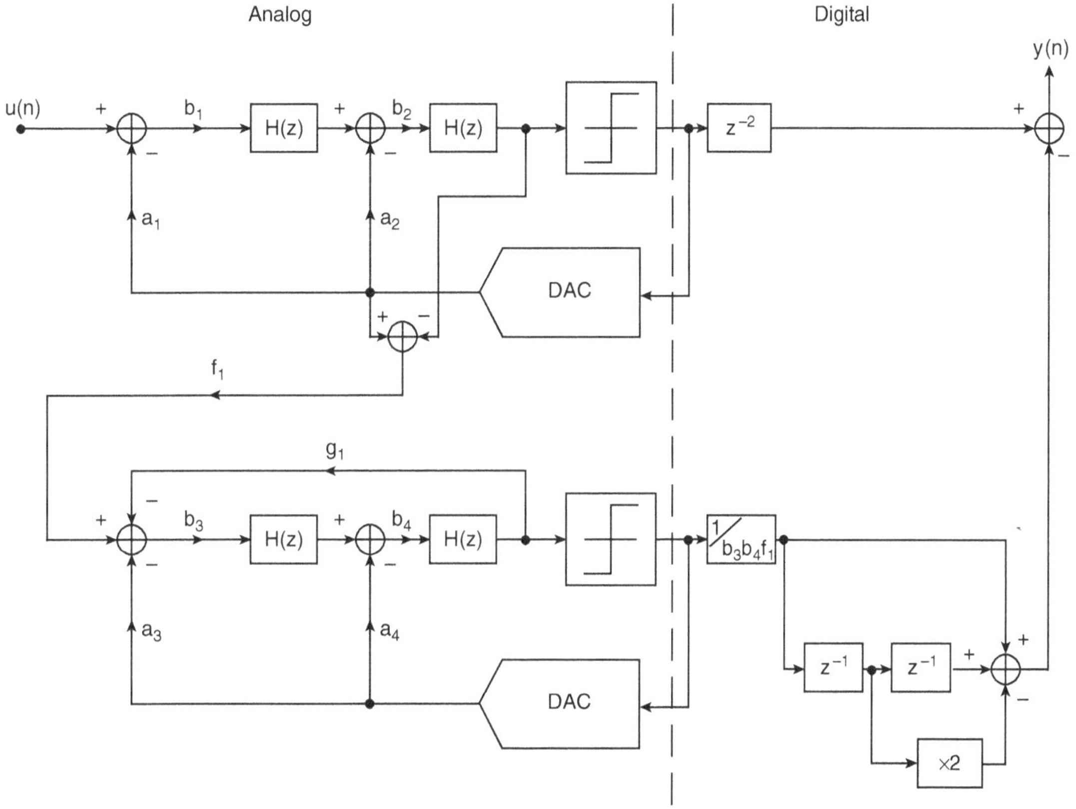

Figure 5.22 2–2 cascaded ΔΣ topology.

by a function of $(1 - z^{-1})^4$. Moreover, it is required that the input signal be delayed by z^{-4}, with a dc gain of one. Assume that each $H(z)$ building block in the schematic is realized by a noninverting and delayed SC integrator—that is, $H(z) = 1/(z - 1)$—which has perfect settling characteristics and an infinite op-amp gain. Also, the quantization noise is assumed to be independent of the input signal.

Solution: Let us start with the first stage in the analog section of the modulator. Utilizing the z-domain analysis, we write the following system function:

$$\{[U(z) - Y_1(z)a_1] \cdot H(z)b_1 - Y_1(z)a_2\} \cdot H(z)b_2 + E_1(z) = Y_1(z) \tag{5.75}$$

where $U(z)$, $Y_1(z)$, and $E_1(z)$ are the z-domain expressions for the system's input signal, the output of the first stage (i.e., the first quantizer's output), and the quantization noise generated by the first stage, respectively. After a few rearrangements, we obtain the expression for $Y_1(z)$ as follows:

$$Y_1(z)=\frac{U(z)\cdot[H(z)]^2 b_1b_2+E_1(z)}{[H(z)]^2 a_1b_1b_2+H(z)a_2b_2+1} \tag{5.76}$$

Similarly, we can write the system function for the second stage of the modulator as follows:

$$\{[E_1(z)f_1-Y_2(z)a_3-Y_2(z)g_1+E_2(z)g_1]\cdot H(z)b_3-Y_2(z)a_4\}\cdot H(z)b_4+E_2(z) \\ =Y_2(z) \tag{5.77}$$

where $Y_2(z)$ and $E_2(z)$ are the z-domain expressions for the output of the second stage (i.e., the second quantizer's output), and the quantization noise generated by the second stage, respectively. Rearranging, we obtain the expression for $Y_2(z)$ as follows:

$$Y_2(z)=\frac{[E_1(z)f_1+E_2(z)g_1]\cdot[H(z)]^2 b_3b_4+E_2(z)}{[H(z)]^2 b_3b_4(a_3+g_1)+H(z)a_4b_4+1} \tag{5.78}$$

Thus, based on Figure 5.22, we can write the expression for the final output, $Y(z)$, as follows:

$$Y(z)=Y_1(z)\cdot z^{-2}-Y_2(z)\cdot\frac{1}{f_1b_3b_4}\cdot(1-z^{-1})^2 \tag{5.79}$$

Substituting $Y_1(z)$ and $Y_2(z)$ in Equation (5.79) with the expressions given by Equation (5.76) and Equation (5.78), respectively, we obtain the following:

$$Y(z)=\frac{U(z)\cdot[H(z)]^2 b_1b_2+E_1(z)}{[H(z)]^2 a_1b_1b_2+H(z)a_2b_2+1}\cdot z^{-2} \\ -\frac{[E_1(z)f_1+E_2(z)g_1]\cdot[H(z)]^2 b_3b_4+E_2(z)}{[H(z)]^2 b_3b_4(a_3+g_1)+H(z)a_4b_4+1}\cdot\frac{1}{f_1b_3b_4}\cdot(1-z^{-1})^2 \tag{5.80}$$

To cancel the quantization noise of the first stage, $E_1(z)$, from the final output $Y(z)$, we should design the modulator so that the following condition is met:

$$\frac{z^{-2}}{[H(z)]^2 a_1b_1b_2+H(z)a_2b_2+1}=\frac{(1-z^{-1})^2\cdot[H(z)]^2}{[H(z)]^2 b_1b_2(a_3+g_1)+H(z)a_4b_4+1} \tag{5.81}$$

Equating the system coefficients of z between the denominators in the preceding equation, we obtain the following conditions:

$$\begin{cases} a_2b_2=a_4b_4 \\ a_1b_1b_2=(a_3+g_1)b_3b_4 \end{cases} \tag{5.82}$$

For the system to modulate the quantization noise from the second stage by a function of $(1-z^{-1})^4$, the following condition needs to be met based on Equation (5.80):

$$\frac{\left\{[H(z)]^2 b_3 b_4 g_1 + 1\right\} \cdot \dfrac{1}{f_1 b_3 b_4} \cdot (1-z^{-1})^2}{[H(z)]^2 b_3 b_4 (a_3 + g_1) + H(z) a_4 b_4 + 1} = (1-z^{-1})^4 \tag{5.83}$$

After some rearrangements, we obtain the following conditions:

$$\begin{cases} f_1 b_3 b_4 = 1 \\ (a_3 + g_1) b_3 b_4 = 1 \\ a_4 b_4 = 2 \\ g_1 = 0 \end{cases} \tag{5.84}$$

Finally, for the input signal, we specify the following:

$$\frac{[H(z)]^2 b_1 b_2 \cdot z^{-2}}{[H(z)]^2 a_1 b_1 b_2 + H(z) a_2 b_2 + 1} = z^{-4} \tag{5.85}$$

Thus, we obtain more conditions:

$$\begin{cases} b_1 b_2 = 1 \\ a_1 = 1 \\ a_2 b_2 = 2 \end{cases} \tag{5.86}$$

Incorporating Equations (5.82), (5.84), and (5.86) into one set of necessary conditions, we write the following:

$$\begin{cases} a_1 = 1 \\ b_1 b_2 = 1 \\ a_3 b_3 b_4 = 1 \\ a_2 b_2 = a_4 b_4 = 2 \\ f_1 = a_3 \\ g_1 = 0 \end{cases} \tag{5.87}$$

Similar to the 2–2 modulator, the 2–1–1 cascaded modulator is constructed from a cascade of one second-order and two first-order modulators. In an ideal 2–1–1 modulator, the quantization noise errors of the first and second stages are eliminated by using the digital error cancellation logic, and the quantization noise of the third stage (i.e., a first-order modulator) is shaped by the response of a fourth-order $NTF(z)$.

Nevertheless, cascaded $\Delta\Sigma$ modulators have a common drawback: *quantization noise leakage*. That is, the imperfections of the analog circuitry such as the finite op-amp gain and capacitor mismatch errors result in an incomplete cancellation of the quantization noise that is fed into the last stage.

In a cascaded ΔΣ modulator, the capacitor mismatch or finite op-amp gain introduces gain errors to the integrators and to the feedback DACs. The gain errors are less important to the 1-bit feedback DACs since they are always linear (they become much more pronounced when multibit DACs are used).

However, the gain errors in the integrators are more problematic. They cause the modulator's noise transfer function to deviate from its desired form, thereby weakening the effect of digital error cancellation logic and causing a noticeable error leakage in the modulator's final output. Take the 2-1-1 modulator as an example. Assuming that all op-amps have the same dc gain A V/V, that the maximum capacitor match is Δ, and that the input signal is a sinusoid with a peak-to-peak amplitude of 2 V (i.e., its value ranges between –1 V and +1 V), it can be shown that the consequent in-band quantization noise leakage power is given by [74]

$$E_l^2 = \frac{1}{6A^4\mathrm{OSR}} + \frac{\pi^2}{2A^2\mathrm{OSR}^3} + \frac{2\pi^4}{15\mathrm{OSR}^5}\left(\frac{2}{A} + 3\Delta\right)^2 + \frac{2\pi^6}{21\mathrm{OSR}^7}\left(\frac{3}{2A} + 6\Delta\right)^2 \qquad (5.88)$$

Quantitatively speaking, if we assume that the maximum capacitor mismatch is equal to 0.2%, that an OSR of 32 is employed, and that the charge injection and op-amp offset errors are neglected, then it can be found from the preceding formula that a dc gain of approximately 10,000 V/V, or, equivalently, 80 dB will be required to maintain a 113-dB SNR, which is the nominal SNR of a fourth-order modulator according to Equation (5.71).

As a final note, the cascaded (or multistage) modulator cannot rely on the loop filters to achieve a high SNR because they are only first- or second-order modulators. Rather, the achievable SNR depends on how effectively the quantization noise errors from the first few stages can be canceled in the digital domain. In comparison, for the high-order single-stage modulator, the high SNR is achieved by using the high-order loop filter that is constructed from a cascade of op-amps, which can easily provide a high dc gain to effectively suppress the in-band quantization noise power [75]. However, as mentioned earlier, the cascaded modulator normally has a robust stability whereas the high-order single-stage modulator is likely to become unstable.

ΔΣ Modulators with Multibit Quantization

A straightforward solution to the noise leakage problem of a cascade ΔΣ modulator is to use a fine or multibit quantizer in the first stage. Using a multibit quantizer significantly reduces the in-band noise power, thereby alleviating the error leakage at the modulator's output.

Another significant benefit of using a multibit quantizer is the improved SNR versus OSR figure. Today, the fast-growing market for applications such as broadband video processing and digital subscriber line (DSL) grows the demand for delta-sigma ADCs that can work across a wide signal bandwidth (from 0.5 to 5 MHz) while providing a high resolution (14 to 16 bits) [4][5][6][7]. In such cases, high OSRs (e. g., OSR > 16) are often not desirable because the resultant high clock sampling rates make the ADCs power hungry.

Furthermore, given a fixed loop filter (i.e., L is a constant), the lower the OSR, the less effective the noise shaping will be, thereby decreasing the SNR. Recalling Equation (5.71), we can find that besides OSR and L, the last parameter capable of improving the SNR is M (i.e., the number of bits in the quantizer). Theoretically speaking, by increasing M, we can decrease the OSR without losing much SNR; hence, the SNR versus OSR figure can be improved.

Nevertheless, like many other circuit design scenarios, there is a price to pay. The price is the nonlinearity error of the multibit DAC in the feedback path due to the capacitor (if a thermometer-coded SC DAC is used) or current-source (if a thermometer-coded current-steering DAC is used) mismatch. As shown in Figure 5.1, a 1-bit ADC needs a 1-bit feedback DAC to extract the quantization noise, and similarly a multibit ADC needs a multibit DAC. As mentioned earlier, the two input levels of a 1-bit DAC can always be connected through a single straight line (regardless of the slope value); hence, the DAC is always linear. By contrast, a multibit DAC has more than two levels. Thus, its linearity is not guaranteed.

As mentioned in Section 5.3, the minimum relative capacitor matching error in standard CMOS technologies is about 0.02%, which is equivalent to a maximum ENOB of 12 bits for an SC converter whose accuracy performance depends on the capacitance ratios (without calibrations). Thus, if the multibit DAC in the feedback path of a multibit ADC is realized as an SC DAC, then its accuracy cannot be better than 12 bits unless calibrations are carried out. Very often, the feedback DAC's nonlinearity error is directly inserted into the first integrator of the $\Delta\Sigma$ modulator, which means the DAC error is in essence a second (or auxiliary) input; hence, both the DAC error and the main input signal are processed by the same $STF(z)$, which is typically a pure delay given by (z^{-L}) [64]. In other words, the $\Delta\Sigma$ modulator's output is contaminated by the unsuppressed (or unshaped) DAC error; consequently, the $\Delta\Sigma$ modulator's maximum accuracy is limited to about 12 bits as well. Apparently, this is not comparable to our envisioned number, which in this case is at least 14 bits.

As we learned in Section 5.3, there are many calibration/correction (either in the analog or digital domain) techniques to resolve the capacitor-mismatch issues. Unfortunately, most of them are not applicable (except for the correlation calibration) to ΔΣ ADCs because of the fundamental difference between the Nyquist-rate and oversampling converters—that is, Nyquist-rate converters are memoryless, while oversampling ones have memory.

One of the most widely adopted capacitor mismatch error correction techniques appropriate for the needs of linearizing the multibit ΔΣ DAC is called *dynamic element matching* (DEM) [76]. It is proposed that the nonlinearity errors (i.e., harmonics in the output) induced by the multibit feedback DAC can be averaged or smoothed by randomizing the selection of capacitor branches (or *unit elements*) in the SC DAC, which is similar to the one shown in Figure 5.3(b). However, the randomization tends to raise the noise floor, thereby degrading the SNR [64]. Additionally, the hardware complexity and the total power consumption are increased due to design of the digital pseudo-random-signal generator and the multiplexers.

Another popular correction technique is called *data-weighted averaging* (DWA) [77]. In the proposed scheme, the DAC unit elements are selected in a sequential fashion. It uses a pointer to record where the previous unit element selection ends, and it starts the next new selection from that point. In addition, the number of selected elements during each selection is decided by the input word. For example, if the input reads a "3," then the next three new unit elements in the sequence will be selected. Next, if the input is a "2," then the subsequent two new unit elements will be selected. This DWA technique has the advantage of shaping the DAC mismatch errors with a first-order high-pass transfer function, hence the name *mismatch shaping* [77]. However, for certain input dc values or frequencies, some tones will still show up in the spectrum [64]. Improved but more sophisticated DWA algorithms, such as bidirectional DWA [4] or partitioned DWA [6], have been developed to remove the tones.

At the time of this writing, the most flexible unit-element selection scheme is the tree-structure algorithm [78]. In the proposed scheme, a tree-structure digital encoder is placed in front of the DAC's input. The encoder consists of digital logic cells called the *switching blocks*, which are organized in a tree structure. Each switching block contains a swapping cell and a first-order random sequence generator. For calibrating an M-bit DAC, $2^M - 1$ switching blocks are needed. Compared to the DEM technique, the hardware cost for generating the randomization is reduced in this configuration because the complex random signal generator is now disintegrated and incorporated into each of the switching blocks.

As a final note, although multibit quantizers have been seen mostly in cascaded modulators so far, they can also be employed to build single-stage modulators. A practical example is a multibit audio ΔΣ DAC that was built based on the third-order DQIR topology shown in Figure 5.19 [79].

Note that there are many ΔΣ modulator topologies other than the ones presented here. For example, the band-pass ΔΣ modulator is a subject of many ongoing research activities in the community of analog and RF circuit designers at the time of this writing, mainly due to its great potential in wireless communications. Very often, a band-pass ΔΣ modulator can be transformed from a low-pass ΔΣ modulator by simply replacing z in the latter's transfer function with $(-z^2)$; consequently, the fundamental active building block in a band-pass ΔΣ modulator is the resonator, corresponding to the integrator in a low-pass modulator. For the sake of brevity, the design of band-pass ΔΣ modulators is not detailed in this book. The interested reader is referred to the literature for more information.

Appendix 5.1

This appendix presents the exemplary simulation of a fourth-order 1-bit audio ΔΣ ADC. The target specifications of the ADC are listed in Table 5.A. Note that the term "simplified 4th-order DFFIR" means the topology used here is a simplified version of the DFFIR topology shown in Figure 5.21. Specifically, all the forward paths (except for the input path that leads to the first integrator) are removed. The source codes are as follows.

Table 5.A. Target specifications of the fourth-order 1-bit audio ΔΣ ADC.

OSR	128
Signal bandwidth	0 to 20 kHz
Clock frequency	5.12 MHz
SNR	>120 dB
No. of decimation stages	5
Modulator topology	Simplified 4th-order DFFIR

PART I

```
% Use MATLAB codes to derive the fourth-order NTF(z) and STF(z)%
% MATLAB functions in Schreier's toolbox [68] are utilized%

Clear all;
fprintf(1,'NTF Synthesis- Fourth-order modulator')
order=4;
r=128;
opt=1;      %Enable zero optimization to minimize the in-band noise
power%
H_inf=2.0; %Lee's rules%
f0=0;
H= synthesizeNTF(order,r,opt,H_inf,f0);
[num1,den1]=tfdata(H,'v') %The original NTF(z) in 'zpk' format%

figure(1);
subplot(221);
plotPZ(H); %Plot the NTF(z)'s zeros and poles%
title('NTF poles and zeros diagram');
f = [linspace(0,0.75/r,100) linspace(0.75/r,0.5,100)];
z = exp(j*2*pi*f);
magH = dbv(evalTF(H,z));
subplot(222);
plot(f,magH); %Plot the NTF(z)'s magnitude response%
axis([0 0.5 -150 10]);
xlabel('Normalized frequency (1\rightarrow f_s)');
ylabel('dB');
title('NTF Magnitude Response');
grid on;

[snr_pred,amp] = predictSNR(H,r);
[snr,amp] = simulateSNR(H,r);
subplot(223);
plot(amp,snr_pred,amp,snr,'o'); %Predict the best SNR performance%
grid on;
figureMagic([-120 0], 10, 1, [0 150], 10, 1);
xlabel('Input Level, dB');
ylabel('SNR dB');
```

```
title('SNR curve');
s=sprintf('peak SNR = %4.1fdB\n(OSR=%d)',max(snr),r);
text(-49,15,s);
[a,g,b,c]=realizeNTF(H,'CIFB') %Realize NTF(z) with the simplified
DFFIR (or CIFB [68])%
b=[b(1) zeros(1,length(b)-1)];
ABCD = stuffABCD(a,g,b,c,'CIFB');
[H,G]=calculateTF(ABCD);
[num2 den2]=tfdata(H,'v') %Check whether the realized transfer
function matches NTF(z)%

subplot(224); %Plot the frequency responses of NTF(z) and STF(z)%

f=logspace(-3,0,200);
z=exp(2i*pi*f);
magG=dbv(evalTF(G,z));
magH=dbv(evalTF(H,z));
semilogx(f,magG,'-',f,magH,'-');
grid on

sigma_H=dbv(rmsGain(H,0,0.5/r));
hold on;
semilogx([.001 0.5/r], sigma_H*[1 1]);
plot([.001 0.5/r], sigma_H*[1 1],'o');
text( 0.001, sigma_H+6, sprintf('rms gain = %5.0fdB',sigma_H)); %RMS
figureMagic([1e-3 0.5], [], [], [-150 10], 10, 2);
xlabel('Normailized f');
ylabel('dB');
title('Frequency response of NTF and STF');
grid on;
[ABCDs,umax]=scaleABCD(ABCD); %Scale the system coefficients to the
%
[as,gs,bs,cs] = mapABCD(ABCDs ,'CIFB') %Assign the system
coefficients to circuitry%
```

PART II

```
/*SWITCAP program*/
/*The capacitances are obtained based on kT/C noise requirements
[64]*/
```

```
TITLE:  1-BIT 4TH-ORDER DELTA SIGMA MODULATOR
TIMING;
PERIOD 1.95E-7
CLOCK PHI1 1 (0 1/2);
CLOCK PHI2 1 (1/2 1);
END;

SUBCKT (N1 N2 N3 N4)
DIFAMP (P1);
E1 (N1 0 N3 N4) P1;
E2 (0 N2 N3 N4) P1;
END;

CIRCUIT;
CP1 (3 5) 3.17;
CP2 (7 9) 24.18;
CP3 (11 13) 0.1;
CP4 (15 17) 0.6;
CP5 (19 21) 0.1;
CP6 (23 25) 0.3;
CP7 (27 29) 0.06;
CP8 (31 33) 0.15;
CP9 (5 35) 3.17;
CP10 (13 37) 0.12;
CP11 (21 39) 0.1;
CP12 (29 41) 0.04;
CP13 (5 45) 0.15;
CP14 (21 46) 0.06;
CN1 (4 6) 3.17;
CN2 (8 10) 24.18;
CN3 (12 14) 0.1;
CN4 (16 18) 0.6;
CN5 (20 22) 0.1;
CN6 (24 26) 0.3;
CN7 (28 30) 0.06;
CN8 (32 34) 0.15;
CN9 (6 36) 3.17;
CN10 (14 38) 0.12;
```

```
CN11 (22 40) 0.1;
CN12 (30 42) 0.06;
SP1 (1 3) PHI1;
SP2 (3 0) PHI2;
SP3 (5 0) PHI1;
SP4 (5 7) PHI2;
SP5 (9 11) PHI1;
SP6 (11 0) PHI2;
SP7 (13 0) PHI1;
SP8 (13 15) PHI2;
SP9 (17 19) PHI1;
SP10 (19 0) PHI2;
SP11 (21 0) PHI1;
SP12 (21 23) PHI2;
SP13 (25 27) PHI1;
SP14 (27 0) PHI2;
SP15 (29 0) PHI1;
SP16 (29 31) PHI2;
SP17 (35 0) PHI1;
SP18 (35 43) PHI2;
SP19 (37 0) PHI1;
SP20 (37 43) PHI2;
SP21 (39 0) PHI1;
SP22 (39 43) PHI2;
SP23 (41 0) PHI1;
SP24 (41 43) PHI2;
SN1 (2 4) PHI1;
SN2 (4 0) PHI2;
SN3 (6 0) PHI1;
SN4 (6 8) PHI2;
SN5 (10 12) PHI1;
SN6 (12 0) PHI2;
SN7 (14 0) PHI1;
SN8 (14 16) PHI2;
SN9 (18 20) PHI1;
SN10 (20 0) PHI2;
SN11 (22 0) PHI1;
SN12 (22 24) PHI2;
```

```
SN13 (26 28) PHI1;
SN14 (28 0) PHI2;
SN15 (30 0) PHI1;
SN16 (30 32) PHI2;
SN17 (36 0) PHI1;
SN18 (36 44) PHI2;
SN19 (38 0) PHI1;
SN20 (38 44) PHI2;
SN21 (40 0) PHI1;
SN22 (40 44) PHI2;
SN23 (42 0) PHI1;
SN24 (42 44) PHI2;
SP25 (45 0) PHI1;
SP26 (45 17) PHI2;
SN25 (46 0) PHI1;
SN26 (46 33) PHI2;

SREFP1 (43 47) CMPNEG;
SREFN1 (48 43) CMPPOS;
SREFP2 (47 44) CMPPOS;
SREFN2 (48 44) CMPNEG;
YCMP1 CMPLAT (33 34 PHI1 CMPPOS);
YG1 NOT (CMPPOS CMPNEG);
X1 (9 10 8 7) DIFAMP (1E4);
X2 (17 18 16 15) DIFAMP (1E4);    /*Use fully differential op-amps*/
X3 (25 26 24 23) DIFAMP (1E4);
X4 (33 34 32 31) DIFAMP (1E4);

VREFP (47 0);
VREFN (48 0);
VINP (1 0);
VINN (2 0);
END;
ANALYZE NTRAN;   /*Perform transient analysis with a sinusoidal
input Vpp = 2 V*/
TIME 65535/2- 98306 1;
SET VREFP DC -2;
SET VREFN DC +2;
```

```
SET VINP COSINE 0.0 1.0 20000 0 0 -90;
SET VINN COSINE 0.0 -1.0 20000 0 0 -90;
PRINT V(1) V(43);
END;
END;
```

PART III

```
%Use MATLAB codes to perform FFT analysis on the result from
SWITCAP transient analysis%
close all;
clear all;
load delsig.out;    %Load the result from SWITCAP transient
analysis%
tt=delsig(:,1);
vv=delsig(:,3);
r=128;
N=65536;
fB=ceil(N/(2*r));
f=170;
u=0.5*sin(2*pi*f/N*[0:N-1]);
v=vv(1:65536);
t=tt(1:65536);
spec=fft((v)'.*hann(N))/(N/4);
figure(1);
plot(linspace(0,1,N/2),dbv(spec(1:N/2))); %Plot the result of the FFT
analysis%
axis([0 1 -150 0]);
ylabel('dBFS');
snr=calculateSNR(spec(1:fB),f);
s=sprintf('SNR=%4.1fdB\n',snr);
text(0.5,-110,s);
```

PART IV

```
%Use MATLAB codes to simulate the behavior of the five-stage
digital decimation filter%
clear all;
close all;

load delsig1.out; %Load the result from SWITCAP transient analysis%
tt=delsig1(:,1);
```

```
vv=delsig1(:,3);
v=vv(1:65536);
r=128;
N=65536;
fB=ceil(N/(2*r));
f=floor(2*fB/3);

%Configure the five cascaded delayed discrete-time integrators (or
averaging filter)%

a = ((32)^5)*[1 -5 10 -10 5 -1];
b = [1 zeros(1,31) -5 zeros(1,31) 10 zeros(1,31) -10...zeros(1,31) 5
zeros(1,31) -1];
figure(1);
[H,w]=freqz(b,a,65536);
plot(linspace(0.00001,1,65536),dbv(H));
axis([0.00001 1 -220 10]);
x=filter(b, a, v);
spec=fft((x)'.*hann(N))/(N/4);
figure(2);
subplot(221);
plot(linspace(0,1,N/2),dbv(spec(1:N/2)));
axis([0 1 -220 0]);
grid on;
ylabel('dBFS');
snr=calculateSNR(spec(1:fB),f);
s=sprintf('SNR=%4.1fdB\n',snr);
text(0.5,-100,s);

%Configure the five-tap FIR filter following the integrators%

b1=firls(45,[0 1/128 1/127 1],[1 1 0 0]);
y1=resample(x,1,2,b1,10000); %Down-sample the digital output by a
factor of 32%
spec=fft((y1)'.*hann(N/2))/(N/8);
subplot(222); %Plot the FFT curve of the averaging filter's output%
plot(linspace(0,1,N/4),dbv(spec(1:N/4)));
axis([0 1 -220 0]);
```

```
grid on;
ylabel('dBFS');
snr=calculateSNR(spec(1:fB),f);
s=sprintf('SNR=%4.1fdB\n',snr);
text(0.5,-100,s);
b2=firls(41,[0 1/128 1/127 1],[1 1 0 0]);
y2=resample(y1,1,2,b2,10000);
spec=fft((y2)'.*hann(N/4))/(N/16);
subplot(223);
plot(linspace(0,1,N/8),dbv(spec(1:N/8)));
axis([0 1 -210 0]);
grid on;
ylabel('dBFS');
snr=calculateSNR(spec(1:fB),f);
s=sprintf('SNR=%4.1fdB\n',snr);
text(0.5,-100,s);
b3=fir2(45,[0 1/128 1/127 1],[1 1 0 0]);
a3=[1 zeros(1,46)];
y3=filter(b3,a3,y2);
spec=fft((y3)'.*hann(N/4))/(N/16);
subplot(224); %Plot the FFT curve of modulator's final output%
plot(linspace(0,1,N/8),dbv(spec(1:N/8)));
axis([0 1 -220 0]);
grid on;
ylabel('dBFS');
snr=calculateSNR(spec(1:fB),f);
s=sprintf('SNR=%4.1fdB\n',snr);
text(0.5,-100,s);
```

PART V

```
*Use SPICE code  to calculate the minimum slew rate of the first
op-amp.

vclk phi1 0 pulse  (-1.5 1.5v 0 0.06n 0.06n 96n 195n)
vclk0 phi2 0 pulse  (-1.5 1.5v  98n  0.06n 0.06n 96n 195n)
g1 vi 1 VCR pwl(1) phi1 0 -1 200meg 1 100
g2 1 0  VCR pwl(1) phi2 0 -1 200meg 1 100
g3 3 0  VCR pwl(1) phi1 0 -1 200meg 1 100
g4 3 5  VCR pwl(1) phi2 0 -1 200meg 1 100
```

```
c1 1 3 3.17p
c2 5 7 24.18p
.subckt slew vip vin out
gm 0 vo  vip vin 1m max=1m min=-1m
r vo 0 5meg
c vo 0 6.2p
eb out 0 vo 0 1
.ends
x_1 0 5 7 slew
vin vi 0 sin( 0 1 20e4 0 0 0)
.tran 10n 30u
.op
.options post
.print v(7)
.END
```

References

[1] P. R. Gray and R. Meyer, "Future directions of silicon ICs for RF personal communications," *Proceedings of the IEEE Custom Integrated Circuits Conference*, pp. 83–90, May 1995.

[2] A. A. Abidi et al., "The future of CMOS wireless transceivers," *IEEE Solid-State Circuits Conference, Digest of Technical Papers*, pp. 118–119, February 1997.

[3] I. Oppermann et al. (Ed.), *UWB: Theory and applications*, John Wiley & Sons, New York, 2004.

[4] I. Fujimori et al., "A 90-dB SNR 2.5-MHz output-rate ADC using cascaded multibit delta-sigma modulation at 8x oversampling," *IEEE Journal of Solid-State Circuits*, Vol. 35, pp. 1820–1828, July 1999.

[5] Y. Geerts et al., "A 3.3-V, 15-bit, delta-sigma ADC with a signal bandwidth of 1.1 MHz for ADSL applications," *IEEE Journal of Solid-State Circuits*, Vol. 34, pp. 927–936, July 1999.

[6] K. Vleugels et al., "A 2.5-V sigma-delta modulator for wideband communication applications," *IEEE Journal of Solid-State Circuits*, Vol. 36, pp. 1887–1898, December 2001.

[7] X. Wang, U. Moon, M. Liu, and G. C. Temes, "Digital correlation technique for the estimation and correction of DAC errors in multibit MASH ΔΣ ADCs,"

Proceedings of IEEE International Symposium on Circuits and Systems, Vol. IV, pp. 691–694, Phoenix, AZ, May 2002.

[8] S. Tewksbury et al., "Terminology related to the performance of S/H, A/D, and D/A circuits," *IEEE Trans. on Circuits & Systems*, Vol. CAS-25, No. 7, pp. 419–426, July 1978.

[9] R. van de Plassche, *Integrated analog-to-digital and digital-to-analog data converters*, Kluwer Academic Publisher, Berlin, Germany, 1994.

[10] B. Razavi, *Principles of data conversion system design*, IEEE Press, Piscataway, NJ, 1995.

[11] D. A. Johns and K. Martin, *Analog integrated circuits design*, John Wiley & Sons, New York, 1997.

[12] IEEE Std. 746–1984, *IEEE standard for performance measurements of A/D and D/A converters for PCM television video circuits*, IEEE Press, New York, 1984.

[13] IEEE Std. 1241–2000, *IEEE standard for terminology and test methods for analog-to-digital converters*, IEEE Press, New York, 2001.

[14] P. Holloway, "A trimless 16-bit digital potentiometer," *IEEE Solid-State Circuits Conference, Digest of Technical Papers*, pp. 66–67, February 1984.

[15] A. Abrial et al., "A 27-MHz digital-to-analog video processor," *IEEE Journal of Solid-State Circuits*, Vol. SC-23, No. 6, pp. 1358–1369, December 1988.

[16] T. Miki et al., "An 80-MHz 8-bit CMOS D/A converter," *IEEE Journal of Solid-State Circuits*, Vol. SC-21, No. 6, pp. 983–988, December 1986.

[17] H. Schouwenaars, D. Groeneveld, and H. Termeer, "A low-power stereo 16-bit CMOS D/A converter for digital audio," *IEEE Journal of Solid-State Circuits*, Vol. SC-23, No. 6, pp. 1290–1297, December 1988.

[18] D. Groeneveld et al., "A self-calibration technique for monolithic high-resolution D/A converters," *IEEE Journal of Solid-State Circuits*, Vol. SC-24, No. 6, pp. 1517–1522, December 1989.

[19] D. Mercer and L. Singer, "12-bit 125-MS/s CMOS D/A designed for spectral performance," *Proceedings of International Symposium on Low Power Electronics and Design*, Vol. 2, No. 1, pp. 243–246, August 1996.

[20] R. Hester et al., "CODEC for echo-canceling full-rate ADSL modems," *IEEE Solid-State Circuits Conference, Digest of Technical Papers*, pp. 242–243, February 1999.

[21] A. Bugeja et al., "A self-trimming 14-b 100-MS/s CMOS DAC," *IEEE Journal of Solid-State Circuits*, Vol. 35, No. 12, pp. 1841–1852, December 2000.

[22] K. Martin et al., "A differential switched-capacitor amplifier," *IEEE Journal of Solid-State Circuits*, Vol. SC-22, pp. 104–106, February 1987.

[23] H. Matsumoto and K. Watanabe, "Spike-free switched-capacitor circuits," *Electronics Letters*, Vol. 23, pp. 428–429, April 1987.

[24] H. Matsumoto and K. Watanabe, "Switched-capacitor algorithmic digital-to-analog converters," *IEEE Trans. on Circuits & Systems*, Vol. CAS-33, pp. 721–724, 1986.

[25] C. C. Shih and P. R. Gray, "Reference refreshing cyclic analog-to-digital and digital-to-analog converters," *IEEE Journal of Solid-State Circuits,* Vol. 21, pp. 544–554, August 1986.

[26] P. Li et al., "A ratio-independent algorithmic analog-to-digital conversion technique," *IEEE Journal of Solid-State Circuits*, Vol. 19, pp. 828–836, December 1984.

[27] H. Onodera et al., "A cyclic A/D converter that does not require ratio-matched components," *IEEE Journal of Solid-State Circuits,* Vol. 23, pp. 152–158, February 1988.

[28] S. Y. Chin and C. Y. Wu, "A CMOS ratio-independent and gain-insensitive algorithmic analog-to-digital converter," *IEEE Journal of Solid-State Circuits*, Vol. 31, pp. 1201–1207, August 1996.

[29] Y. Huang, *Design of high-performance switched-capacitor circuits in the presence of component imperfections*, Ph.D. dissertation, Oregon State University, Corvallis, OR, 1997.

[30] B.-S. Song et al., "A 12-b 1-Msample/s capacitor error-averaging pipelined A/D converter," *IEEE Journal of Solid-State Circuits*, Vol. SC-23, pp. 1324–1333, December 1988.

[31] H. S. Lee et al., "A self-calibrating 15 bit CMOS A/D converter," *IEEE Journal of Solid-State Circuits*, Vol. SC-19, pp. 813–819, December 1984.

[32] Y. Lin, B. Kim, and P. R. Gray, "A 13-b 2.5-MHz self-calibrated pipelined A/D converter in 3-μm CMOS," *IEEE Journal of Solid-State Circuits*, Vol. 26, pp. 628–636, April 1991.

[33] U. Moon, J. Silva, J. Steensgaard, and G. C. Temes, "Switched-capacitor DAC with analogue mismatch correction," *Electronics Letters*, Vol. 35, pp. 1903–1904, October 1999.

[34] B. Razavi and B. A. Wooley, "Design techniques for high-speed high-resolution comparators," *IEEE Journal of Solid-State Circuits*, Vol. SC-27, pp. 1916–1926, December 1992.

[35] R. van de Plassche and P. Baltus, "An 8 b 100 MHz folding ADC," *IEEE Solid-State Circuits Conference, Digest of Technical Papers*, pp. 222–223, February 1988.

[36] K. Kusumoto et al., "A 10 b 20 MHz 30 mW pipelined interpolating CMOS ADC," *IEEE Journal of Solid-State Circuits*, Vol. 28, pp. 1200–1206, December 1993.

[37] M. Steyaert et al., "A 100-MHz 8-bit CMOS interpolating A/D converter," *IEEE Custom Integrated Circuits Conference*, pp. 28.1.1–28.1.4, May 1993.

[38] R. E. J. van de Grift et al., "An 8-bit video ADC incorporating folding and interpolation techniques," *IEEE Journal of Solid-State Circuits*, Vol. SC-22, pp. 944–953, December 1987.

[39] I. Mehr and D. Dalton, "A 500-MSamples/s, 6-b Nyquist-rate ADC for disk-drive read-channel application," *IEEE Journal of Solid-State Circuits*, Vol. 34, pp. 912–920, July 1999.

[40] K. Nagaraj et al., "A dual-mode 700-Msamples/s 6-bit 200-MSamples/s 7-bit A/D converter in a 0.25-μm digital CMOS process," *IEEE Journal of Solid-State Circuits*, Vol. 35, pp. 1760–1768, December 2000.

[41] M. Choi and A. A. Abidi, "A 6-b 1.3-Gsamples/s A/D converter in 0.35-μm CMOS," *IEEE Journal of Solid-State Circuits*, Vol. 36, pp. 1847–1858, December 2001.

[42] X. Jiang et al., "A 2 GS/s 6 b ADC in 0.18 μm CMOS," *IEEE Solid-State Circuits Conference, Digest of Technical Papers*, pp. 322–323, February 2003.

[43] K. Poulton et al., "A 20 GS/s 8 b ADC with a 1 MB memory in 0.18 μm CMOS," *IEEE Solid-State Circuits Conference, Digest of Technical Papers*, pp. 318–319, February 2003.

[44] R. Taft et al., "A 1.8 V 1.6 GS/s 8 b self-calibrating folding ADC with 7.26 ENOB at Nyquist frequency," *IEEE Solid-State Circuits Conference, Digest of Technical Papers*, pp. 252–253, February 2004.

[45] T. Shimizu et al., "A 10-bit 20-MHz two-step parallel A/D converter with internal S/H," *IEEE Journal of Solid-State Circuits*, Vol. 24, pp. 13–20, February 1989.

[46] Y. Wang and B. Razavi, "An 8-bit 150-MHz CMOS A/D converter," *IEEE Journal of Solid-State Circuits*, Vol. 35, pp. 308–317, March 2000.

[47] S. Limotyrakis et al., "A 150-MS/s 8-b 71-mW CMOS time-interleaved ADC," *IEEE Journal of Solid-State Circuits*, Vol. 40, pp. 1057–1067, May 2005.

[48] H. Pan et al., "A 3.3-V 12-b 50-MS/s A/D converter in 0.6-μm CMOS with over 80-dB SFDR," *IEEE Journal of Solid-State Circuits*, Vol. 35, pp. 1769–1780, December 2000.

[49] H. Pan and A. A. Abidi, "Spectral spurs due to quantization in Nyquist ADCs," *IEEE Trans. on Circuits & Systems—I: Regular Papers*, Vol. 51, pp. 1422–1439, August 2004.

[50] S. H. Lewis and P. R. Gray, "A pipelined 5-Msample/s 9-bit analog-to-digital converter," *IEEE Journal of Solid-State Circuits*, Vol. SC-22, pp. 954–961, December 1987.

[51] A. N. Karanicolas et al., "A 15-b 1-Msample/s digitally self-calibrated pipelined ADC," *IEEE Journal of Solid-State Circuits*, Vol. 28, pp. 1207–1215, December 1993.

[52] D. Cline, *Noise, speed, and power trade-offs in pipelined analog-to-digital converters*, Ph.D. dissertation, UC Berkeley, Berkeley, CA, 1995.

[53] S. Lee and B. S. Song, "Digital-domain calibration of multi-step analog-to-digital converter," *IEEE Journal of Solid-State Circuits*, Vol. 27, pp. 1679–1688, December 1992.

[54] B. Murmann et al., "A 12-b 75-MS/s pipelined ADC using open-loop residue amplification," *IEEE Journal of Solid-State Circuits*, Vol. 38, pp. 2040–2050, December 2003.

[55] E. Siragusa et al., "A digitally enhanced 1.8 V 15 b 40 Ms/s CMOS pipelined ADC," *IEEE Solid-State Circuits Conference, Digest of Technical Papers*, pp. 452–453, February 2004.

[56] H. C. Liu et al., "A 15-b 40-MS/s CMOS pipelined analog-to-digital converter with digital background calibration," *IEEE Journal of Solid-State Circuits*, Vol. 40, pp. 1047–1055, May 2005.

[57] W. Yang et al., "A 3-V 340-mW 14-b 75-Msample/s CMOS ADC with 85-dB SFDR at Nyquist input," *IEEE Journal of Solid-State Circuits*, Vol. 36, pp. 1931–1936, December 2001.

[58] A. Kitagawa et al., "A 10 b 3MSamples/s CMOS cyclic ADC," *IEEE Solid-State Circuits Conference, Digest of Technical Papers*, pp. 280–281, February 1995.

[59] J. L. McCreary et al., "All-MOS charge redistribution A/D conversion technique—Part I," *IEEE Journal of Solid-State Circuits*, Vol. SC-10, pp. 371–379, December 1975.

[60] G. C. Temes, "Accurate linear data conversion using inaccurate nonlinear analog components," Research Seminar, Oregon State University, Corvallis, OR, January 2003.

[61] C. C. Enz and G. C. Temes, "Circuit techniques for reducing the effects of op-amp imperfections: autozeroing, correlated double sampling, and chopper stabilization," *Proceedings of the IEEE*, Vol. 84, pp. 1584–1614, November 1996.

[62] J. Robert et al., "A 16-bit low-voltage CMOS A/D converters," *IEEE Journal of Solid-State Circuits*, Vol. SC-22, pp. 157–163, April 1987.

[63] W. C. Black et al., "Time-interleaved converter arrays," *IEEE Journal of Solid-State Circuits*, Vol. SC-15, pp. 1022–1029, December 1980.

[64] S. R. Norsworthy, R. Schreier, and G. C. Temes, *Delta-sigma data converters—Theory, design and simulation*, IEEE Press, New York, 1997.

[65] B. E. Boser and B. A. Wooley, "The design of sigma-delta modulation analog-to-digital converters," *IEEE Journal of Solid-State Circuits*, Vol. SC-23, pp. 1298–1308, December 1988.

[66] J. Silva et al., "Wideband low-distortion delta-sigma ADC topology," *Electronics Letters*, Vol. 37, pp. 737–738, June 2001.

[67] W. L. Lee, *A novel higher order interpolative modulator topology for high resolution oversampling A/D converters*, Master's thesis, MIT, Cambridge, MA, 1987.

[68] R. Schreier, *Delta-sigma Toolbox in MATLAB*, [Online]. Available at ftp://ftp.mathworks.com/pub/contrib/v5/control/delsig.

[69] SWITCAP2 (v1.1) Manual, [Online]. Available at www.cisl.Columbia.edu/projects/switcap.

[70] R. Schreier, "An empirical study of high-order single-bit delta-sigma modulators," *IEEE Trans. on Circuits & Systems*, Vol. 40, No. 8, pp. 461–466, August 1993.

[71] N. S. Sooch et al., "18-bit stereo D/A converter with integrated digital and analog filter," *Proceedings of Ninth AES International Convention*, October 1991.

[72] P. Ferguson et al., "An 18 b 20 kHz dual sigma-delta A/D converter," *IEEE Solid-State Circuits Conference, Digest of Technical Papers*, pp. 68–292, February 1991.

[73] A. Feldman et al., "A 13-bit, 1.4-MS/s sigma-delta modulator for RF baseband channel applications," *IEEE Journal of Solid-State Circuits*, Vol. 33, No. 10, pp. 1462–1469, October 1998.

[74] T. Sun, *Compensation techniques for cascaded delta-sigma A/D converters and high-performance switched-capacitor circuits*, Ph.D. dissertation, Oregon State University, Corvallis, OR, 1998.

[75] A. Abidi, "On the operation of cascode gain stages," *IEEE Journal of Solid-State Circuits*, Vol. 23, No. 6, pp. 1434–1437, December 1988.

[76] L. R. Carley, "A noise-shaping coder topology for 15+ bit converters," *IEEE Journal of Solid-State Circuits*, Vol. SC-24, No. 2, pp. 267–273, April 1989.

[77] R. T. Baird and T. S. Fiez, "Linearity enhancement of multi-bit ΔΣ A/D and D/A converters using data weighted averaging," *IEEE Trans. on Circuits & Systems*, Vol. 42, No. 12, pp. 753–762, December 1995.

[78] J. Grilo et al., "A 12-mW ADC delta-sigma modulator with 80 dB of dynamic range integrated in a single-chip blue-tooth transceiver," *IEEE Journal of Solid-State Circuits*, Vol. 37, No. 3, pp. 271–277, March 2002.

[79] I. Fujimori et al., "A multibit delta-sigma audio DAC with 120-dB dynamic range," *IEEE Journal of Solid-State Circuits*, Vol. 35, No. 8, pp. 1066–1073, August 2000.

CHAPTER 6

Switched-Capacitor DC-DC Converters

6.1 Introduction

The development of switched-capacitor (SC) direct-current-to-direct-current (DC-DC) converters is motivated by the ever-increasing demands for a small-form-factor (i.e., small-size and light-weight), high-conversion-efficiency, and low-cost power management system, which is the best candidate suitable to meet the needs of continuously shrinking portable electronic devices such as MP3 players, cellular phones, PDAs, and so on.

Modern portable electronic devices must be powered by batteries (standard LiIon, Ni/H, solar-cell, fuel-cell, and so on). When being used for the first time, a battery is able to supply a full-range dc voltage. As time passes by, the battery voltage drops at a certain rate, which is determined by the type of battery and the load current requirement.

Strictly speaking, the battery voltage does not decrease with time linearly. In fact, it normally decreases at a relatively slow rate before the operation time reaches a threshold point. Once the threshold point is passed, the battery voltage drops abruptly. For example, the nominal output voltage of a standard AA-type LiIon battery decreases from 4.05 V to 3.45 V within the first 4 hours of usage time if a 200 mA load current is required. Once the first 4 hours are over, the battery voltage begins dropping with a much sharper slope, and its value is reduced from 3.45 V to 2.25 V within the next 30 minutes. Similar arguments apply to other battery types such as the solar-cell battery and the Ni/H battery. Such variations in the battery voltage will cause undesirable effects on the electronic device, therefore a DC-DC converter is always needed to establish and maintain a constant supply voltage in the presence of a time-varying battery voltage.

Conventional *switched-mode power supply* (SMPS) circuits use inductors built from magnetic coils to process energy and stabilize the output voltage. In some cases, transformers are utilized either in combination with or in place of inductors, depending on the conversion efficiency requirement and the output ripple limit. Power conversion processed by inductors or transformers is typically very efficient and has been adopted in many practical designs (particularly for delivering a load current over 0.5 A) since the 1930s [1].

However, most low-cost and space-constrained mobile equipment requires integration of power supplies in CMOS ICs, which remains a challenge mainly due to the bulky dimensions of the off-chip (or *external*) inductor, and the unacceptable electromagnetic interference (EMI). Although the compact on-chip *spiral inductor* [2] has been invented to facilitate the construction of fully integrated radio-frequency (RF) filters and voltage-controlled oscillators (VCO), the resultant quality factor (Q) is relatively low (typically less than 16) as compared to that of the conventional off-chip inductor, for Q is roughly proportional to the square of the linear dimensions of the inductor [2]. A low quality factor leads to a high power loss, thus the maximum achievable conversion efficiency of a power converter using spiral inductors is typically much lower than that of a converter using external inductors. At the time of this writing, the use of the on-chip inductor is considered less attractive to the applications of high-efficiency monolithic DC-DC converters.

As an alternative to the inductive DC-DC converter, the switched-capacitor (SC) DC-DC converter substitutes the magnetic coils with a few capacitors and an array of switches, making it possible to fabricate the entire converter in a single chip. In the SC DC-DC converter, the switching array is responsible for charging and discharging the capacitors so that a desired output supply voltage is provided to power the electronic device. Besides high integration, the advantages of SC DC-DC converters include low fabrication cost, high switching frequency, medium-to-high conversion efficiency, and the reduced *voltage-mode* electromagnetic interference (EMI).

On the other hand, SC DC-DC converters suffer from signal-dependent current spikes mainly due to the *nonzero switch on-resistance* (R_{on}). In an ideal case, when a perfect switch with zero on-resistance is used to charge a capacitor in an SC DC-DC converter, the electric charge flows onto the capacitor instantaneously (while the voltage across the capacitor changes slowly), and the resulting current waveform consists of a sequence of impulse functions. But for a real switch with a nonzero on-resistance, the charge cannot be transferred instantaneously; hence, the output current waveform contains finite-pulse-width current spikes instead of zero-pulse-width impulses. Additionally, the value of R_{on} depends on the input signal, thereby introducing signal-dependent distortions to the converter. In a word, SC DC-DC converters suffer from *current-mode* EMI in the form of tones or harmonics.

Moreover, R_{on} dissipates energy, hence reducing the achievable conversion efficiency. To minimize the power loss due to R_{on}, the MOS transistors, which are used to implement the switches in an SC DC-DC converter, must be designed to have large (W/L) ratios, where W is the effective gate width and L is the effective gate length. This in a sense limits the degree of compactness of an SC DC-DC converter.

Finally, if the capacitors of the converter are implemented on chip, the parasitic capacitances will also introduce a power loss, thereby further reducing the conversion efficiency. Advanced process technologies that are appropriate for fabricating on-chip capacitors with low parasitic capacitances are needed to achieve a high conversion efficiency. However, this typically incurs a high cost; as a result, many existing commercial SC DC-DC converters continue to use off-chip capacitors to maintain a high conversion efficiency without losing too much integration (since the required capacitance values are comparatively large, typically on the order of 0.1 μF, the actual difference in linear dimensions between an on-chip realization and its off-chip counterpart is not significant).

Types of SC DC-DC Converters

There are three basic types of SC DC-DC converters. The first two are the step-down (or *buck*) converter and the step-up (or *boost*) converter. As the names suggest, a step-down converter processes the input voltage with a gain less than or equal to one, while a step-up converter has a voltage gain greater than or equal to one. The third type is called the step-down-step-up (or *buck-boost*) converter, which is a combination of the step-down and the step-up gain configurations.

Applications of SC DC-DC Converters

The SC DC-DC converter is often referred to as a *charge-pump* (sometimes also called the *push-pull converter*) for it uses capacitors to store and transfer energy in the form of an electric charge. SC DC-DC converters have been widely used in many applications, including dynamic random access memory (DRAM) circuits [3], electrically-erasable programmable read-only memory (EEPROM) circuits [4], and phase-locked loop (PLL) circuits [5]. Since the late 1990s, they have come into wide use for low-voltage SC circuits such as CMOS data converters.

Chapter Outline

Section 6.2 discusses the design of a widely adopted open-loop SC step-up DC-DC converter called the *Dickson charge-pump*. Section 6.3 introduces cross-coupled SC step-up DC-DC converters, and Section 6.4 provides an overview of SC step-down DC-DC converters. Finally, Section 6.5 introduces multiple-gain SC step-down-step-up DC-DC converters.

6.2 Dickson Charge-Pump

Conventional Dickson Charge-Pump

Probably the most widely used step-up SC DC-DC converter in DRAM, EEPROM, and flash memory applications is the *Dickson charge-pump* [6]. A five-stage Dickson charge-pump is shown in Figure 6.1. In the circuit, six NMOS transistors and five large *pump capacitors* are used as the diodes and the energy processing devices, respectively. Here, we assume that all six NMOS transistors are identical, that the time taken to charge each pump capacitor is shorter than the duration of Φ_1 or Φ_2 (i.e., $R_{on}C_i < 0.5T_{clk}$), and that transistor body effects and parasitic capacitances are negligible.

The circuit operates as follows. When Φ_1 goes low (i.e., $\Phi_1 = 0$), the voltage on the top plate (note that the term *top plate* used in this chapter does not necessarily mean the top plate of an on-chip *double-poly capacitor*) of the capacitor C_1 is set to $(V_{dd} - V_{thn})$, where V_{dd} is the power supply voltage and V_{thn} is the threshold voltage of an NMOS transistor. When Φ_1 goes high (i.e., $\Phi_1 = V_{dd}$), the voltage on the top plate

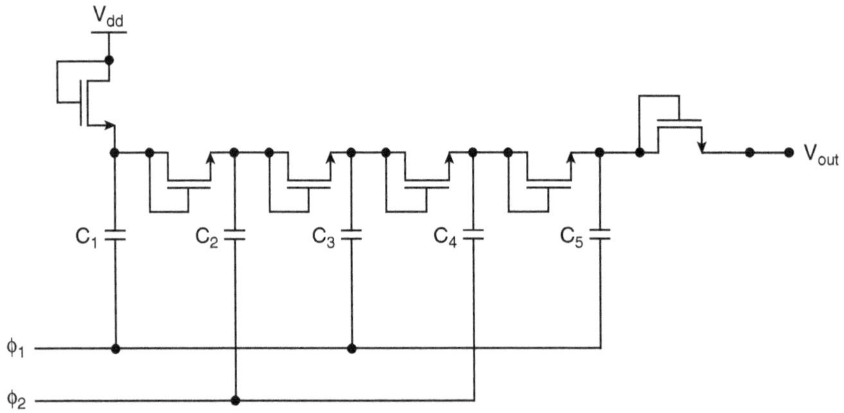

Figure 6.1 Conventional Dickson charge-pump.

of C_1 is set to $(2V_{dd} - V_{thn})$, for the voltage across C_1 does not change instantaneously. As a result, the diode-connected NMOS transistor between C_1 and C_2 is turned on, and the top plate of C_2 is thus charged to $(2V_{dd} - 2V_{thn})$. Next, Φ_1 goes low and Φ_2 goes high (i.e., $\Phi_2 = V_{dd}$), the top plates of C_2 and C_3 are boosted to $(3V_{dd} - 2V_{thn})$ and $(3V_{dd} - 3V_{thn})$, respectively, whereas the top plate of C_1 is pulled back down to $(V_{dd} - V_{thn})$. The process continues through the succeeding stages, and eventually, the top-plate voltage of C_5 is interchanging between $(5V_{dd} - 5V_{thn})$ and $(6V_{dd} - 5V_{thn})$. After the last diode (i.e., NMOS transistor), the output voltage V_{out} swings between $(5V_{dd} - 6V_{thn})$ and $(6V_{dd} - 6V_{thn})$. For example, if it is assumed that $V_{dd} = 1.5\,V$ and $V_{thn} = 0.5\,V$, then theoretically V_{out} would swing between 4.5 V and 6 V.

In practice, there is a parasitic capacitance (C_{pi}) between the top plate of each pump capacitor C_i (i = 1, 2, 3, 4, 5) and ground. Taking these parasitic capacitances into account, we can find the general expression for the nominal output voltage of an N-stage Dickson charge-pump circuit, which is given by

$$V_{out} = V_{dd} + \sum_{i=1}^{N}\left(\frac{V_{dd}C_i}{C_i + C_{pi}} - \frac{I_{li}T_{clk}}{C_i + C_{pi}} - V_{thn}\right) \tag{6.1}$$

The expression in parentheses represents the *voltage pumping gain* of the ith power stage, which is equivalent to the maximum voltage difference between the top plates of C_i and C_{i-1} ($i \geq 1$). The second item of this expression in parentheses is the voltage variation at the output due to the SC configuration of the ith power stage (also known as *ripple*) [7][8]. I_{li} is the load current of the ith power stage. It is instructive to mention that the magnitude of this ripple is inversely proportional to the clock frequency (or *switching frequency*). That is, the higher is the clock frequency, the smoother the charge-pump's output voltage will be.

To simplify the analysis, we assume that all pump capacitors in the foregoing circuit have the same capacitance value (i.e., $C_i = C$) and the same parasitic capacitance values (i.e., $C_{pi} = C_p$), and that all power stages have the same load current I_l. So the foregoing equation can be modified to

$$V_{out} = V_{dd} + N\left(\frac{V_{dd}C}{C + C_p} - \frac{I_l T_{clk}}{C + C_p} - V_{thn}\right) \tag{6.2}$$

As this equation indicates, the inherent dc voltage drop in an N-stage charge-pump is approximately NV_{thn}. Given a load current at the output terminal I_l, the total current provided by V_{dd}, I_{in}, flows through clock drivers into the charge-pump circuit and it can be approximately expressed as

$$I_{in} \cong (N+1) \cdot I_l \tag{6.3}$$

From Equations (6.2) and (6.3), we can obtain the approximate expression of the charge-pump's conversion efficiency, which is as follows:

$$\eta = \frac{V_{out} I_l}{V_{dd} I_{in}} \cong \frac{1+\dfrac{NC}{C+C_p}}{N+1} - \left(\frac{N}{N+1} \cdot \frac{I_l T_{clk}}{C+C_p} \cdot \frac{1}{V_{dd}}\right) - \left(\frac{N}{N+1} \cdot \frac{V_{thn}}{V_{dd}}\right) \tag{6.4}$$

The second item in the right part of the foregoing is the efficiency loss due to output ripples, which is determined by the number of power stages, load current, switching frequency, parasitic capacitance, and power supply voltage.

The last item of Equation (6.4) reflects a major drawback of the Dickson charge-pump—that is, the threshold voltage drop across each diode-connected transistor significantly reduces the conversion efficiency. As the number of stage N increases, the conversion efficiency decreases. Additionally, if it is assumed that V_{dd} is close to V_{thn} (e.g., $V_{dd} \leq 1.2\,\mathrm{V}$), which is often the case for ultra-low-power applications, then the conventional Dickson charge-pump circuit cannot function properly since its conversion efficiency would be near zero.

Improved Dickson Charge-Pumps

To compensate the inherent transistor threshold voltage drops, various modified Dickson charge-pump circuits have been reported. One of the most noteworthy techniques is based on the *static charge transfer switches*, which were reported by Wu et al. [7]. A four-stage modified Dickson charge-pump circuit is shown in Figure 6.2. As the schematic shows, 10 NMOS transistors and five capacitors are used to build the charge-pump. At this moment, we assume that transistor body effects are negligible to simplify the analysis.

The circuit operates as follows. After the initial startup, the power supply voltage V_{dd} is applied to the drain terminals of two coupled NMOS devices, with the one on top being a diode-connected NMOS transistor and the other an NMOS charge transfer switch whose operation is controlled by the top-plate voltage of capacitor C_2. In the beginning, the NMOS charge transfer switch is turned off, and the top plate of C_1 is charged to $(V_{dd} - V_{thn})$. Similar to the circuit presented earlier, when Φ_1 goes high (i.e., $\Phi_1 = V_{dd}$), the top plate of C_1 is charged to $(2V_{dd} - V_{thn})$, and the top plate of C_2 is charged to $(2V_{dd} - 2V_{thn})$. Next, Φ_2 goes high (i.e., $\Phi_2 = V_{dd}$), and the top plate of C_2 is charged to $(3V_{dd} - 2V_{thn})$.

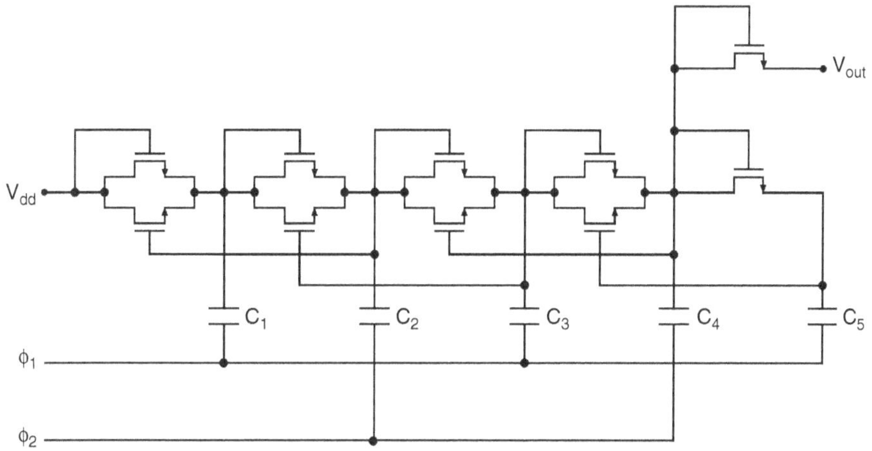

Figure 6.2 Modified Dickson charge-pump.

As the top-plate voltage of C_2 toggles between $(2V_{dd} - 2V_{thn})$ and $(3V_{dd} - 2V_{thn})$, the first NMOS charge transfer switch (from the left) is turned on, and it operates in the active region since its gate-to-source voltage $V_{gs} \approx V_{dd} > V_{thn}$ (assuming $V_{dd} > V_{thn}$). Consequently, the top-plate voltage of C_1 is charged to V_{dd} via this NMOS charge transfer switch. In a similar fashion, the top plates of the capacitors in the succeeding stages will be respectively charged to $2V_{dd}$, $3V_{dd}$, and $4V_{dd}$. In other words, the diode-connected NMOS transistors work as the *preconditioning* or *startup* devices, whereas the NMOS charge transfer switches are responsible for the effective voltage boosting. Additionally, it can be shown that the average charge transported by each NMOS charge transfer switch is constant, hence the name *static charge transfer switch.*

Based on the foregoing observation and Equation (6.2), we can write the expression for the nominal output voltage of an N-stage Dickson charge-pump circuit (before the last diode) that employs static charge transfer switches, which is given by

$$V_{out} = V_{dd} + N\left(\frac{V_{dd}C}{C+C_p} - \frac{I_l T_{clk}}{C+C_p}\right) \tag{6.5}$$

As the equation indicates, this modified charge-pump configuration helps remove the V_{thn}-related item in both Equations (6.2) and (6.4). Therefore, theoretically the modified charge-pump is more efficient than the conventional one shown in Figure 6.1.

However, a major drawback of the preceding charge-pump is that the NMOS charge transfer switch cannot be completely turned off after the desired charge transfer is accomplished, resulting in an undesirable *reverse charge leakage*, which reduces the voltage pumping gain. For example, in the circuit shown in Figure 6.2, after the top plates of C_1 and C_2 are charged to V_{dd} and $2V_{dd}$, respectively, the first NMOS switch (from the left) cannot be completely turned off unless the condition $V_{gs} < V_{thn}$ is satisfied. But here $V_{gs} = V_{dd}$, and the foregoing condition is usually *not* possible to meet. As a result, a portion of the supply voltage has to be diverted to cover the on-voltage drop across this particular NMOS charge transfer switch, thereby reducing the voltage pumping gain as well as the conversion efficiency.

An improved design proposed by the same researchers two years later [8] claims to be capable of solving this problem by simply adding a pair of auxiliary NMOS and PMOS transistors into each power stage. A four-stage example is shown in Figure 6.3.

Let us restart from the point when the top plates of C_1 and C_2 are charged to V_{dd} and $2V_{dd}$, respectively. Consider the auxiliary PMOS transistor in the first power stage (from the left), its drain-to-gate voltage (V_{dg}) is approximately V_{dd} at this moment. Because $V_{sd} = V_{sg} - V_{dg} \approx V_{sg} - V_{dd}$, the auxiliary PMOS transistor will stay in the active region as long as $V_{dd} > |V_{thp}|$, which is true. As the auxiliary PMOS transistor is turned on, the gate of the NMOS charge transfer switch is charged to approximately $2V_{dd}$. In the meanwhile, the auxiliary NMOS transistor is turned off because its V_{gs} is approximately zero.

Next, Φ_1 goes high and Φ_2 goes low; the auxiliary PMOS transistor is turned off while the auxiliary NMOS transistor is turned on, for both transistors' gates are now

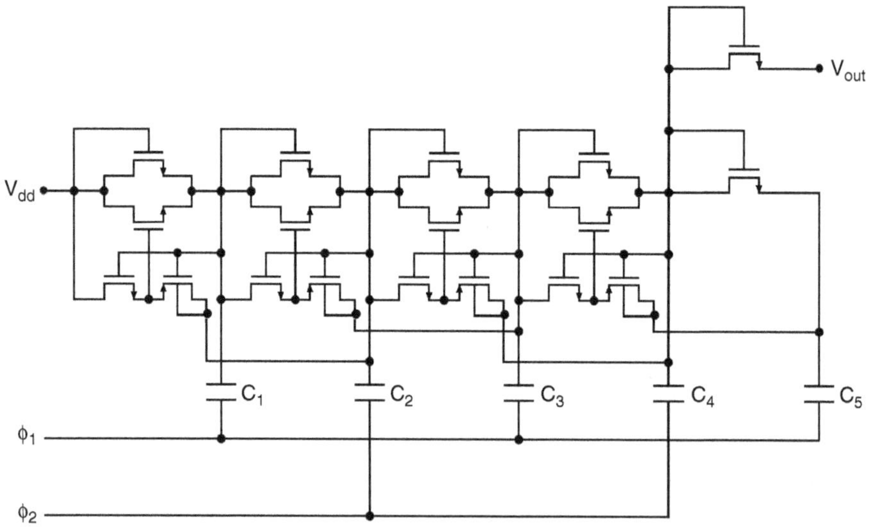

Figure 6.3 Improved Dickson charge-pump with auxiliary transistors.

charged to $2V_{dd}$. In the meanwhile, the gate voltage of the NMOS charge transfer switch remains at V_{dd}, whereas its source voltage is changed to $2V_{dd}$, resulting in a negative V_{gs} of $(-V_{dd})$. Therefore, the NMOS charge transfer switch is *completely* shut down and the reverse charge leakage is avoided.

However, note that thus far we have not considered the impact of the *body effect* or *substrate effect* on the operation of a Dickson charge-pump circuit. Body effect is important especially when both NMOS and PMOS transistors are implemented in a CMOS process that has a high substrate doping density.

The body effect phenomenon can be described qualitatively as follows. In an NMOS transistor, when the source-to-body voltage (V_{sb}) is increased, the amount of active electrons in the channel beneath the gate is reduced (consider the body as the auxiliary gate of the transistor). In effect, the average channel current I_{ds} is reduced, and the effective threshold voltage V_{thn} is increased, for it has become more difficult to turn on the NMOS transistor. By contrast, in a PMOS transistor, the body effect is likely to occur when the body-to-source voltage (V_{bs}) is large.

Now let us revisit the charge-pump circuit shown in Figure 6.3. Note that during the charge transfer mode, the fourth NMOS charge transfer switch (from the left) may not be turned *on* completely due to the body effect. The analysis is as follows. First, from previous discussions we know that to turn on the NMOS charge transfer switch, the following condition must be met: $V_{dd} > V_{thn}$. Second, in accounting for body effects, we find that the realistic threshold voltage of an NMOS transistor is given by

$$V_{thn} = V_{thn\_0} + \gamma\left(\sqrt{V_{sb} + 2|\phi_F|} - \sqrt{2|\phi_F|}\right) \tag{6.6}$$

Here, V_{thn_0} is the threshold voltage when the NMOS transistor's source and body are shorted (i.e., $V_{sb} = 0$), and γ is the *body effect coefficient* with a unit of $V^{1/2}$. Φ_F is called *Fermi potential* and is approximated as 0.35 V for typical CMOS processes [9]. We then assume that $V_{dd} = 1.8$ V, $\gamma = 0.6$, $V_{thn\_0} = 0.7$ V, and that the required output voltage from this four-stage charge-pump is set to about 7.0 V (the maximum achievable gain of an N-stage charge-pump is usually less than 4 due to parasitic capacitances)—that is, $V_{sb} \approx 7.0$ V (the substrate of the NMOS switch is tied to ground to prevent latch-up [9]). Substituting these values into Equation (6.6), we can calculate the value of V_{thn}, which is about 1.863 V (i.e., $V_{dd} < V_{thn}$).

Therefore, in this situation, the NMOS charge transfer switch of the fourth power stage cannot be turned on successfully. This indicates that this fourth power stage will not function. In fact, experiments have shown that the maximum achievable conversion efficiency of this charge-pump topology drops abruptly when the number of power stages exceeds five.

This performance limitation due to the body effect can be alleviated by using a cross-coupled voltage doubler, which is the topic of Section 6.3, to boost the amplitude of the clock voltage that is used to drive the final power stage via C_5 [8]. Alternatively, this problematic NMOS charge transfer switch and the output NMOS diodes can be substituted by their PMOS counterparts to compensate the body effect [10]. The reference offers a detailed description of the circuitry.

Furthermore, it is of practical interest to determine the minimum total capacitance that will be needed to build a Dickson charge-pump, given the specified power supply voltage (V_{dd}), load current (I_l), and clock frequency (f_{clk}). This total capacitance value can then be used to estimate the total silicon area and power consumption of the charge-pump circuit. Equation (6.5) shows that the total capacitance can be expressed as

$$C_{total} = NC = \frac{(C + C_p)(V_{out} - V_{dd})}{CV_{dd} - I_l T_{clk}} C \tag{6.7}$$

After differentiating the preceding expression with respect to C, we obtain the optimum value of C required for minimizing the total capacitance value [10], which is expressed as follow:

$$C_{opt} = \frac{I_l}{V_{dd} f_{clk}} + \sqrt{\left(\frac{I_l}{V_{dd} f_{clk}}\right)^2 + \frac{I_l C_p}{V_{dd} f_{clk}}} \tag{6.8}$$

To get a rough feel for the number, assume that in a Dickson charge-pump that is powered by a 2-V battery voltage, a load current of 100 mA is required when the charge-pump is operating at a clock frequency of 1 MHz. In addition, assume that the parasitic capacitance C_p is equal to $0.05C$. Then the optimum value of C can be calculated based on the two foregoing formulae, and the result turns out to be approximately 0.1025 μF. For a four-stage converter (i.e., $N = 4$), the minimum total capacitance value is thus about 0.41 μF.

6.3 Cross-Coupled SC Step-Up DC-DC Converters

Although it is straightforward to build a step-up DC-DC converter based on the Dickson charge-pump topology, the resultant conversion efficiency is not adequate to many applications, especially when the power supply voltage V_{dd} is very low (e.g., below 1.2 V). Additionally, from Equation (6.8) we find that a low-supply-voltage and high-load-current Dickson charge-pump circuit typically occupies a large silicon area, which is not desirable to the design of space-constrained devices such as hearing aids.

As the alternative to Dickson charge-pump circuits, cross-coupled SC DC-DC converters are more appropriate for battery-driven portable applications that require a higher conversion efficiency and a smaller silicon area. One of the earlier high-efficiency cross-coupled charge-pump topologies was configured as a voltage doubler (×2), which consists of four NMOS transistors, four PMOS transistors, and three large capacitors [3]. Recently, this topology found its use in low-supply-voltage CMOS data converters for minimizing the switch on-resistance [11]. One of the several possible transistor-level realizations of this charge-pump topology is shown in Figure 6.4 [12].

In this circuit, all NMOS transistors share the common substrate, which is connected to ground, while all PMOS transistors are constructed in a common N-well tied to a bias voltage V_{bias}, which is either greater than or equal to the desired output voltage value ($2V_{dd}$). The main purpose of such a configuration is to ensure that the PN junction between the source (in the NMOS case) or the N-well (in the PMOS case) and the substrate is reversely biased, avoiding latch-up problems during startup [9].

Figure 6.4 Cross-coupled charge-pump configured as a voltage doubler.

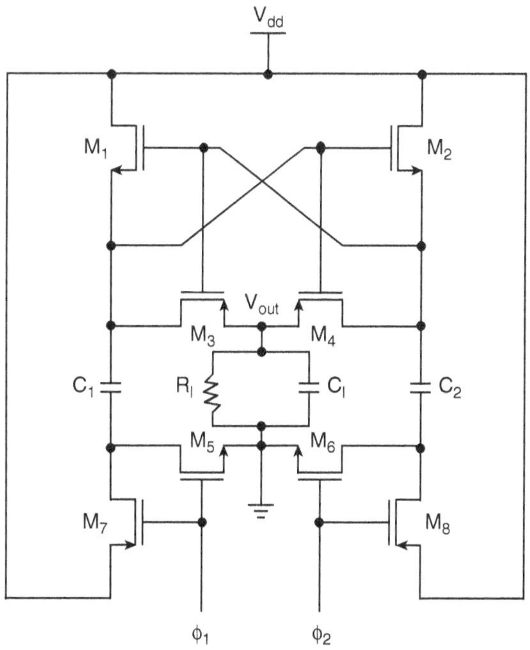

The circuit operates as follows. First, let us start with the half circuit consisting of transistors M_1, M_3, M_5, and M_7. When Φ_1 goes high (i.e., $\Phi_1 = V_{dd}$), both M_1 and M_5 are turned on (since Φ_1 and Φ_2 are not overlapping, M_1 and M_2 are never on at the same time), thus the capacitor C_1 is charged to V_{dd}. Next, Φ_1 goes low (i.e., $\Phi_1 = 0$) and Φ_2 goes high (i.e., $\Phi_2 = V_{dd}$), thereby M_1 and M_5 are turned off while M_3 and M_7 are turned on. Ideally, V_{out} would be charged to $2V_{dd}$; hence the name *voltage doubler.*

The other half of the charge-pump circuit, which consists of transistors M_2, M_4, M_6, and M_8, is the opposite-phase version of the preceding half circuit, meaning the circuit shown in Figure 6.4 is composed of two identical voltage doublers operating in antiphase. This configuration has two major advantages. First, the effective switching frequency is doubled (similar to the *double sampling* concept that we discussed in Chapter 3), which means the output filter constructed from R_l and C_l is effectively clocked at $2f_{clk}$, where f_{clk} is the system clock frequency. This feature helps reduce the output ripples and increase the conversion efficiency. Second, the cross-coupling configuration eliminates the need for diode-connected transistors that are required in the Dickson charge-pump; therefore the threshold voltage drop is no longer a concern.

However, one of the major design issues with respect to the cross-coupled charge-pump topology is that parasitic capacitances have a direct impact on the conversion efficiency. As a matter of fact, the power loss during the conversion cycles is mainly due to the charging and discharging of the internal parasitic capacitors.

As developed in [12], the conversion efficiency of the charge-pump shown in Figure 6.4 can be approximately expressed as (assuming that the pump capacitors $C_l = C_2 = C$)

$$\eta \cong \frac{1}{1 + \alpha \cdot C f_{clk} \dfrac{(R_l + R_0)^2}{2R_l} + 2C f_{clk} \dfrac{R_0^2}{R_l}} \tag{6.9}$$

where $\alpha \cdot C$ represents the equivalent parasitic capacitance associated with each pump capacitor. The equation indicates that the conversion efficiency is inversely proportional to the parasitic factor α. The value of α is determined by the type of integrated capacitors used. It can be as large as 50% if poly-metal capacitors are used. For double-poly capacitors, the value of α is typically between 5% and 10%. In the cases of thin oxide CMOS capacitors, the value of α ranges between 10% and 15%.

R_0 in Equation (6.9) is the equivalent output resistance that is composed of the simulated SC resistance due to the switching of internal capacitors, and the switch on-resistance (it does not include the resistive load R_l) [12][13]. To the first-order approximation, R_0 can be expressed as

$$R_0 \cong \begin{cases} \dfrac{1}{2f_{clk}(1+\alpha)C} & \text{iff} \quad f_{clk} < f_{cutoff} = \dfrac{1}{2[R_{on}\cdot(1+\alpha)C + t_{sw}]} \\ R_{on} & \text{iff} \quad f_{clk} \geq f_{cutoff} = \dfrac{1}{2[R_{on}\cdot(1+\alpha)C + t_{sw}]} \end{cases} \tag{6.10}$$

where t_{sw} is the switching delay that includes the nonoverlapping interval (i.e., when both clock phases are at 0 V). Note that $2f_{clk}$ instead of f_{clk} is used in one of the foregoing expressions due to the double sampling effect mentioned earlier.

From Equations (6.9) and (6.10), we can find that as long as the clock frequency f_{clk} is below the cutoff frequency f_{cutoff}, the conversion efficiency increases with f_{clk}. However, once f_{clk} exceeds the cutoff frequency limit, the conversion efficiency decreases as f_{clk} continues increasing. In other words, the practical frequency range of the cross-coupled charge-pump circuit is limited by f_{cutoff}.

The value of f_{cutoff} greatly depends on the type of capacitors and the fabrication technology. A good reference point is at about 10 MHz; it is for a cross-coupled voltage doubler that employs double-poly integrated capacitors fabricated in a standard 0.5 μm CMOS technology. In such a case, depending on the resistive load (R_l), the maximum achievable conversion efficiency ranges between 70% and 80%.

Based on the cross-coupled voltage doubler discussed earlier, a voltage tripler (×3) [14] and a voltage quadrupler (×4) [15] have been reported. Alternatively, the cross-coupled voltage doubler shown in Figure 6.4 can be used in combination with a two-stage Dickson charge-pump to construct a hybrid voltage quadrupler [16].

Before we wrap up the discussion of SC step-up DC-DC converters, note that there are many other practical circuit topologies beyond what have been investigated here. For instance, an interesting topology called the *multiple-lift SC converter* was reported in [17]. The two-lift (×2) version of this topology (called the *H-bridge*) has been adopted in the design of many commercial SC voltage doublers. By simply repeating the lower-order converters (e.g., two-lift or three-lift), one can build higher-order converters up to 128-lift (×128) in a modular fashion. Another noteworthy example is the *pseudo-4-phase charge-pump* reported by Lee et al. [18], which is particularly suitable for ultra-high-speed applications such as flash memories implemented in submicron digital CMOS technology.

6.4 SC Step-Down DC-DC Converters

As the concept of low-power design becomes increasingly popular, many state-of-the-art handheld device including MP3 players and cellular phones are designed to operate at 1.5 V, 1.2 V, or even lower voltages (with 1 mA to 200 mA total load current). However, most standard batteries have a full-range dc voltage of at least 2.5 V, and consequently, a step-down DC-DC converter is usually required to convert such a high battery input voltage into a low constant output voltage.

A widely used conventional SC step-down DC-DC converter (with three power stages) is shown in Figure 6.5. Note that M_1–M_6 are typically implemented as *Schottky diodes*, which prevent charge leakages while maintain the circuit's high speed.

The circuit operates as follows. When Φ_1 goes high and Φ_2 goes low, the pump capacitors C_1–C_3 are charged in series by V_{dd} through the Φ_1 switch (near the input), M_2, and M_5. If it is assumed that all pump capacitors have the same capacitance values (i.e., $C_1 = C_2 = C_3 = C$), then it can be shown that the charge that flows into each capacitor is given by $V_{dd}C/3$. Next, Φ_1 goes low and Φ_2 goes high, the pump capacitors are connected in parallel between V_{out} and ground—that is, they are discharged simultaneously to the load through the Φ_2 switch close to the output. The ideal output voltage is thus given by

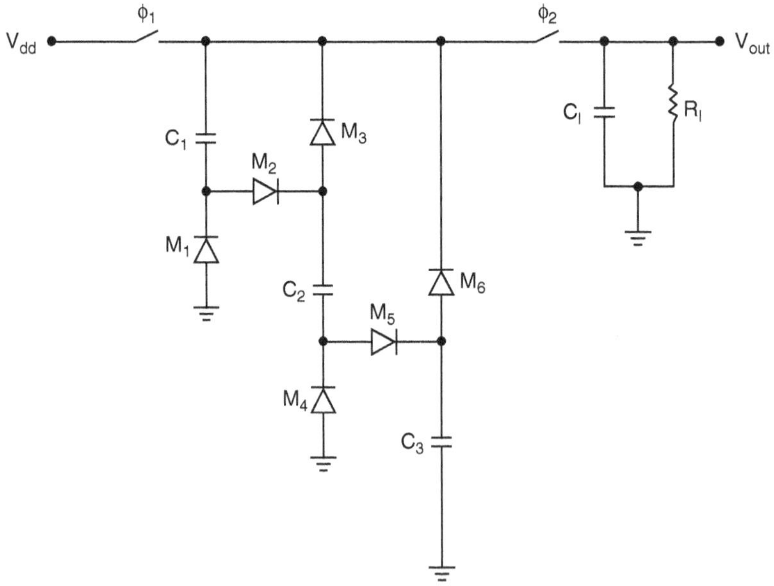

Figure 6.5 Conventional SC step-down DC-DC converter.

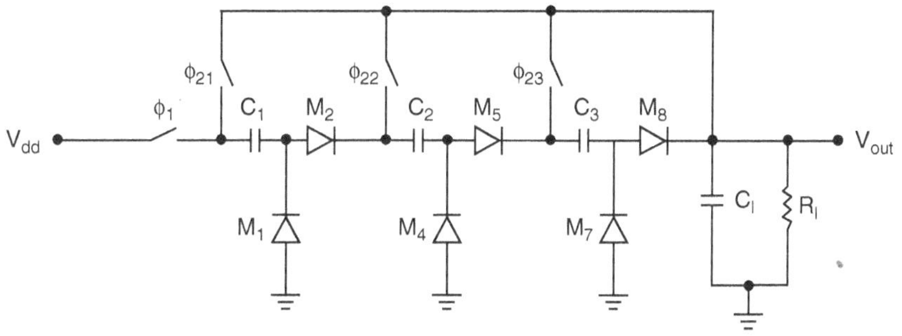

Figure 6.6 Improved SC step-down DC-DC converter.

$$V_{out\_ideal} = \frac{V_{dd}}{3} \tag{6.11}$$

An alternative topology with a reduced component count was reported by Harris and Ngo [19]. A four-stage example is shown in Figure 6.6. Note that when Φ_1 goes high, the load capacitor C_l joins the three serially connected pump capacitors via the diode M_8.

The circuit operates as follows. When Φ_1 goes high, the pump capacitors C_1–C_3 and the load capacitor C_l are charged in series by V_{dd} through the Φ_1 switch, M_2, M_5, and M_8. Theoretically speaking, if all four capacitors are equally sized (i.e., $C_1 = C_2 = C_3 = C_l = C$), then the charge sent into each capacitor is given by $V_{dd}C/4$. Next, Φ_1 goes low. If we assume that Φ_{21}, Φ_{22}, and Φ_{23} switches are turned on simultaneously, then the ideal output voltage is $V_{dd}/4$.

In practice, it is common to use a load capacitor C_l several times larger than the pump capacitor to effectively suppress the ripples in the converter's output signal. The larger the load capacitor, the greater the RC time constant of the output low-pass filter will be, and the high-frequency components in the output signal (including ripples) can be reduced more effectively. For example, assume that the capacitance value of all four pump capacitors in Figure 6.6 is equal to 9.6 μF and the load capacitance value is equal to 86.4 μF (i.e., $C_l = 9C$). For f_{clk} = 100 kHz, R_l = 3 Ω, and V_{dd} = 55 V, the measured output is about 11.98 V with a maximum ripple of only 96 mV [19].

To minimize ripples, we can control the output switches (Φ_{21}, Φ_{22}, Φ_{23}) so that they operate in a *time-interleaving* manner; this is a popular technique used in many low-voltage high-current switched-mode power supplies (SMPS) to stabilize the output voltage [20]. Specifically, for the circuit shown in Figure 6.6, when Φ_2

goes high, C_1 is discharged to the load through the Φ_{21} switch. But the remaining capacitors (C_2 and C_3) are not discharged, hence their voltages are intact. Next, the Φ_{21} switch is turned off and the Φ_{22} switch is turned on, and C_2 is discharged to the load while C_1 and C_3 are not. Finally, Φ_{23} is on and only C_3 is being discharged to the load. We will now look into how this interleaving configuration helps further reduce output ripples.

Consider an $(N-1)$-stage converter, if the Φ_{21}, Φ_{22}, Φ_{23}, till Φ_{2N} switches in Figure 6.6 are turned on simultaneously, then input/output voltage waveforms such as those illustrated in Figure 6.7(a) will result. As the figure shows, all pump capacitors have identical charging/discharging patterns (see V_{ci}), and there is a large ripple in the output voltage.

On the other hand, if Φ_{21}, Φ_{22}, Φ_{23}, till Φ_{2N} switches are turned on one by one, then we obtain the voltage waveforms shown in Figure 6.7(b). In contrast to the previous noninterleaving situation, the pump capacitors are being discharged sequentially during different phases. Also, the previously mentioned ripple in the output voltage is now *chopped* into pieces, resulting in a smaller average ripple.

As we saw earlier, in an SC DC-DC converter, the major cause of output ripples is the switching of internal pump capacitors. In addition, as Equations (6.1) and (6.2) indicate, given a fixed load current value, the average ripple magnitude (V_{ripple}) is proportional to the product of $T_{clk}/(C + C_p)$, where T_{clk} is the *effective* switching clock period, and $(C + C_p)$ is the equivalent capacitance per stage. It is well known that an M-step interleaving configuration is capable of reducing T_{clk} by a factor of M. In the circuit shown in Figure 6.6, M is equal to 3; thus, the maximum value of V_{ripple} is reduced by a factor of 3 (e.g., 96 mV/3 = 32 mV).

Furthermore, Chung et al. [21] proposed an interesting closed-loop SC step-down DC-DC converter. The basic idea is to form a feedback loop from the converter's output terminal, through an RC network, to the input transistors. In effect, the input current is controlled by the output voltage through a negative feedback, thereby further reducing output ripples.

6.5 Multiple-Gain SC DC-DC Converters

For handheld electronic devices that are powered by rechargeable batteries, it is desirable to have the capability of controlling multiple dc voltage gains through a single DC-DC converter. For instance, most GSM/EDGE dual-mode cellular phones use 3.6-V 700-mA Lithium Ion rechargeable batteries. In such cases, the battery voltage typically drops from 3.6 V down to 0.8 V (i.e., the end-of-life threshold

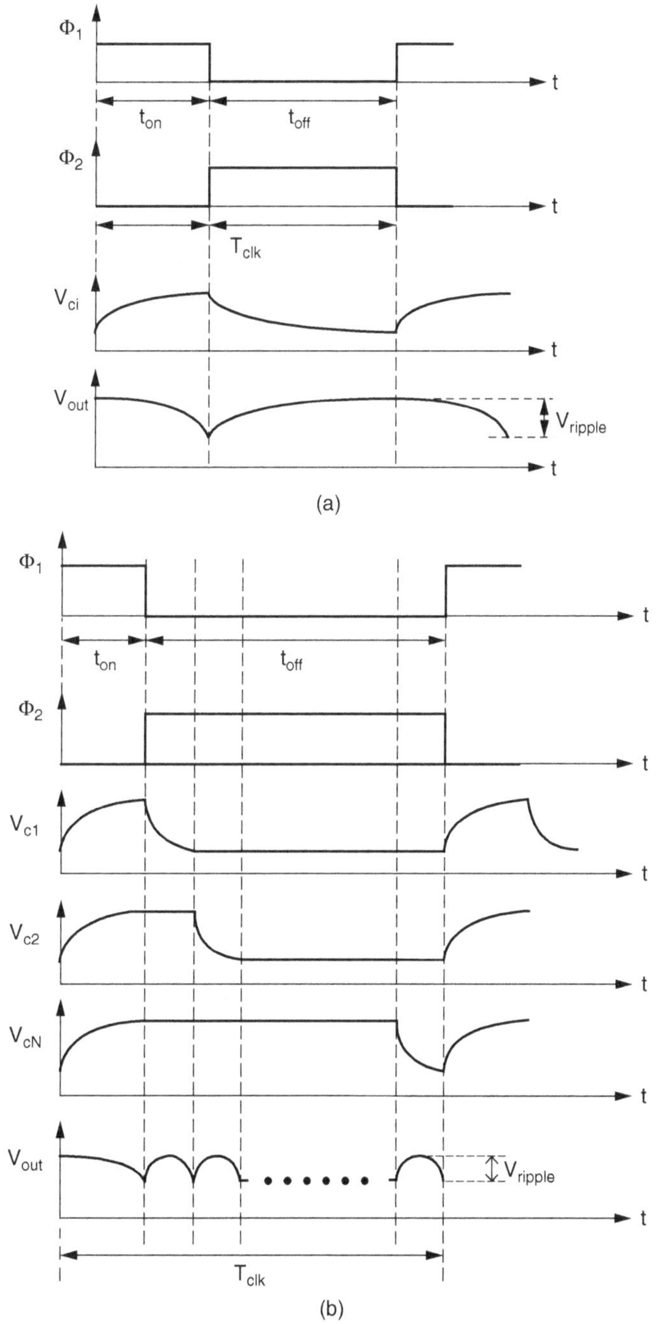

Figure 6.7 (a) Voltage waveforms without interleaving. (b) Voltage waveforms with interleaving.

voltage) as the battery continues to operate without recharging, whereas it climbs up to about 4.2 V when the battery is being recharged. However, for the sake of performance, the cellular phone normally requires a constant and stable dc voltage between the highest and the lowest battery voltages, regardless of the recharging process. As a result, the designed DC-DC converter must be capable of providing either step-up or step-down gains, depending on the status of the battery voltage. Additionally, the number of different gains (or *conversion ratios*) should be maximized in order to minimize the conversion inaccuracies and maximize battery life.

A practical multiple-gain SC DC-DC converter is shown in Figure 6.8 [22]. The circuit consists of three equal-size pump capacitors C_1–C_3 (on the order of 0.1 μF) and 12 switches. The clock scheme is slightly complicated and it is described as follows: Φ_1, Φ_2, and Φ_3 are three nonoverlapping clock phases, and $(\Phi_1 + \Phi_2 + \Phi_3)$ is equivalent to one clock cycle, T_{clk}. The rest of the clock phases are related to Φ_1, Φ_2, and Φ_3 in the following manner:

$$\begin{cases} \phi_4 = \overline{\phi_2}, \phi_5 = \overline{\phi_3}, \phi_6 = \overline{\phi_1}, \\ \phi_7 = \overline{A}\cdot B\cdot\phi_1 + A\cdot\overline{B}\cdot\phi_2 + A\cdot B\cdot\phi_3, \\ \phi_8 = A\cdot B\cdot\phi_1 + \overline{A}\cdot B\cdot\phi_2 + A\cdot\overline{B}\cdot\phi_3, \\ \phi_9 = A\cdot\overline{B}\cdot\phi_1 + A\cdot B\cdot\phi_2 + \overline{A}\cdot B\cdot\phi_3, \\ \phi_{10} = \overline{C}\cdot D\cdot\phi_1 + C\cdot\overline{D}\cdot\phi_2 + C\cdot D\cdot\phi_3, \\ \phi_{11} = C\cdot D\cdot\phi_1 + \overline{C}\cdot D\cdot\phi_2 + C\cdot\overline{D}\cdot\phi_3, \\ \phi_{12} = C\cdot\overline{D}\cdot\phi_1 + C\cdot D\cdot\phi_2 + \overline{C}\cdot D\cdot\phi_3. \end{cases} \tag{6.12}$$

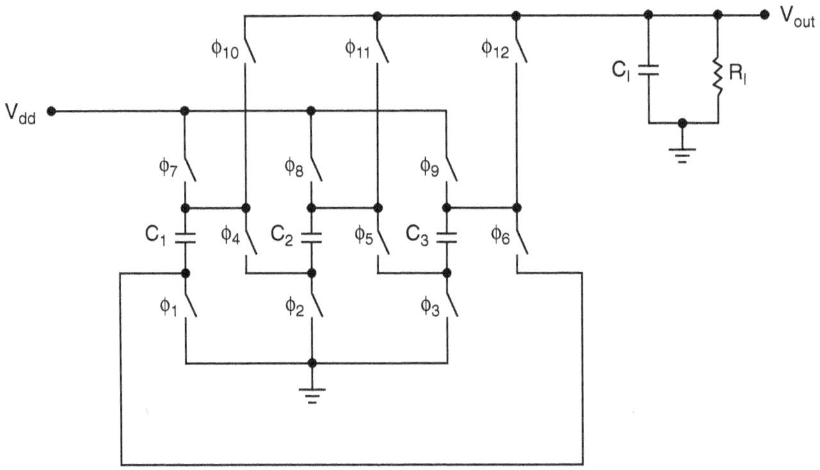

Figure 6.8 Multiple-gain SC DC-DC converter.

where AB and CD are 2-bit digital (binary) control words. Note that 18 three-input AND gates, 6 three-input OR gates, and 7 inverters are needed to realize the clock control circuit specified in the preceding expression.

For the sake of brevity, the gain configuration of (AB = 11, CD = 10) is used here as an example to explain the implementation of gain settings through the SC array. When AB = 11 and CD = 10, the clock phases are arranged as follows:

$$\begin{cases} \phi_4 = \overline{\phi_2}, \phi_5 = \overline{\phi_3}, \phi_6 = \overline{\phi_1}, \\ \phi_7 = \phi_3, \phi_8 = \phi_1, \phi_9 = \phi_2, \\ \phi_{10} = \phi_2, \phi_{11} = \phi_3, \phi_{12} = \phi_1. \end{cases} \tag{6.13}$$

Thus, when Φ_1 is on, there are two charge transfer paths through the SC array, which are

$$\begin{cases} V_{dd} \rightarrow \phi_8 \rightarrow C_2 \rightarrow \phi_4 \rightarrow C_1 \rightarrow \phi_1 \rightarrow ground, \\ V_{dd} \rightarrow \phi_8 \rightarrow \phi_5 \rightarrow C_3 \rightarrow \phi_{12} \rightarrow V_{out}. \end{cases} \tag{6.14}$$

The first path is called the *input path* while the second is called the *output path.* As the schematic shows, C_1 and C_2 are connected in series between V_{dd} and ground through the input path, whereas C_3 is connected between V_{dd} and V_{out} through the output path. Based on the charge reservation principle, we can calculate the ratio of V_{out} to V_{dd} (i.e., the conversion ratio), which is equal to 3/2.

When Φ_2 is on, there are two new charge transfer paths that are different from the preceding, and they travel through the SC array as well. They are expressed as follows:

$$\begin{cases} V_{dd} \rightarrow \phi_9 \rightarrow C_3 \rightarrow \phi_5 \rightarrow C_2 \rightarrow \phi_2 \rightarrow ground, \\ V_{dd} \rightarrow \phi_9 \rightarrow \phi_6 \rightarrow C_1 \rightarrow \phi_{10} \rightarrow V_{out}. \end{cases} \tag{6.15}$$

The resulting conversion ratio is also equal to 3/2.

Similarly, it can be found that when Φ_3 is on, the conversion ratio is equal to 3/2 as well. Therefore, the combination of AB = 11 and CD = 10 sets the overall conversion ratio to 3/2. The practical combinations of AB, CD, and the conversion ratio are listed in Table 6.1. As the table indicates, only 8 out of 16 possible cases are listed. This is because among the unlisted cases, some are equivalences of Case 4 or Case 8, while the remainders cut off the input paths altogether. Case 8, which is a special setting with a cutoff output path, is included here because it may be useful for facilitating a fast power recovery once the *power-saving* or *sleep mode* is terminated [22]. However, for low-voltage and high-current applications, this configuration is prone to latch-up problems and may cause damage to the devices.

Table 6.1 Gain setting by AB and CD.

Case #	AB	CD	Conversion ratio
1	10	01	1/3
2	11	01	1/2
3	10	11	2/3
4	01	01	1
5	11	10	3/2
6	01	11	2
7	01	10	3
8	01	00	0

A similar multiple-gain topology was presented in [23]. In the reported circuit, three pump capacitors and 20 switches are used to construct an SC array that is capable of providing seven different *nonzero* conversion ratios: 1/2, 2/3, 3/4, 1, 4/3, 3/2, and 2.

In addition, from the preceding development we can find that the more are the gain options, the more pump capacitors and switches will be needed. Theoretically speaking, with M pump capacitors and sufficient switches, one may realize up to $(2^M - 1)$ different gain values.

Finally, note that voltage regulation is essential to an SC converter (particularly to a multiple-gain DC-DC converter) because deviations of the converter's output voltage from its desired value will result in significant losses in efficiency.

One of the widely used techniques to regulate voltage in an SC DC-DC converter is called the *pulse-frequency modulation* (PFM) [1]. In this regulation scheme, the operation of the converter is controlled by a PFM control loop, which typically consists of a reference voltage generator, a comparator, a clock signal generator (or oscillator), and a digital gate.

The PFM control loop operates as follows. The desired voltage ($V_{desired}$) is generated by the reference voltage generator, and it is compared with the output voltage (V_{out}) by the comparator. If V_{out} is less than $V_{desired}$, then the comparator outputs 1, which opens the digital gate, and the converter is activated by the clock signal. As a result, the SC array is clocked to boost the output voltage. On the other hand, if V_{out} is greater than $V_{desired}$, then the gate is closed and the converter is idle. As a result, the SC array stops delivering charge to the output, thereby reducing the output voltage.

Moreover, in such a PFM-controlled DC-DC converter, for a given load impedance and a fixed conversion ratio, the duration of the idle mode will be decreased as the load current increases, because the converter must boost the output voltage to keep up with the increasing product of the load current and impedance. However, under the same conditions, the duration of the active mode will be decreased once the input battery voltage reaches a point when the resulting output voltage is equal to or higher than the desired voltage ($V_{desired}$). In an intuitive sense, we can say that the switching frequency, which is reflected by the rate at which the idle mode and the active mode interchange and has a direct effect on the output ripples and the conversion efficiency according to Equations (6.1) and (6.2), depends on both the input battery voltage and the (output) load current. This is a major drawback of PFM voltage regulation, because the resulting output has comparatively strong tones or harmonics in its frequency spectrum. Since the characteristics of these tones are affected by both input and output parameters, it is difficult to remove them by linear filtering.

Rao et al. [24] reported a possible solution to this chaotic distortion problem, which involves incorporating a *dithered* first-order delta-sigma modulator with the PFM control loop to transform the chaotic tones into pseudo-white noise, which is easier to filter out.

References

[1] N. Mohan et al., *Power electronics converters, applications, and design* (2nd ed.), John Wiley & Sons, New York, 1995.

[2] C. P. Yue and S. S. Wong, "On-chip spiral inductors with patterned ground shields for Si-based RF ICs," *IEEE Journal of Solid-State Circuits*, Vol. 33, No. 5, pp. 743–752, May 1998.

[3] Y. Nakagome et al., "An experimental 1.5-V 64 Mb DRAM," *IEEE Journal of Solid-State Circuits*, Vol. 26, No. 4, pp. 465–472, April 1991.

[4] K. Sawada, Y. Sugawara, and S. Masui, "An on-chip high-voltage generator circuit for EEPROM's with a power supply voltage below 2 V," *IEEE Symposium on VLSI Circuits, Digest of Technical Papers*, pp. 75–76, June 1995.

[5] U. Rohde and T. Bucher, *Communication receivers: Principles and design*, McGraw-Hill, New York, 1988.

[6] J. Dickson, "On-chip high-voltage generation in MNOS integrated circuits using an improved voltage multiplier technique," *IEEE Journal of Solid-State Circuits*, Vol. SC-11, No. 3, pp. 374–378, June 1976.

[7] J. Wu, Y. Chang, and K. Chang, "1.2 V CMOS switched-capacitor circuits," *IEEE International Solid-State Circuits Conference, Digest of Technical Papers*, pp. 388–389, February 1996.

[8] J. Wu and K. Chang, "MOS charge-pumps for low-voltage operation," *IEEE Journal of Solid-State Circuits*, Vol. 33, No. 4, pp. 592–597, April 1998.

[9] D. A. Johns and K. Martin, *Analog integrated circuits design*, John Wiley & Sons, New York, 1997.

[10] S. C. Lee et al., "A low-ripple switched-capacitor DC-DC up converter for low-voltage applications," *Proceedings of the 2nd IEEE Asia Pacific Conference on Integrated Circuits (ASIC)*, pp. 13–16, August 2000.

[11] T. Cho and P. R. Gray, "A 10 b, 20 Msamples/s, 35 mW pipeline A/D converter," *IEEE Journal of Solid-State Circuit*s, Vol. 30, No. 3, pp. 166–172, March 1995.

[12] P. Favrat, P. Deval, and M. Declercq, "A high-efficiency CMOS voltage doubler," *IEEE Journal of Solid-State Circuits*, Vol. 33, No. 3, pp. 410–416, March 1998.

[13] G. van Steenwijk et al., "Analysis and design of a charge-pump circuit for high output current applications," *Proceedings of IEEE European Solid-State Circuits Conference*, Vol. 3, pp. 118–121, September 1993.

[14] D. Maksimovic and S. Dhar, "Switched-capacitor dc-dc converters for low-power on-chip applications," *Proceedings of IEEE Power Electronics Specialist Conference*, Vol. 1, pp. 54–59, August 1999.

[15] T. Ying, W. H. Ki, and M. Chan, "Area-efficient CMOS charge-pumps for LCD drivers," *IEEE Journal of Solid-State Circuits*, Vol. 38, No. 10, pp. 1721–1725, October 2003.

[16] C.-C. Wang and J.-C. Wu, "Efficiency improvement in charge-pump circuits," *IEEE Journal of Solid-State Circuits*, Vol. 32, No. 4, pp. 852–860, April 1998.

[17] F. Luo, H. Ye, and M. H. Rashid, "Multiple-lift push-pull switched-capacitor Luo-converters," *Proceedings of IEEE Power Electronics Specialist Conference*, Vol. 2, pp. 415–420, June 2002.

[18] J. Lee et al., "High-performance 1-Gb NAND flash memory with 0.12-μm technology," *IEEE Journal of Solid-State Circuits*, Vol. 37, No. 11, pp. 1502–1509, November 2002.

[19] W. S. Harris and K. D. T. Ngo, "Power switched-capacitor DC-DC converter: Analysis and design," *IEEE Trans. on Aerospace and Electronic Systems*, Vol. 33, No. 2, pp. 386–395, April 1997.

[20] C. Chang and M. Knights, "Interleaving technique in distributed power conversion systems," *IEEE Trans. on Circuit and Systems*, Vol. 42, No. 5, pp. 245–251, May 1995.

[21] H. Chung, S. Hui, and S. Tang, "Development of a multistage current-controlled switched-capacitor step-down DC-DC converter with continuous input current," *IEEE Trans. on Circuit and Systems*, Vol. 47, No. 7, pp. 1017–1026, July 2000.

[22] N. Hara, I. Oota, and F. Ueno, "A new ring type switched-capacitor DC-DC converter with low inrush current and low ripple," *Proceedings of IEEE Power Electronics Specialist Conference*, Vol. 2, pp. 1536–1542, May 1998.

[23] J. Kotowski et al., "Capacitor DC-DC converter with PFM and gain hopping," U.S. Patent 6055168, April 2000.

[24] A. Rao, W. McIntyre, U. Moon, and G. C. Temes, "A noise-shaped switched-capacitor DC-DC voltage regulator," *Proceedings of IEEE European Solid-State Circuits Conference*, Vol. 2, pp. 375–378, September 2002.

CHAPTER 7

Advanced Switched-Capacitor Circuit Techniques

7.1 Introduction

The design of linear active circuits typically involves multiple tradeoffs among supply voltage, accuracy, power, speed, and other factors. However, an *all-round* design that meets all the expectations is impracticable as we are living in a wonderful and yet imperfect world. Depending on the application's requirements and the technology available, circuit designers may choose to optimize either one of these fundamental design aspects.

This chapter investigates two major challenges that are of immediate relevance to modern switched-capacitor (SC) circuits. One is to design high-performance SC circuits in the presence of a low power supply voltage ($V_{dd} < 1.5\,\text{V}$), and the other is to reduce the effect of the imperfections (or nonidealities) of operational amplifiers (op-amps) on SC circuits.

Chapter Outline

Section 7.2 presents a number of low-voltage SC circuit techniques such as *clock boosting*, *bootstrapped switch*, and *switched op-amp*. Then Section 7.3 explores two accuracy-enhancement techniques suitable to desensitize SC circuits from op-amp imperfections, namely *autozeroing* and *correlated double sampling*.

7.2 Low-Voltage SC Circuits Techniques

The Low-Voltage Challenge

Since the late 1990s, the brisk market for small-form-factor portable electronic products—including cellular phones, MP3 players, hearing aids, and handheld medical testing devices—has driven the fast advancement of modern fine-line-width submicron CMOS technologies. As CMOS device dimensions continue to shrink for achieving higher integration densities, it is almost essential that the allowable power supply voltage (V_{dd}) be scaled down proportionally to guarantee the device's long-term reliability.

As the Semiconductor Industry Association predicted [1], along with the continual downscaling of deep-submicron CMOS technologies, the nominal power supply voltage for most high-performance *digital* CMOS ICs will plummet to as low as 0.4 V by the year 2016. Also, as discussed in Chapter 1, the cutoff frequency (f_t) of an MOS transistor is increased as its effective gate length (L) is reduced. Moreover, it is known that the operating power dissipation in a digital IC primarily depends on V_{dd}, and the amount of power reduction is proportional to the square of V_{dd}.

Although the foregoing technological trend promises smaller, faster, and more power-efficient digital signal processing (DSP) integrated systems in the near future, the low supply voltage remains a fundamental restraint for the design of *analog* CMOS circuits. This is mainly due to the fact that the threshold voltage of the MOS transistor (V_{th}), which is governed by the intrinsic process breakdown voltage and the stress limit of the thin gate oxide, cannot be scaled down proportionally with V_{dd} or with transistor geometries.

In addition, the aggressive downscaling of the effective gate length (L) may result in short-channel effects such as *velocity saturation*, in which case the relationship between the drain current (I_d) and the gate-source voltage (V_{gs}) is after an incrementally *linear* pattern as opposed to the *square-law* rule introduced in Chapter 1.

Consequently, under the low-supply-voltage and small-geometry conditions, the majority of the classical analog design rules are no longer applicable, and the realization of high-performance analog circuits faces two major challenges. The first challenge is to design the type of op-amp that can provide a high voltage gain and a high output swing in the presence of a low supply voltage, while dissipating a minimum amount of power. The design of such high-performance op-amps is an active research topic at the time of this writing, and some state-of-the-art prototypes have been reported [2][3][4][5].

The second primary challenge is to drive the *floating switch* (i.e., the switch that is never connected to ground or the virtual ground) when the supply voltage drops to about the same as or less than the sum of the absolute values of the PMOS and NMOS threshold voltages. The floating-switch problem is a primary concern to the design of

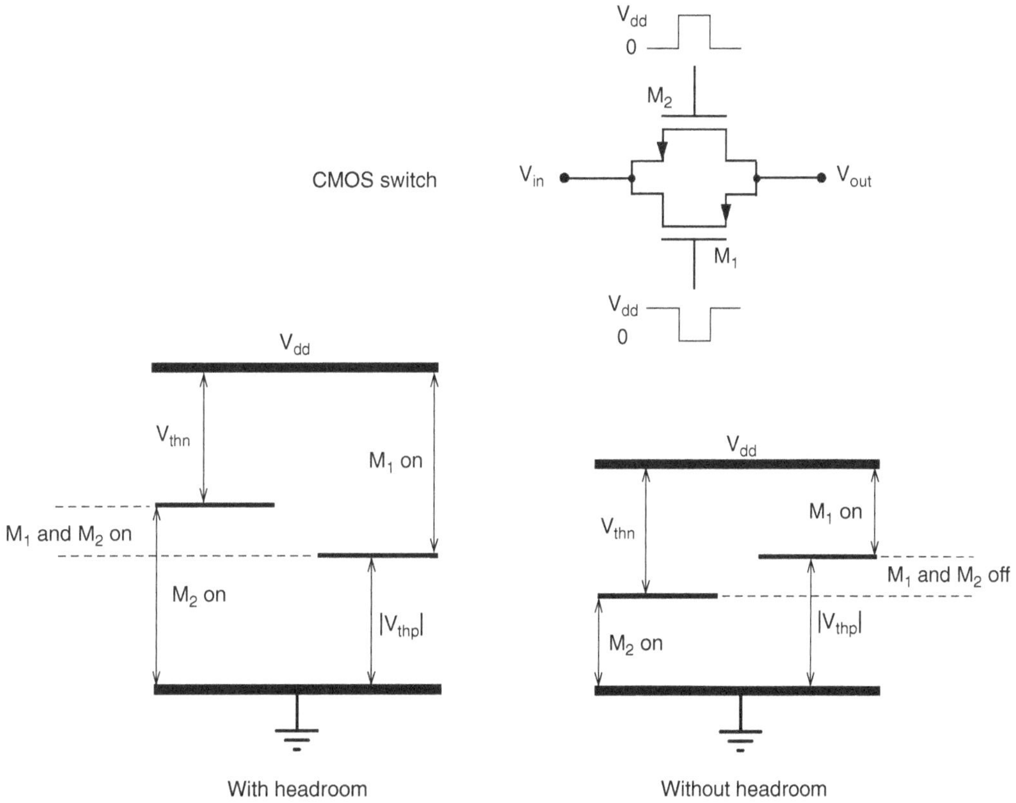

Figure 7.1 CMOS switch problem in low-voltage applications.

low-voltage SC circuits and may be best understood with the help of Figure 7.1. As the diagram shows, a CMOS transmission gate or CMOS switch is composed of a PMOS transistor (M_1) and an NMOS transistor (M_2). Two complementary clock signals (alternating between 0 V and V_{dd}) are used to drive the transistors. As shown in the lower-left part of the diagram, M_1 rather than M_2 is turned on by an input signal that has a magnitude between $|V_{thp}|$ and V_{dd}, while M_2 rather than M_1 is turned on by an input signal whose magnitude is between 0 V and $(V_{dd} - V_{thn})$. For an input between $|V_{thp}|$ and $(V_{dd} - V_{thn})$, both M_1 and M_2 are turned on. In low-voltage applications, it is preferred that the level of the input signal (V_{in}) reside at about halfway between 0 V and V_{dd}, to obtain a *rail-to-rail* output swing. To successfully pass V_{in} through to the output terminal, we need to make sure the following conditions are satisfied:

$$V_{dd} \geq V_{thn} + |V_{thp}| \tag{7.1}$$

The result of subtracting $(V_{thn} + |V_{thp}|)$ from V_{dd} is commonly referred to as the *headroom*. Apparently, the headroom decreases with V_{dd}. In addition, we have known that the switch on-resistance is inversely proportional to the value of ($V_{dd} - V_{in} - V_{thn}$). Thus, as V_{dd} is reduced, the switch on-resistance is increased; so is the magnitude of the signal-dependent error being injected to the output.

Once V_{dd} is reduced to below ($V_{thn} + |V_{thp}|$), no headroom remains and the two transistors can never be on simultaneously, as shown in the lower-right part of the diagram. In this situation, to maintain the input-to-output connection, we may choose either an NMOS switch or a PMOS switch, depending on the magnitude of the input signal. Specifically, if the input level is close to ground, then an NMOS transistor should be used to realize the switch. By contrast, if the input level is near V_{dd}, then a PMOS transistor should be used.

What's more, it can be found that the *maximum allowable input voltage range* (sometimes also called the *input dynamic range*) is limited between 0 V and ($V_{dd} - V_{thn}$) in the NMOS case or between $|V_{thp}|$ and V_{dd} in the PMOS case. In either case, the range is narrower as compared to the full voltage swing (i.e., from 0 V to V_{dd}) in the CMOS case.

In low-voltage SC circuits, particularly those with sub-1 V power supplies, a non-full-swing input dynamic range is often the *show-stopper.* For example, consider a standard 0.25 μm CMOS process with the following threshold voltage values: V_{thn} = 0.45 V and V_{thp} = −0.5 V. If it is assumed that V_{dd} is equal to 0.8 V and the input is biased at 0.4 V, then the switch (be it an NMOS, PMOS, or CMOS switch) will never be turned on because the input signal level has fallen into the "dead zone" between 0.35 V and 0.5 V.

Also, a narrower input dynamic range usually results in a lower *signal-to-noise-ratio* (SNR), because the desired signal power is reduced whereas the total noise power remains intact. It is known that the SNR performance of a basic SC circuit such as an integrator is limited primarily by the *kT/C noise* (or *sampling noise*). Therefore, to retain the SNR of a low-voltage SC integrator, we may use large sampling capacitors to suppress the *kT/C* noise power. However, the resultant large $R_{on}C$ time constant places a constraint on the maximum achievable speed of the circuit. Additionally, the increase in the total capacitor area results in a higher power dissipation.

The foregoing treatment represents a practical example of the multidimensional tradeoff optimization that involves supply voltage, accuracy, speed, and power. If an ideal *class-B* amplifier (or *push-pull* amplifier) is used to build the SC integrator, which consumes zero power when the input voltage is not changing (i.e., in the standby mode), then it can be shown that the average power dissipation is given by [6]

$$P = 4kT(DR)f_N \frac{\overline{V_{in}}}{V_{dd}} \quad (7.2)$$

where DR represents the dynamic range (i.e., accuracy), f_N is the Nyquist signal bandwidth (i.e., speed), and V_{dd} is the power supply voltage. Thus, in contrast to a digital CMOS IC whose power dissipation decreases in proportion to the square of V_{dd}, an analog CMOS IC may actually consume more power as the supply voltage is lowered.

Special CMOS process techniques such as the multiple-layer masking [7] can be used to reduce the threshold voltage of the MOS transistor. Also, the floating gate MOSFET technique [8] and the bulk-driven transistor technique [9] were proposed to alleviate the threshold voltage limitations. However, at the time of this writing, none of these implementations have been readily put into mass production using standard CMOS technologies, because they typically require extra fabrication steps, which lead to an increased process complexity and hence a higher cost. As the result, innovative low-voltage analog circuit design approaches that can take effect in standard CMOS devices are preferred.

Clock Boosting and Switch Bootstrapping

To ensure a rail-to-rail input/output connection without applying special process steps for reducing the threshold voltages, we may adopt the *clock boosting* (sometimes also called the *gate voltage boosting*) approach to increase the NMOS switch's gate-source voltage (use NMOS switch as an example). The basic idea is to double the clock voltage on the gate of the NMOS floating switch. The cross-coupled voltage doubler [10] that we discussed in Chapter 6 can be used here, and its SC implementation is shown in Figure 7.2 [11].

The circuit operates as follows. The cross-couple configuration consisting of two NMOS transistors, M_1 and M_2, allows the capacitors, C_1 and C_2, to be charged alternately by the power supply voltage V_{dd}. As shown in the lower part of the schematic, an input clock signal with a swing of V_{dd} is applied to C_1, and it is passed on to C_2 through an inverter. When the input clock signal is low, the voltage at the top plate of C_2 is boosted from V_{dd} to about $2V_{dd}$, and the PMOS transistor M_4 conducts a boosted voltage (about $2V_{dd}$) to the gate of the floating switch (in gray shade).

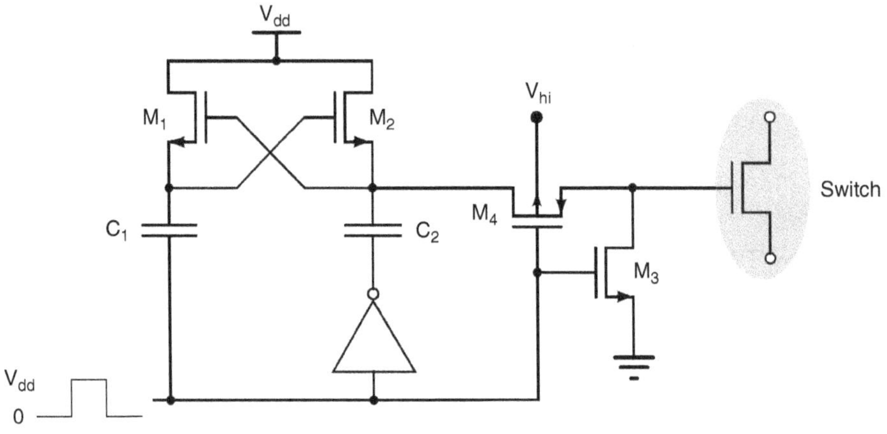

Figure 7.2 Clock boosting circuit.

When the input clock signal is high, the NMOS transistor M_3 is on, turning off the floating switch by pulling its gate voltage down to ground. To avoid latch-up, the N-well of the PMOS transistor M_4 needs to be tied to a high-voltage V_{hi}, which is typically equal to or greater than $2V_{dd}$.

The clock boosting circuit in Figure 7.2 was applied to a CMOS pipelined ADC reported by Cho and Gray [11]. However, generally speaking, the clock boosting technique is not suitable for deep-submicron CMOS technologies primarily due to the limitations of *gate-oxide breakdown*, *gate-induced drain leakage*, *hot-electron effect*, and *punch-through* [12][13]. The critical terminal voltages of the switch such as V_{gs}, V_{gd}, and V_{ds} should be kept below V_{dd} in a given technology to ensure the long-term device reliability. However, when a voltage doubler is used, the switch's gate voltage is always boosted to about $2V_{dd}$, regardless of the input signal; hence, V_{gs} cannot remain constant unless the input stays still. Moreover, V_{gs} may become as large as about $2V_{dd}$ in the presence of a small V_{in}, which is rather dangerous. Therefore, the clock boosting circuit is not the most reliable candidate to meet the needs of low-voltage deep-submicron CMOS devices.

To avoid the long-term reliability pitfalls, an alternative approach called the *bootstrapped switch* (perhaps because the switch can support itself and does not need a clock booster) was reported in [12]. The conceptual diagram of a bootstrapped switch is shown in Figure 7.3. The basic idea is that an auxiliary circuit is used to provide a constant V_{gs} (for all levels of V_{in}), which has a maximum value of V_{dd}, thereby significantly reducing the possibility of device failure.

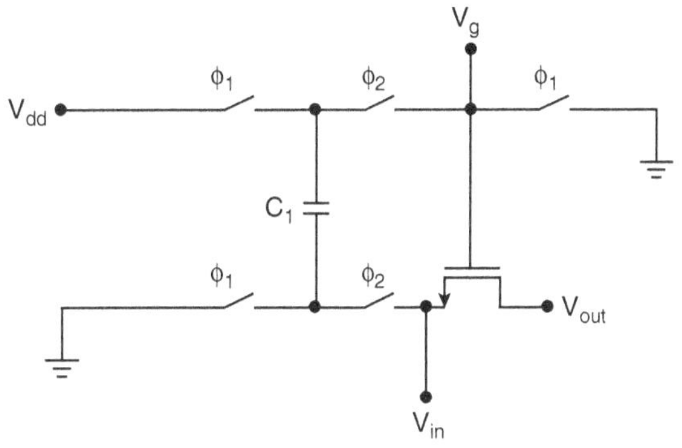

Figure 7.3 Conceptual diagram of a bootstrapped switch.

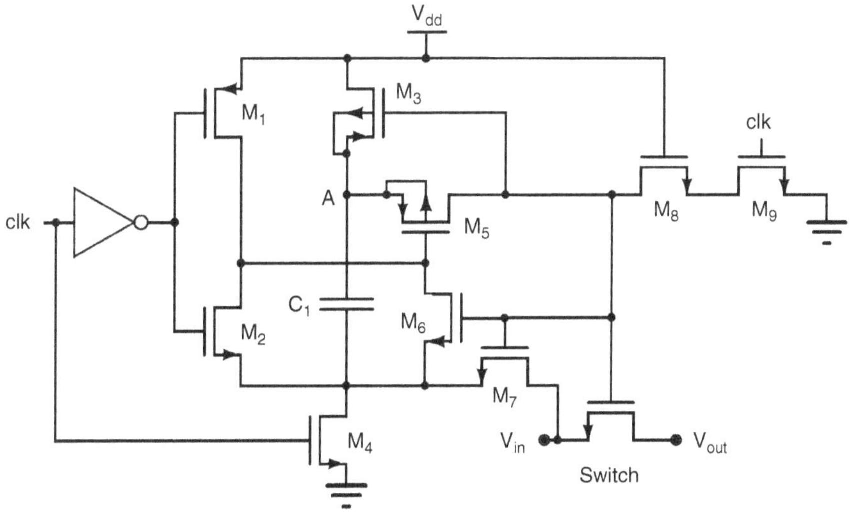

Figure 7.4 Transistor-level implementation of a bootstrapped switch.

The circuit operates as follows. When Φ_1 is on, the gate voltage (V_g) is connected to ground, so the switch is shut off. In the meanwhile, the capacitor C_1 is charged by V_{dd}. Next, Φ_2 is on and the voltage across C_1 builds up a step between the gate and source terminals, resulting in a V_g given by ($V_{in} + V_{dd}$). In effect, V_{gs} is always equal to V_{dd}, regardless of the input.

Several bootstrapped switch configurations can be found in the literature [12][14][15][16]. One of the transistor-level implementations is shown in Figure 7.4 [14].

The circuit operates as follows. When the clock signal *clk* goes high, the gate of the switch is discharged to ground through transistors M_8 and M_9. In the meanwhile, capacitor C_1 is charged by V_{dd} through transistors M_3 and M_4. The switch and C_1 are isolated from each other by transistors M_3 and M_5. When *clk* goes low, so does the gate voltage of the PMOS transistor M_5 (note that M_1 and M_2 form an inverter). At the same time, both M_6 and M_7 are turned on, allowing the gate voltage (V_g) to track the input signal V_{in} with an offset of V_{dd} (M_7 also protects V_g from the input loading). As a result, the charge stored on C_1 flows onto the gate of the switch through M_5, and V_g becomes the sum of V_{in} and V_{dd}.

Note that the bodies (or bulks) of M_3 and M_5 are connected to node A, which usually has the highest voltage in the circuit, in order to avoid latch-up [12][14]. M_6 helps M_2 pull down the gate voltage of M_6 when *clk* goes low, thereby increasing the circuit's speed. Also, it keeps the source-to-gate voltage (V_{sg}) of M_5 from exceeding V_{dd} for the sake of the long-time reliability. Finally, M_8 is useful for preventing the critical voltages (e.g., V_{gs} and V_{gd}) of M_9 from exceeding V_{dd} when *clk* goes high, and in practice the length of M_8 is often made long to avoid *punching-through* from M_5 to M_9 [13].

Thus, the circuit in Figure 7.4 provides a constant and stable V_{gs}, allowing the switch to conduct input signals within the full supply voltage range (i.e., between 0 V and V_{dd}). Nevertheless, the *effectiveness* of conduction, which is typically quantified by the switch's on-conductance (g_{ds}), drops as the input signal level increases, due to the *body effect*. As discussed in Chapter 6, the body effect tends to cause both the effective threshold voltage of the switch and the signal-dependent switch on-resistance (R_{on}) to increase.

To alleviate the body effect on a bootstrapped switch, we may permanently connect the body of the NMOS switch to its source. However, this arrangement is not applicable to some fabrication processes. Alternatively, we can replace the NMOS switch with a CMOS switch as shown in Figure 7.5.

As the schematic shows, M_3 and M_4 form the main switch, while M_1 and M_2 forms the auxiliary switch. When *clk* goes low, both M_1 and M_3 are shut off, and the body of M_3 is tied to the highest voltage in the circuit (i.e., V_{dd}) through the PMOS transistor M_5, in order to prevent latch-up. When the clock signal *clk* goes high, both the main and auxiliary switches are conducting, and the body of the PMOS transistor M_3 is connected to its source rather than to V_{dd}. As a result, its body-to-source voltage (V_{bs}) is constantly set to zero, and the body effect is thus removed. Also, its on-resistance is significantly lowered.

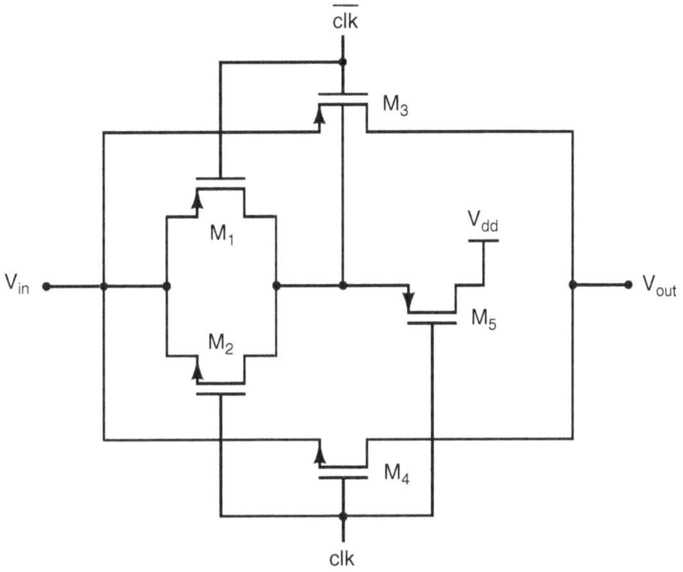

Figure 7.5 CMOS switch configuration for bootstrapping.

In [16], a current source is used to drive a replica switch with a constant on-conductance (g_{ds}), and an op-amp loop is used to force the source of the replica switch to track the input signal, thereby allowing the main switch to maintain a constant V_{gs}, and also, a fixed g_{ds}. As a result, the distortion caused by the input-dependent on-resistance (R_{on}) is greatly suppressed. However, due to the use of active devices such as the op-amp, this bootstrapping circuit dissipates more power as compared to the aforementioned passive configurations, given the same power supply voltage.

Switched Op-Amp

The essence of *switched-op-amp* (SOA) technique is to avoid the floating-switch problem by simply eliminating the switch itself [17]. To understand this, let us take a look at the circuit shown in Figure 7.6.

As we can see, the schematic shows a generic noninverting SC integrator (clock phases are not shown), followed by a second noninverting SC integrator, whose op-amp and integrating capacitor are not shown. It can be found that S_{f1} and S_{f2} are the floating switches of the first and the second integrator, respectively. When S_{f2} is eliminated (i.e., shorted), for the second integrator to retain its functionality, the output path of the op-amp (as shown) must be turned on/off during the sampling/integrating phase by the switch S_{o1} (and additional switches if any). Similarly, when S_{f1} is eliminated, the op-amp in the previous stage (not shown) should be able to be turned on and off alternately. In other words, the op-amp should be made *switchable*, hence the name *switched op-amp.* A simple switched-op-amp circuit is shown in Figure 7.7 [17].

As the schematic shows, this circuit is basically a classical differential-input, single-ended-output two-stage op-amp with two additional transistors, M_6 and S_{o1}.

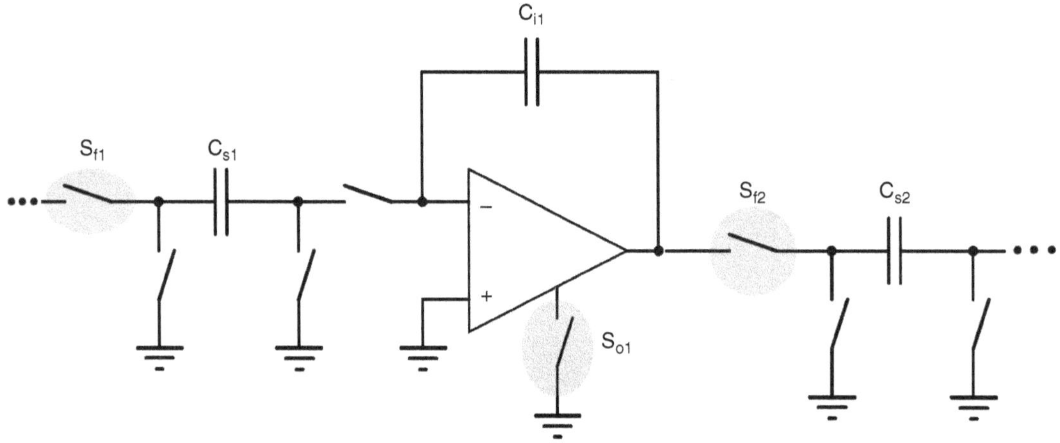

Figure 7.6 Conceptual diagram of switched-op-amp circuit.

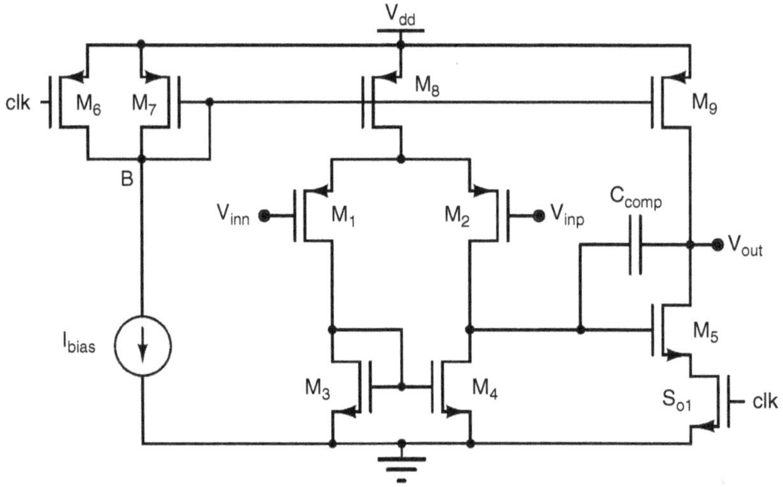

Figure 7.7 Simple switched op-amp.

The transistor S_{o1} is used to implement the switch of the op-amp shown in Figure 7.6. When *clk* goes low, S_{o1} cuts off the output current path consisting of M_5 and M_9, preventing the stored charge on the compensation capacitor C_{comp} from leaking away; thus, it will take a shorter time to recharge the op-amp once *clk* goes high again. Meanwhile, M_6 pulls the nodal voltage at node B up to about V_{dd}, shutting off all three current paths driven by M_7, M_8, and M_9. Also, it speeds up the operation of turning off the op-amp. The off op-amp provides high output impedance (seen from the subsequent stage), and its output node is typically connected to a fixed voltage reference (e.g., ground or virtual ground). When *clk* goes high, S_{o1} is on while M_6 is off, and the op-amp is turned on.

However, as intuition suggests, the speed of turning on/off an op-amp is not likely to be comparable to that of turning on/off a switch, which is typically constructed from only one or two MOSFETs. In practice, this intuition is proven to be correct. For a typical SOA-based analog signal processing (ASP) application, the op-amps shall be idle during one half the clock cycle (e.g., within the integrating phase). Thus, the overall system suffers from speed limitation due to the transient time required for powering up/down all of its op-amps. Moreover, due to the low-voltage and low-power constraints, the maximum achievable slew rate and gain-bandwidth product of each op-amp are further hampered, limiting the accuracy (i.e., dynamic range and linearity) of the overall SOA circuit. As a result, the design of high-speed and accurate SOA-type SC circuits remains a challenge.

An interesting technique called *unity-gain-reset-op-amp* [18] was reported to tackle the aforementioned issues. In the proposed configuration, the op-amp's output is fed back to its negative input terminal when the subsequent integrator is operating in the integrating phase. Thus, the op-amp is never completely turned off, which eliminates the settling time required in the SOA circuits. Nevertheless, this technique requires additional building blocks to avoid the potential forward-biased PN junction problem (i.e., latch-up) [18]. Finally, it might be worthwhile to make an experimental chip that incorporates this reset-op-amp technique in a different CMOS process such as the silicon-on-insulator (SOI) process, which is immune to latch-up problems.

At the time of this writing, the SOA technique has been mostly adopted to meet the needs of applications that require modest frequency-accuracy products, including SC filters [19][20], delta-sigma ($\Delta\Sigma$) modulators [2][3][21], and pipelined ADCs [4].

7.3 Accuracy-Enhancement Techniques for SC Circuits

The Imperfect Op-Amp

In active analog circuits, the op-amp is perhaps the most important building block. For example, in an SC integrator, the primary use of an op-amp is to create a perfect virtual ground (i.e., a node with a very high impedance and a constant potential) at its negative input terminal, ensuring a *lossless* charge transfer at all times. In other words, ideally no electric charge is absorbed by the op-amp through the virtual ground. This feature is very important to the active SC circuit that processes signals in the charge domain. However, in reality, a lossless charge transfer around the op-amp is not possible mainly due to op-amp imperfections including the *dc offset voltage*, *finite op-amp gain*, and *finite op-amp bandwidth*.

A common effect of these imperfections is that the voltage magnitude at the op-amp's negative input terminal is deviated from its desired value (i.e., 0 V), meaning that the virtual ground is degraded. In comparison with the dc offset, the finite op-amp gain and the finite op-amp bandwidth alter the virtual ground in a more complex fashion. Specifically, for a closed-loop op-amp with a finite gain A_0, the absolute magnitude error at its negative input is approximated as $(-V_{out}/A_0)$. It can be shown that the actual transfer function of a standard noninverting and delaying SC integrator (see Chapter 3) in the presence of a finite-gain (A_0) op-amp is given by [22][23]

$$H(\omega) = \frac{\left(\frac{C_s}{C_i}\right)e^{-j(\omega T)/2} \Big/ j2\sin(\omega T/2)}{\left[1+\frac{1}{A_0}\left(1+\frac{C_s}{2C_i}\right)-j\frac{(C_s/C_i)}{2A_0\tan\left(\frac{\omega T}{2}\right)}\right]} \tag{7.3}$$

where C_s is the sampling capacitor, C_i is the integrating capacitor, and T is the sampling clock period. The expression in the numerator is the ideal transfer function (i.e., with infinite op-amp gain).

If it is further assumed that this op-amp has a finite bandwidth of f_t, then the op-amp gain is given by

$$A_0(\omega) \cong \frac{2\pi \cdot f_t}{j\omega} \tag{7.4}$$

Substituting A_0 in Equation (7.3) with the previous expression, we obtain the integrator's realistic transfer function (approximated to the first order):

$$H(\omega) \cong \frac{\left(\dfrac{C_s}{C_i}\right) e^{-j(\omega T/2)}}{j2\sin(\omega T/2)} \cdot \left[1 - e^{-k} \cdot \frac{C_s}{C_s + C_i}\right] \tag{7.5}$$

where k is given by

$$k = \pi \cdot f_t T \cdot \frac{C_i}{C_s + C_i} \tag{7.6}$$

The interested reader is referred to the references [22][23] for the rigorous proof of the foregoing transfer functions.

Moreover, low-frequency *flicker noise* (also called *1/f noise*) and *thermal noise* further alter this voltage. In effect, the input-referred offset voltage (V_{off}) of an op-amp in a CMOS technology typically ranges from 5 to 20 mV [24], which becomes more pronounced in low-voltage applications, where the inherent signal swing is reduced.

In addition to the lowered supply voltage, the continual shrinking of device dimensions in deep-submicron CMOS technologies, which has caused a significant reduction in the intrinsic dc gain of a MOSFET (usually lower than 20 dB), degrades the effectiveness of conventional approaches (e.g., cascoding) to achieve high op-amp dc gains. Consequently, the effect of finite op-amp gain becomes even more significant.

Autozeroing

The basic idea of the *autozeroing* technique is to store the low-frequency random noise (e.g., flicker noise) and the dc offset voltage using one or more capacitors and then subtract them from the signal at either the input or output of the op-amp [25]. Therefore, the autozeroing process requires at least two clock phases: a sampling phase and a cancellation (or compensation) phase. During the sampling phase, the dc offset and the flicker noise are sampled and stored on the capacitor(s), while during the cancellation phase these stored errors are subtracted from the signal.

Razavi and Wooley [25] reported one of the simpler autozeroing methods used to reduce the effect of the op-amp dc offset in an SC comparator. In the proposed scheme, two different comparator configurations are provided. One is called the *input offset storage* (or the *closed-loop autozeroing*), and its basic configuration is shown in Figure 7.8(a). The other is called the *output offset storage* (or the *open-loop autozeroing*), and its basic configuration is shown in Figure 7.8(b).

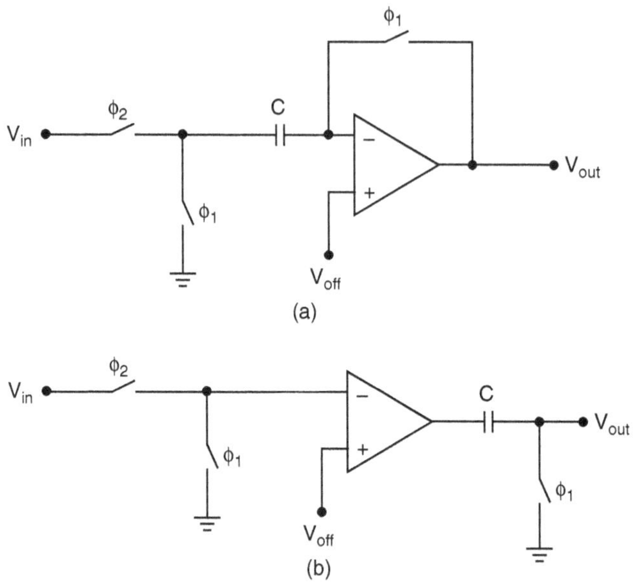

Figure 7.8 Razavi's autozeroing techniques. (a) Input offset storage. (b) Output offset storage.

The previous chapters explored several circuits such as S&Hs and data converters that adopt the input offset storage method. Here, the description of its operation is repeated. When Φ_2 is on, ideally the input capacitor C is charged to $(V_{in} - V_{off})$. Next, Φ_1 is on, and the output voltage is thus given by $V_{in} - V_{off} + V_{off} = V_{in}$.

Note that the accuracy of this offset cancellation is determined by the open-loop dc gain of the op-amp (A_0). Specifically, in the circuit shown in Figure 7.8(a), when Φ_1 is on, the op-amp is included in a unity-gain feedback loop, thus the voltage magnitude at its negative input terminal is given by

$$(V_{off} - V_n)A_0 = V_n \Rightarrow V_n = \frac{A_0}{A_0+1}V_{off} \tag{7.7}$$

Next, Φ_2 is on, and C is charged to $(V_{in} - V_n)$. When Φ_1 is on again, the actual output voltage is given by

$$V_{out} = V_{in} - V_n + V_{off} + \varepsilon = V_{in} + \left(\frac{1}{A_0+1}V_{off} + \varepsilon\right) \tag{7.8}$$

The expression in parentheses is called the *residual offset error.* ε is the additional error voltage introduced by the charge injection when the Φ_2 switch is turned off, and it can be effectively reduced by using a fully differential configuration. As the preceding equation shows, the residual offset error increases as A_0 decreases.

In the output offset storage configuration shown in Figure 7.8(b), an open-loop op-amp is employed as a buffer. When Φ_1 is on, the offset voltage is amplified, and the result is sampled by the output capacitor C. Next, Φ_2 is on, $(V_{in} - V_{off})$ is amplified, and the result is sampled by C. So ideally the output voltage on the left plate of C is also given by $V_{in} - V_{off} + V_{off} = V_{in}$.

As compared to its input offset storage counterpart, the output offset storage configuration is typically faster due to the use of an open-loop op-amp. However, the value of A_0 has to be rather small (usually less than 10 V/V), otherwise the op-amp will be easily saturated by the offset voltage.

Generally speaking, the input offset storage configuration is a better choice when accuracy is the fundamental aspect (e.g., in precision amplifiers, integrating ADCs, or MDACs), whereas the output offset storage configuration is more appropriate for applications that require high speed and low power dissipation (e.g., in comparators or ring oscillators). In addition, both configurations have the same added benefit of reducing the flicker noise and thermal noise that reside in the low-frequency signal band.

A survey of the literature will show a few examples of basic SC building blocks such as the integrator [26] and the amplifier [26][27] that use the autozeroing technique to reduce the op-amp's offset and low-frequency noise.

Correlated Double-Sampling

Although the autozeroing technique is very effective in reducing the effect of the dc offset and the flicker noise, it does not alleviate the SC circuit's dependence on the op-amp dc gain (A_0), as reflected by Equation (7.8). This finite op-amp gain issue is especially problematic in low-voltage applications that require a high accuracy.

The *correlated double-sampling* (CDS) technique can be considered a generalization of the autozeroing technique. In a typical CDS configuration, after the first sampling acquires the amplifier's offset and noise, a second sampling is carried out during the compensation phase to acquire the *instantaneous* value of the magnitude error at the amplifier's negative input terminal. In the aforementioned autozeroing configuration, this magnitude error is assumed to be a dc signal (i.e., constant), which is not applicable to the situation where the effect of finite op-amp gain needs to be taken into account. The CDS scheme requires two sampling operations in each clock cycle. In effect, the correlation properties of adjacent signal samples are exploited and utilized to desensitize the circuit's accuracy from the amplifier's dc gain [28].

The CDS technique can be roughly categorized into three groups: *offset-compensating* CDS configurations where only the dc offset and in-band random noise are eliminated (i.e., autozeroing), *gain and offset-compensating* CDS configurations where the effect of finite op-amp gain is reduced in addition to offset and noise, and *predictive* CDS configurations with gain and offset compensation where prediction is incorporated into each CDS operation to provide a preliminary approximation of the finite gain error that will occur during the next clock interval.

A gain and offset-compensated SC amplifier is shown in Figure 7.9 [29]. Although a single-ended configuration is shown here for simplicity, in practice the fully differential version [30] is usually adopted to minimize the charge injection errors. Here, we assume that the op-amp has a finite gain of A_0 and an input dc offset of V_{off}.

By inspection, we recall that this circuit was used in the DAC shown in Figure 5.3(b). As mentioned before, a small deglitching capacitor is often connected between the output (V_{out}) and the left-hand side of C_1, which creates a feedback path to reduce spikes during the intervals when no clock phase is on.

The circuit operates as follows. When Φ_1 is on, C_1 samples the voltage difference between the circuit's input (V_{in}) and the op-amp's negative input terminal (V_n), while C_2 is charged to the instantaneous signal voltage value at V_n, and C_3 is the feedback capacitor. Next, Φ_2 is on and the charge stored on C_1 is transferred onto C_2, whereas

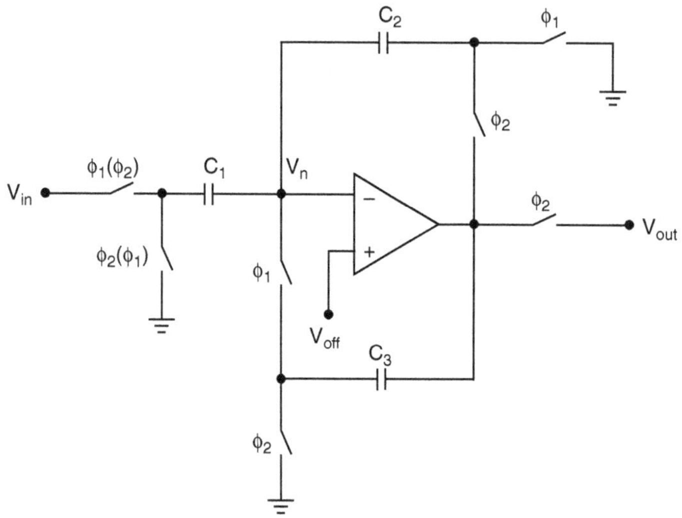

Figure 7.9 Offset and gain-compensated SC amplifier.

C_3 samples the instantaneous value of V_{out} and holds it till Φ_1 is turned on again, which is equivalent to placing an S&H after the amplifier's output.

Although mathematically interesting, the derivation of the circuit's transfer function based on charge equations is rather tedious and only the result [29][30] is presented here:

$$H(z) = \frac{V_{out}(z)}{V_{in}(z)} = \frac{C_1}{C_2} \frac{az^{-1/2}}{1 - bz^{-1}} \tag{7.9}$$

where a and b are respectively given by

$$\begin{cases} a = \dfrac{1}{1 + \dfrac{C_1 + C_2}{C_2 A_0}} \cdot \left(1 - \dfrac{C_1 + C_2}{A_0 C_3 + C_1 + C_2 + C_3}\right), \\ b = \dfrac{\dfrac{C_1 + C_2}{A_0} \cdot \left(C_2 + C_3 + \dfrac{C_1 C_2}{A_0}\right)}{\left(C_2 + \dfrac{C_1 + C_2}{A_0}\right) \cdot \left(C_3 + \dfrac{C_1 + C_2 + C_3}{A_0}\right)} \end{cases} \tag{7.10}$$

As the preceding indicates, this circuit is a noninverting and integrating SC amplifier with a delay of one half clock cycle. Note that the op-amp dc offset voltage (V_{off}) has no effect on the preceding transfer function.

In low-frequency applications such as audio, z is approximately equal to unity, and the previous transfer function can be simplified to

$$H(1) = \frac{V_{out}}{V_{in}} = \frac{C_1}{C_2} \frac{a}{1 - b} = \frac{C_1}{C_2}\left(1 - \frac{C_1 + C_2}{C_2 A_0^2}\right) \tag{7.11}$$

The second term in parentheses is the normalized gain error caused by the finite op-amp gain. In comparison with Equation (7.3), we find that the magnitude of the gain error is inversely proportional to the square of A_0 rather than to A_0. Therefore, the SC amplifier's dependence on op-amp gain is significantly reduced.

The inverting and nondelaying version of this SC amplifier can be realized by simply operating the two switches near V_{in} using the clock phases in parentheses (Figure 7.9). As a result, there is a sign inversion from V_{in} to V_{out}. Also, when Φ_2 is on, there is no specified delay between the input and output samples; hence, C_3 is no longer able to fulfill the role as an output S&H. In such a case, the amplifier's output needs to be followed by a dedicated S&H if the subsequent device requires a steady input signal (e.g., an ADC). It can be shown that the low-frequency input-output relation of the inverting amplifier is given by [29]

$$H(1) = -\frac{C_1}{C_2} \cdot \left(1 - \frac{C_1 + C_2}{C_2 A_0^2 + C_1 + C_2}\right) \tag{7.12}$$

Interestingly, the normalized gain error is slightly different from that given by Equation (7.11). Unfortunately, the circuit in Figure 7.9 has a major drawback in that it is not suitable for high-speed/high-frequency applications. This is mainly due to the time taken for the op-amp to catch up with the voltage variations at its negative input terminal (V_n) in the presence of a fast-varying input signal. In fact, all the CDS techniques that we have discussed thus far are referred to as the *narrowband-CDS* techniques since they are effective in suppressing op-amp related errors in the low-frequency range, rather than in the medium or high-frequency range. It has been reported that the finite op-amp bandwidth is not a big concern to SC filters that require low-to-medium quality factors [23]; however, its effect weighs in for higher-frequency devices such as video ADCs.

To overcome this limitation of speed without losing the benefit of accuracy enhancement provided by the CDS configuration, additional capacitors or building blocks can be used to *predict* and *save* the potential error introduced by op-amp imperfections (e.g., offset, finite gain, and finite bandwidth). And then the predicted error voltage can be eliminated by the CDS switching of capacitors. A few examples of basic SC circuits such as amplifiers, unity-gain buffers, and integrators that make use of *predictive-CDS* (or *wideband-CDS*) techniques can be found in the literature [31][32][33].

Figure 7.10 shows an inverting SC amplifier that incorporates the predictive-CDS technique [32]. It shows that there are a total of 12 switches and five capacitors in the circuitry. The main capacitors, C_1 and C_2, are in the amplification path, meaning that they are responsible for amplifying the input signal. The auxiliary capacitors, C_4 and C_5, are in the prediction path, meaning that they are used to predict the error introduced by op-amp imperfections. To make the prediction as accuracte as possible, C_4 and C_5 are normally chosen such that $C_4/C_5 = C_1/C_2$. The storage capacitor, C_3, is used to save the predicted error, and its capacitance value is not important [32] but is usually kept small.

The operation of the circuit is as follows. When Φ_2 is on, the prediction path performs a preliminary amplification, and the resultant uncompensated output signal generates an error voltage at the op-amp's negative input terminal, which is given by $(-V_{out}/A_0 + V_{off})$. In the meantime, this error voltage is sampled by the storage capacitor C_3. Next, Φ_1 is on, the amplification path consisting of C_1 and C_2 performs a main amplification to generate a valid output voltage using the left plate of C_3 as the *error-compensated* virtual ground. As the name indicates, the present error voltage at the original virtual ground (i.e., the op-amp's negative input) is compensated by the predicted one that is saved across C_3.

Larson and Temes [31] reported an alternative to the preceding configuration, and its improved version is shown in Figure 7.11 [33]. This is also an inverting amplifier that uses the predictive-CDS technique as the foregoing, but it requires only eight

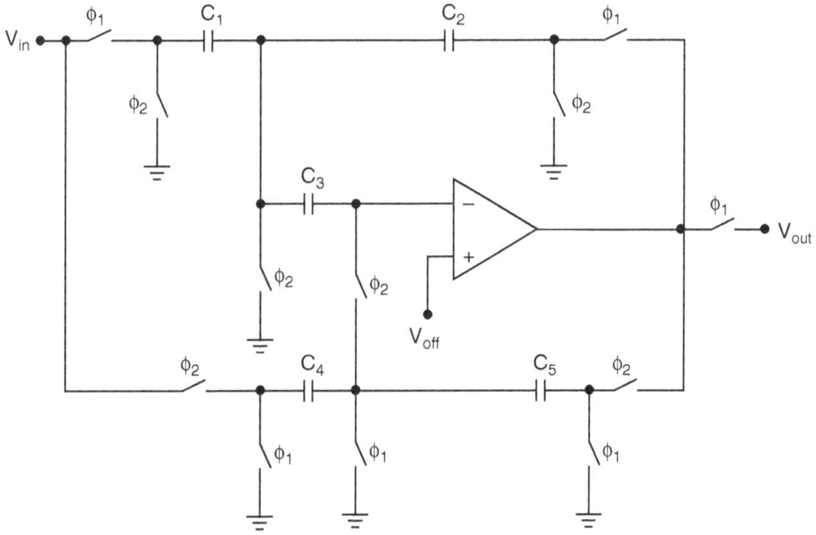

Figure 7.10 SC amplifier using the predictive-CDS technique.

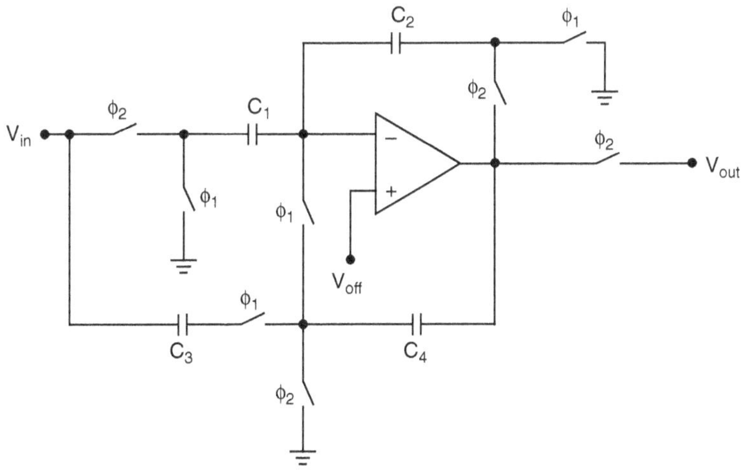

Figure 7.11 Predictive-CDS compensated SC amplifier using four capacitors.

switches and four capacitors besides the op-amp. As before, C_1 and C_2 are the main capacitors, while C_3 and C_4 are the auxiliary capacitors, and $C_3/C_4 = C_1/C_2$.

This circuit can be considered the result of modifying the previous circuit (Figure 7.10). The modification involves two changes. One is to merge the storage capacitor C_3 into C_1, and the other is to simplify the switching configuration in the prediction path. In addition to the area and power saved through the switch sharing and the capacitor reduction, these changes give rise to a more aggressive compensation (especially in the low frequency range) as compared to the previous configuration in Figure 7.10. The high effectiveness of compensation in the low frequency domain may be attributed to the similarity between this circuit and the SC amplifier shown in Figure 7.9, which provides perhaps the most aggressive narrowband gain error compensation.

In summary, assume that the op-amp used in an SC amplifier has a finite dc gain of A_0 and that the realistic gain of the SC amplifier is expressed as:

$$G_{real} = G_{ideal} \cdot (1 - E) \tag{7.13}$$

where E is the normalized gain error. Under the condition of $z = 1$, for an autozeroed SC amplifier that adopts no gain error compensation, the gain error is given by

$$E \cong \frac{1 + |G_{ideal}|}{A_0} \tag{7.14}$$

For the same SC amplifier that incorporates the narrowband-CDS technique (refer to Figure 7.9), the gain error is given by

$$E \cong \frac{1 + |G_{ideal}|}{A_0^2} \tag{7.15}$$

If the SC amplifier uses the wideband-CDS technique (refer to Figures 7.10 and 7.11) to reduce the sensitivity to op-amp imperfections, then the resulting gain error is given by

$$E \cong \frac{(1 + |G_{ideal}|)^2}{A_0^2} \tag{7.16}$$

References

[1] Semiconductor Industry Association, "International technology roadmap for semiconductors: 2002 update," [Online]. Available at www.sematech.org.

[2] A. Baschirotto and R. Castello, "A 1-V CMOS fully differential switched-opamp bandpass sigma–delta modulator," *Proceedings of European Solid-State Circuits Conference (ESSCIRC)*, Vol. 1, pp. 152–155, June 1997.

[3] V. Peluso, P. Vancorenland, A. M. Marques, M. Steyaert, and W. C. Sansen, "A 900-mV low-power ΔΣ A/D converter with 77-dB dynamic range," *IEEE Journal of Solid-State Circuits*, Vol. 33, No. 12, pp. 1887–1897, December 1998.

[4] M. Waltari and K. Halonen, "1-V 9-Bit pipelined switched-opamp ADC," *IEEE Journal of Solid-State Circuits*, Vol. 36, No. 1, pp. 129–134, January 2001.

[5] L. Yao, M. Steyaert, and W. Sansen, "A 0.8-V, 8-μW, CMOS OTA with 50-dB gain and 1.2-MHz GBW in 18-pF load," *Proceedings of European Solid-State Circuits Conference (ESSCIRC)*, Vol. 2, pp. 297–300, June 2003.

[6] S. Rabii and B. A. Wooley, *The design of low-voltage, low-power sigma-delta modulators*, Kluwer, Norwood, MA, 1999.

[7] T. Atachi et al., "A 1.4 V switched capacitor filter," *Proceedings of IEEE Custom Integrated Circuits Conf*erence, Vol. 8, pp. 821–824, May 1990.

[8] J. Ramirez-Angulo, S. Choi, and G. Altamirano, "Low voltage circuits building blocks using multiple input floating gate transistors," *IEEE Trans. on Circuits and Systems—I*, Vol. 42, pp. 971–974, November 1995.

[9] B. Blalock, P. Allen, and G. Rincon-Mora, "Designing 1-V op amps using standard digital CMOS technology," *IEEE Trans. on Circuits and Systems—II*, Vol. 45, pp. 769–780, July 1998.

[10] Y. Nakagome et al., "An experimental 1.5-V 64 Mb DRAM," *IEEE Journal of Solid-State Circuits*, Vol. 26, No. 4, pp. 465–472, April 1991.

[11] T. Cho and P. R. Gray, "A 10 b, 20 Msamples/s, 35 mW pipeline A/D converter," *IEEE Journal of Solid-State Circuit*s, Vol. 30, No. 3, pp. 166–172, March 1995.

[12] A. Abo and P. R. Gray, "A 1.5-V 10-bit 14.3-MS/s CMOS pipeline analog-to-digital converter," *IEEE Journal of Solid-State Circuits*, Vol. 34, No. 5, pp. 599–606, May 1999.

[13] C. Hu, "IC reliability simulation," *IEEE Journal of Solid-State Circuits*, Vol. 27, No. 3, pp. 241–246, March 1992.

[14] M. Dessouky and A. Kaiser, "Very low-voltage digital-audio ΔΣ modulator with 88-dB dynamic range using local switch bootstrapping," *IEEE Journal of Solid-State Circuits*, Vol. 36, No. 3, pp. 349–355, March 2001.

[15] J. Steensgaard, "Bootstrapped low-voltage analog switches," *Proceedings of IEEE International Symposium on Circuits and Systems*, Vol. 2, pp. 29–32, May 1999.

[16] H. Pan et al., "A 3.3-V 12-b 50-MS/s A/D converter in 0.6-μm CMOS with over 80-dB SFDR," *IEEE Journal of Solid-State Circuits*, Vol. 35, pp. 1769–1780, December 2000.

[17] J. Crols and M. Steyaert, "Switched-opamp: An approach to realize full CMOS switched-capacitor circuits at very low power supply voltages," *IEEE Journal of Solid-State Circuits*, Vol. 29, No. 8, pp. 936–924, August 1994.

[18] E. Bidari, M. Keskin, F. Maloberti, U. Moon, J. Steensgaard, and G. C. Temes, "Low-voltage switched-capacitor circuits," *Proceedings of IEEE International Symposium on Circuits and Systems*, Vol. V, pp. 445–448, May 2000.

[19] A. Baschirotto and R. Castello, "A 1-V 1.8-MHz CMOS switched-opamp SC filter with rail-to-rail output swing," *IEEE Journal of Solid-State Circuits*, Vol. 32, pp. 1979–1986, December 1997.

[20] V. S.-L. Cheung, H. C. Luong, and W.-H. Ki, "A 1-V CMOS switched-opamp switched-capacitor pseudo-2-path filter," *IEEE Journal of Solid-State Circuits*, Vol. 36, pp. 14–22, January 2001.

[21] V. S.-L. Cheung, H. C. Luong, and W.-H. Ki, "A 1-V 10.7-MHz switched-opamp bandpass ΣΔ modulator using double-sampling finite-gain-compensation techniques," *IEEE Journal of Solid-State Circuits*, Vol. 37, pp. 1215–1225, October 2002.

[22] G. C. Temes, "Finite amplifier gain and bandwidth effect in switched-capacitor filters," *IEEE Journal of Solid-State Circuits*, Vol. SC-15, pp. 358–361, June 1980.

[23] K. Martin and A. S. Sedra, "Effects of the op amp finite gain and bandwidth on the performance of switched-capacitor filters," *IEEE Trans. Circuits and Systems*, Vol. CAS-28, pp. 822–829, August 1981.

[24] G. C. Temes, "Autozeroing and correlated double sampling techniques," Research Seminar, Oregon State University, Corvallis, OR, May 2002.

[25] B. Razavi and B. A. Wooley, "Design techniques for high-speed, high-resolution comparators," *IEEE Journal of Solid-State Circuits*, Vol. 27, pp. 1916–1926, December 1992.

[26] F. Krummenacher, "Micropower switched capacitor biquadratic cell," *IEEE Journal of Solid-State Circuits*, Vol. SC-17, No. 3, pp. 507–512, June 1982.

[27] R. Gregorian, "High-resolution switched-capacitor D/A converter," *Microelectronic Journal*, No. 12, pp. 10–13, 1981.

[28] C. C. Enz and G. C. Temes, "Circuit techniques for reducing the effects of op-amp imperfections: Autozeroing, correlated double sampling, and chopper stabilization," *Proceedings of the IEEE*, Vol. 84, pp. 1584–1614, November 1996.

[29] K. Haug, G. C. Temes, and K. Martin, "Improved offset-compensation schemes for SC circuits," *Proceedings of IEEE International Symposium on Circuits and Systems*, Vol. 3, pp. 1054–1057, Montreal, Canada, May 1984.

[30] K. Martin et al., "A differential switched-capacitor amplifier," *IEEE Journal of Solid-State Circuits*, Vol. SC-22, No. 1, pp. 104–106, February 1987.

[31] L. E. Larson and G. C. Temes, "SC building blocks with reduced sensitivity to finite amplifier gain, bandwidth and offset voltage," *Proceedings of IEEE International Symposium on Circuits and Systems*, Vol. 1, pp. 334–338, Philadelphia, PA, May 1987.

[32] K. Nagaraj et al., "Switched-capacitor circuits with reduced sensitivity to amplifier gain," *IEEE Trans. Circuits and Systems*, Vol. CAS-34, No, 5, pp. 571–574, May 1987.

[33] H. Yoshizawa, Y. Huang, and G. C. Temes, "Improved SC amplifiers with low sensitivity to op-amp imperfections," *Electron. Letters*, Vol. 33, No. 5, pp. 348–349, February 1997.

CHAPTER 8

Design of SC Delta-Sigma Modulators for Multistandard RF Receivers

8.1 Introduction

Since the mid-1990s, the fast-growing mobile communications market has dramatically increased the number of subscribers to *second-generation* (2G) digital cellular and cordless telephony systems such as the *Global System for Mobile Communications* (GSM) and the *Digital Enhanced Cordless Telecommunications* (DECT), which were originally specified to provide an increased system capacity in comparison with the first-generation analog systems such as the *Advanced Mobile Phone System* (AMPS). Since 1999, the enormous demand for short messaging and mobile multimedia services has been the driving force for revising the existing wireless communications standards and infrastructures. The envisioned outcome is a universal system capable of supporting the *third-generation* (3G) or pre-3G standards, which include the *Enhanced Date rates for GSM Evolution* (EDGE), the *Wideband Code Division Multiple Access* (WCDMA), the *Code Division Multiple Access-2000* (CDMA-2000), and the newest member of the 3G family: *Time Division Synchronous Code Division Multiple Access* (TD-SCDMA). Figure 8.1 sketches the mobile cellular technology roadmap [1].

However, a complete transition from 2G to 3G in a short period of time is not yet feasible, considering the vast volume of existing 2G services and the infrastructure, time, and capital expense needed to achieve competent quality and popularity for a ubiquitous 3G wireless system. To make use of both 2G and 3G standards and services during this transition period, quite a few research efforts have been made to create wireless transceivers that can provide multiband and multistandard performance capabilities [2][3][4][5][6][7].

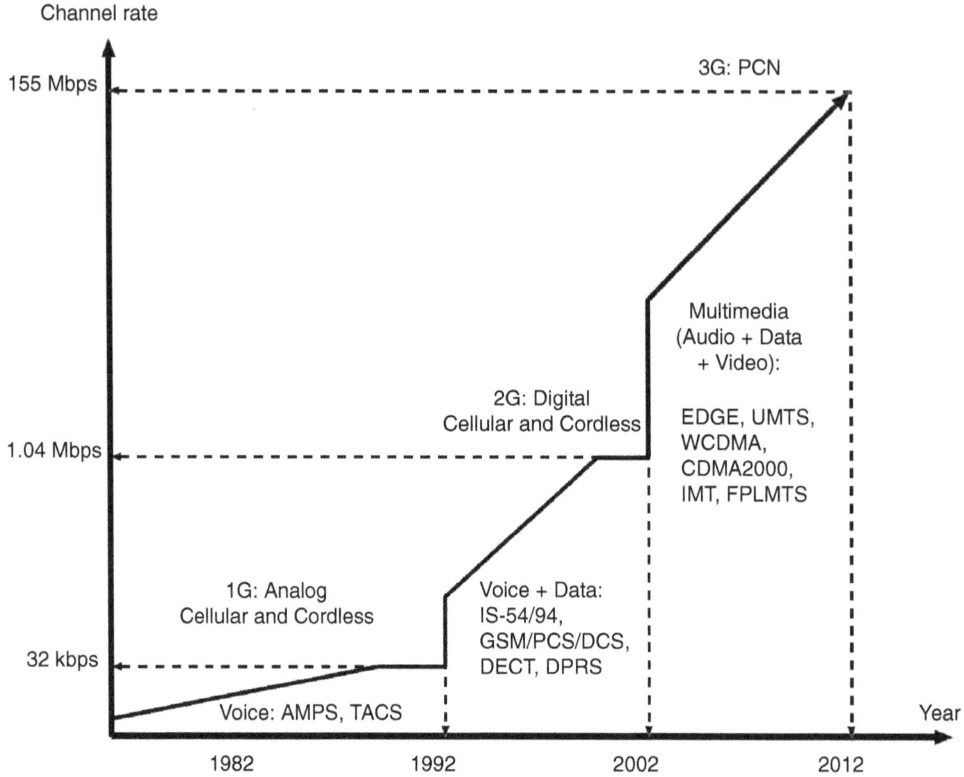

Figure 8.1 Mobile cellular technology road map.

The Multistandard Challenge

Accommodating multiple standards in a monolithic radio-frequency (RF) transceiver tends to significantly increase the circuit complexity in both the RF and baseband portions, resulting in a lower integration and a higher power consumption. The transition from single-standard to multistandard is nontrivial. It requires a complete reconsideration of the RF front-end and baseband circuitries. At the time of this writing, the successful implementation of a high-quility multistandard RF IC is widely regarded as one of the most challenging design tasks.

One notable challenge lies in the design of low-powered and small-sized analog-to-digital converters (ADCs) capable of digitizing the desired small signal in the presence of multiple blockers and carrier frequency components, which have a large signal power. To reduce cost and size, in most modern monolithic RF receivers the conventional high-Q discrete surface acoustic wave (SAW) filters are often substituted with low-Q integrated antialiasing filters. Consequently, the neighboring strong blockers are not attenuated to the satisfaction of the conventional filtering standards, and the desired small signal is often submerged by these insufficiently attenuated blockers. Therefore, the ADC must have a high dynamic range, otherwise the signal will hardly be detected.

Among the plentiful ADC architectures that we have seen so far, pipelined and oversampling delta-sigma (ΔΣ) ADCs are the two best candidates for the analog-to-digital interfaces in a wireless receiver. It is well known that pipelined converters are suitable for applications requiring high conversion rates and medium-to-high resolutions. However, as mentioned in Chapter 5, pipelined converters (among other Nyquist-rate converters) are sensitive to analog component mismatches. As the result, sophisticated calibration/correction circuitries are often required if resolutions higher than 12 bits are desired, which means the circuit complexity as well as power consumption will increase significantly. Although a few 15-bit pipelined ADCs with wideband capabilities have been reported [8][9], they are still inadequate to satisfy the stringent power consumption and size requirements of most modern RF receivers.

In comparison with pipelined ADCs, oversampling ΔΣ ADCs are much less sensitive to analog component mismatch as long as the oversampling-ratio (OSR) is sufficiently high, because it trades more digital signal processing for a relaxed analog circuitry. Hence, given the same resolution and speed, its analog portion typically consumes less power than that of a pipelined ADC.

Additionally, in a ΔΣ ADC, the neighboring blockers and quantization noise are shaped out-of-band in the same fashion, and the decimation filter following the ΔΣ modulator can be used in combination with the digital selective filter and the baseband mixer to attenuate the quantization noise as well as adjacent radio blockers. Furthermore, by choosing different clock sampling rates (i.e., different OSRs), the same ΔΣ ADC architecture can adapt to different signal bandwidth, dynamic range, signal-to-noise ratio (SNR), and linearity requirements that are specified by different wireless standards [10].

ΔΣ Modulator for Multistandard RF Receiver

This chapter presents the top-down design of a pair of quadrature complex third-order ΔΣ modulators for a low-power multistandard RF transceiver [1]. The design of the RF transceiver targets the performance suitable to support three widely adopted wireless communications standards: *Global System for Mobile Communications* (GSM), *Wideband Code Division Multiple Access* (WCDMA) and *Digital Enhanced Cordless Telecommunications* (DECT). As we will see later, the reception

Table 8.1 Summary of each modulator's performance.

Wireless standards	GSM/WCDMA/DECT
Receiver architecture	Digital high-IF
Sampling scheme	IF-sampling
OSR	192 (GSM), 24 (WCDMA) and 64 (DECT)
Dynamic range (dB)	87 (GSM), 69 (WCDMA) and 74 (DECT)
SNDR (dB)	80 (GSM), 55 (WCDMA) and 65 (DECT)
SNR (dB)	83 (GSM), 56 (WCDMA) and 67 (DECT)
IP3 (dBV_{rms})	–27.5(GSM), –19.4 (WCDMA) and –16.8 (DECT)
Total capacitance (modulator)	6.30 pF
Capacitance spread	24 : 1
Reference voltage	1.25 V
Supply voltage	2.5 V
Technology	0.35-μm CMOS Double-Poly

part of the RF transceiver (i.e., the receiver) is implemented based on the high-IF super-heterodyne (or digital high-IF) architecture [11].

In this design, two identical low-pass $\Delta\Sigma$ modulators are used to fulfill the requirements placed by the RF receiver. Table 8.1 summarizes the measurement results of each $\Delta\Sigma$ modulator.

Chapter Outline

This chapter is organized as follows. Section 8.2 provides an overview of different RF receiver architectures and a brief description of IF sampling with a pair of quadrature complex low-pass $\Delta\Sigma$ modulators. Section 8.3 describes the system-level design of the $\Delta\Sigma$ modulators. The results of computer-based behavioral simulations are also presented in this section. Section 8.4 details the circuit implementation of the $\Delta\Sigma$ modulators. Section 8.5 presents the measurement results. Section 8.6 concludes the chapter and offers a few comments for future work.

8.2 Receiver Systems

Figures of Merit

The key figures of merit (FOM) relevant to the design of RF receivers include sensitivity, selectivity, linearity, and dynamic range.

Sensitivity is defined as the lowest possible signal power that an RF receiver can detect in the presence of electronic noise, given a specified *signal-to-noise ratio* (SNR). An alternative representation of sensitivity is called the *minimum detectable signal* (MDS) [12][13].

Selectivity represents the RF receiver's capability of distinguishing a desired small signal in the presence of strong adjacent interferers and blockers. In most cases, the channel band-pass filter (centered at the IF) dictates the receiver's selectivity [12]. It is also dependent on the quality factors of the RF front-end components (e.g., the low-noise amplifier and the first mixer).

Linearity represents the receiver's capability of suppressing the intermodulation products. In practice, the *third-order intercept point* (IP3) measurement is often adopted to test the linearity performance of an RF receiver. In such measurements, two sinusoids with frequencies adjacent to that of the desired signal are fed into the receiver, then the harmonics (second-order and third-order) in the output are measured. The *1-dB compression point*, at which the signal's power gain is 1 dB short from the ideal point, is an alternative figure to IP3 with respect to the linearity performance of an RF receiver.

Dynamic range (DR) is defined as the difference in decibels between the largest possible and smallest possible input voltages that the system is capable of processing. DR can also be interpreted as the ratio of the largest to the smallest possible output signals, given a specified SNR range. The lower limit of dynamic range is normally determined by the sensitivity requirement, while the upper limit depends on the receiver architecture and may vary. The *spurious-free-dynamic range* (SFDR) and the *blocking-dynamic range* (BDR) are the two often seen FOMs in relation to DR.

Conventional Super-Heterodyne Receiver

The conventional *super-heterodyne* receiver shown in Figure 8.2 has been widely used in the wireless industry since its conception in 1917. One of its most significant applications is for handsets for mobile communications systems. In this receiver, the RF signal received by the antenna is filtered by an antialiasing RF filter, and the resultant output is amplified by a low-noise amplifier (LNA). The design of the LNA is critical because in most cases, it dominates the overall noise figure (NF) and dynamic range (DR) of the complete RF receiver. Excellent selectivity and sensitivity can be attained by choosing a relatively high IF frequency, and highly selective (or sharp) RF and IF filters. Additionally, if a high IF is chosen, then the selectivity requirement of the image-rejection (IR) filter will be lessened. The super-heterodyne receiver is insensitive to the dc offset and LO (local-oscillating) leakage thanks to the fact that it employs the *two-step down-conversion* scheme [12][13]. However, due to the difficulties of integrating high-performance (e.g., high-Q) RF, IF, and IR filters in commercially available CMOS processing technologies, they normally have to be realized by using off-chip discrete components, making the receiver inappropriate for size-sensitive applications that typically require a monolithic realization [2].

Zero-IF (Direct-Conversion) Receiver

Figure 8.3 shows the *zero-IF* (or *direct-conversion*) receiver. The zero-IF receiver is the result of eliminating the external IR and IF filters from the foregoing super-heterodyne receiver; thus, it is suitable to meet the small-form-factor requirements. In this receiver, the LO and RF carrier frequency are identical, meaning that the IF frequency is equal to zero (hence the name *zero-IF* receiver). In addition, since the entire RF signal band is directly down-converted to the baseband, the receiver is immune from the image interferences. Despite the promising feature of small-size

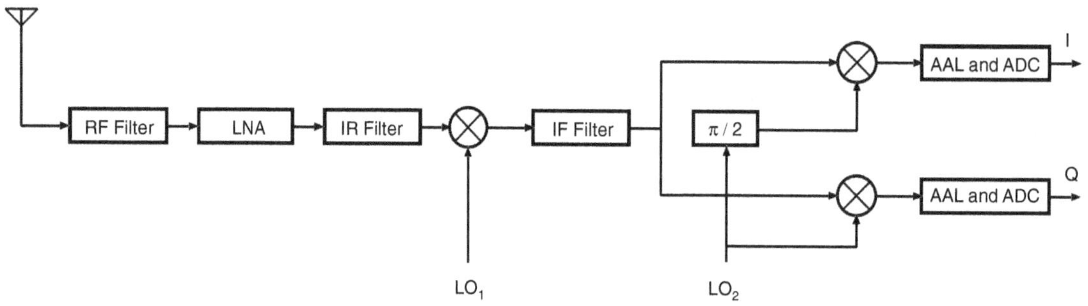

Figure 8.2 Super-heterodyne architecture.

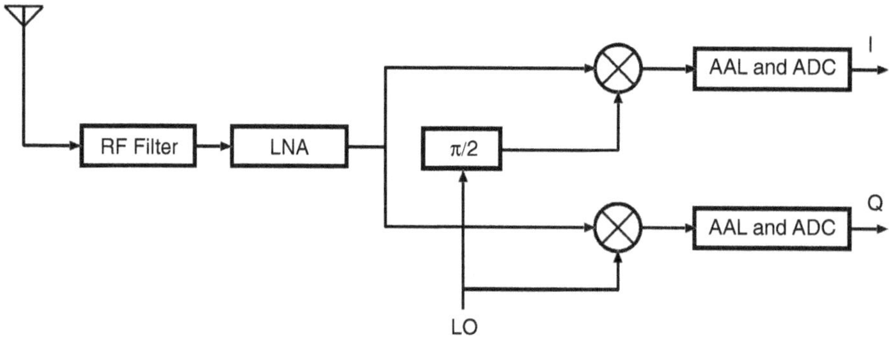

Figure 8.3 Zero-IF (direct-conversion) architecture.

design, the zero-IF receiver has several drawbacks. One is the time-varying dc offset or shift due to the imbalances in phase and amplitude between I and Q paths, which introduces both nonlinearity and phase error to the recovered baseband signal. The other is the self-mixing between the original and leaked LO signals, which may jam the receiver and make it act like an oscillator [14]. Moreover, because the zero-IF receiver adopts a one-step frequency transition, a wideband and low-phase-noise frequency synthesizer is usually required to provide precise LO frequencies for accurate channel selections. Hence, building such a high-performance frequency synthesizer from the low-Q integrated passive components is often considered one of the biggest challenges in the design of a zero-IF receiver. Finally, the dynamic range (DR) requirement of the baseband ADC is nontrivial to meet because the input signal does not experience much prefiltering before it enters into the ADC. In fact, it is known that the zero-IF receiver places the most stringent DR requirement on the ADC in comparison with other standard RF receiver topologies.

Low-IF Receiver

The idea of *on-chip band-pass filtering* perhaps led to the invention of the *low-IF* (or the *single-conversion*) receiver, which is shown in Figure 8.4. In this receiver, the IF is chosen at a relatively low frequency (typically hundreds of kHz) instead of dc. As the result, the dc offset and low-frequency noise (e.g., flicker noise) problems are alleviated. In addition, compared to the zero-IF receiver, the low-IF receiver trades a pair of high-selectivity band-pass filters for the relaxed dynamic range requirements on the baseband ADCs. However, since the low-IF receiver also employs a one-step down-conversion scheme, it still requires a wideband frequency-synthesizer with the precise tuning capability.

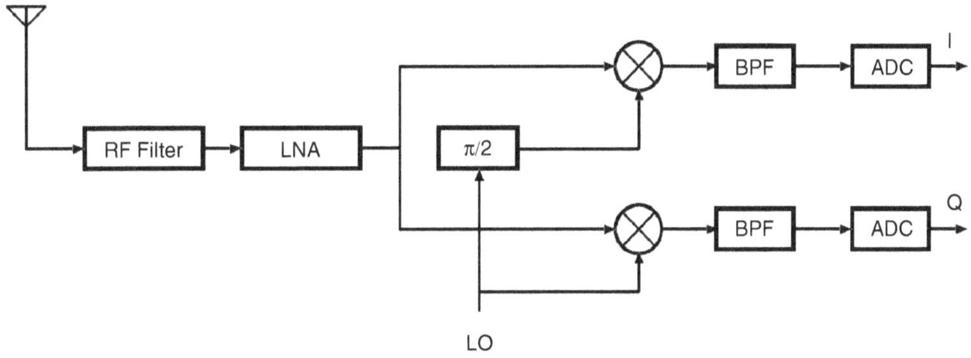

Figure 8.4 Low-IF architecture.

Wideband IF Double-Conversion Receiver

The *wideband IF double-conversion* (WIFDC) receiver shown in Figure 8.5 employs a two-step down-conversion scheme similar to that of the conventional super-heterodyne receiver. The essential difference between the WIFDC and the super-heterodyne receivers is that in a WIFDC receiver, the second LO frequency (rather than the first LO frequency as in the super-heterodyne case) is programmable and can be used to select the desired signal channels. In comparison with the zero-IF receiver, the WIFDC receiver mitigates the errors due to the LO self-mixing by choosing non-zero IFs. Compared to the low-IF receiver, the WIFDC receiver avoids the trouble of designing a wideband frequency-synthesizer because only low (on the kHz level) LO frequencies are generated. Also, this receiver circumvents the need for a pair of high-Q band-pass filters. As the schematic shows, the WIFDC receiver employs a two-step down-conversion scheme, so image interferences are inevitable. To alleviate the imaging problems, four mixers operating at low-medium frequencies (on the order of 100 kHz) are used in the second down-conversion stage [5], resulting in a high power consumption and a complex circuit design.

Digital-IF Receiver

The fast advancement of complementary metal-oxide semiconductor (CMOS) technologies has enabled the successful implementation of smaller and faster digital ICs that consume very little power. What this means to the design of modern RF receivers is that the circuitries performing the IF-to-baseband down-conversion and the

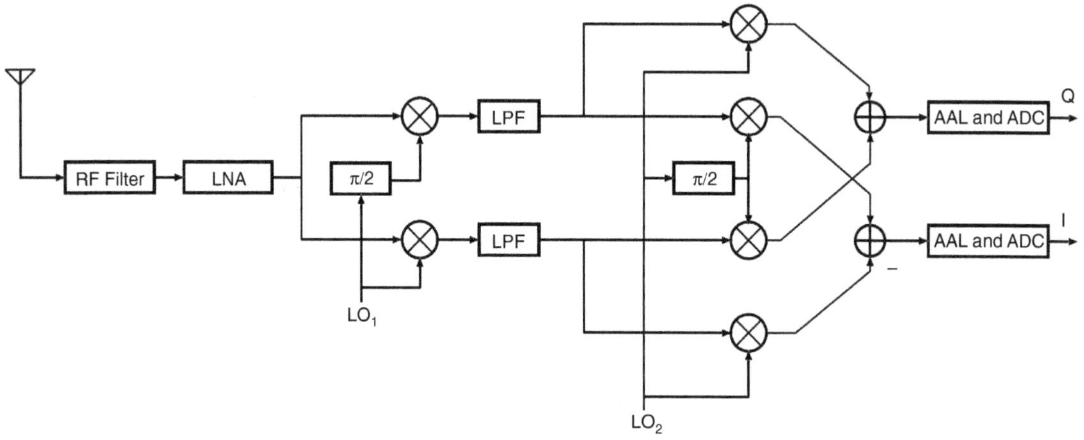

Figure 8.5 Wideband IF double-conversion architecture.

channel selection should be pushed to the digital domain if a low-power and high-speed receiver is envisioned.

An additional benefit of an *early digitization* to an RF receiver is the flexibility or programmability. The closer the analog-to-digital interface is to the antenna, the more conveniently the receiver can be programmed to adapt to different wireless standards (also known as the multistandard adaptability).

To facilitate the early digitization in an RF receiver, the analog-to-digital conversion should be applied at either the RF or the IF. Apparently, the idea of *RF sampling* is impractical in most cases because an ADC with such a wide bandwidth (on the gigahertz level) is nearly impossible to build in a CMOS process. The only appropriate choice is the *IF-sampling* scheme (i.e., the digitization is performed around the IF). A survey of the state-of-the-art design works will show that the IF-sampling scheme has been adopted in super-heterodyne receivers [15][16][17], low-IF receivers [18][19], and zero-IF receivers [6][7].

In the ADCs that require IF-sampling, ΔΣ modulators are often chosen for three reasons: (1) as mentioned earlier, ΔΣ modulators are less sensitive to component mismatch as long as the OSR is sufficiently high; (2) ΔΣ modulators' inherent programmability is a requisite for a multistandard RF receiver; and (3) ΔΣ modulators shape quantization noise and adjacent blockers simultaneously, thereby circumventing the need for highly selective baseband filters.

IF-sampling operations may be performed at either low IFs (on the order of 100 kHz) or high IFs (between 75 MHz to 150 MHz), depending on the application's requirements and the chosen receiver architecture. However, the high-IF sampling is a more appropriate candidate to meet the needs of current and emerging multistandard RF receivers [4][7][11]. It can be found that the high-IF sampling significantly reduces the selectivity requirements for the front-end RF and IR filters, thereby allowing for a smaller-size RF front-end circuitry. In addition, the high-IF sampling alleviates the imaging problems because the IF is at least 10 times higher than the desired signal cutoff frequency.

Recently, an aged receiver architecture called *digital high-IF* has found its renewed use in the mobile communications applications [11][13][16][17][18][19][20]. The general block diagram of a digital high-IF receiver is shown in Figure 8.6.

As the schematic shows, the IF-to-baseband section is split into two conversion paths (I and Q branches), and the signal fed into each path is clocked at half the effective sampling rate f_s (i.e., double sampling). As a result, the power consumption is largely reduced compared to that of a full-rate clocking configuration.

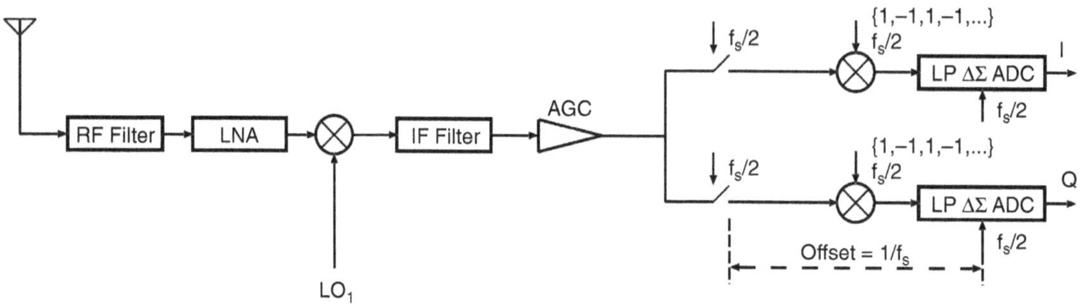

Figure 8.6 Digital high-IF architecture.

However, as we will see later, the double-sampling SC integrator shown in Figure 3.4(c) is not used in this design, because halving the clock rate is performed *outside* the ΔΣ modulators by the samplers and mixers [11]. In other words, standard SC integrators such as those shown in Figures 3.4(a) or (b) can be used in the ΔΣ modulators.

Note that the two-path structure shown in Figure 8.6 is equivalent to a single path built on one band-pass modulator, which is clocked at the full sampling rate f_s and followed by a digital mixer [17][21]. Band-pass ΔΣ modulators are typically immune to low-frequency flicker noise and dc offsets. In addition, because the quadrature mixing is performed digitally after the band-pass ΔΣ modulation, the accuracy of the mixer is independent of analog component imperfections.

However, in comparison with the IF-sampling approach of using a single band-pass modulator, the configuration of two low-pass modulators in a quadrature fashion greatly reduces the circuit complexity because both the modulator order and the effective clock sampling rate are reduced by a factor of 2. Also, the stability of an Lth-order low-pass modulator is easier to control than that of a $(2L)$th-order band-pass modulator [22][23].

In this chapter, two identical low-pass ΔΣ modulators (one is 90 degrees out of phase with respect to the other) are employed to perform IF sampling for the multi-standard RF receiver. Figure 8.7 illustrates the ideal noise transfer functions (NTF) for the GSM and the WCDMA standards, respectively.

The GSM standard requires a signal bandwidth for each channel (or channel bandwidth) equal to 200 kHz, whereas the WCDMA standard typically requires a channel bandwidth as high as 3.84 MHz. Note that in either case (GSM or WCDMA), the NTF shown in the diagram is the result of combining the NTF of the

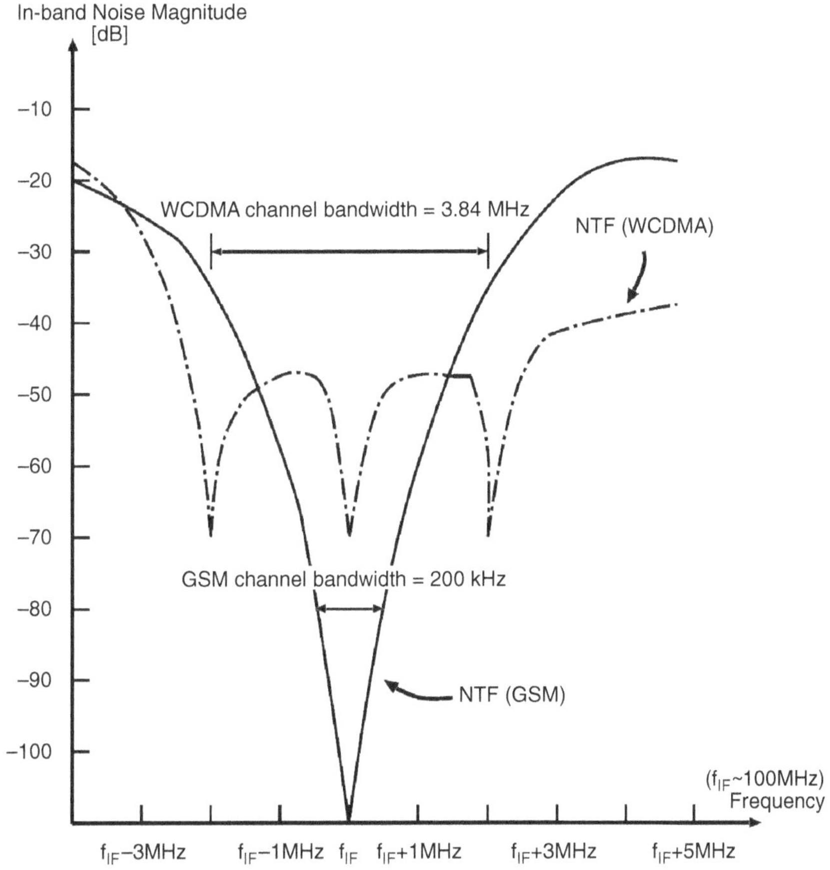

Figure 8.7 IF Sampling: NTFs for GSM and WCDMA.

low-pass modulator in the *I*-branch with its flipped-over version (90 degrees out of phase)—that is, the ideal NTF of the low-pass modulator in the *Q*-branch.

Modulator Specifications

The specifications of the ΔΣ modulators highly depend on the selectivity, sensitivity, linearity, and dynamic range requirements for the multistandard RF receiver. Table 8.2 summarizes the radio specifications of the multistandard RF receiver [24][25][26]. The next logical step is to translate the specifications in the table into the design requirements for the ΔΣ modulators.

The in-band blockers must be taken into account when specifying the dynamic range for each modulator. The bottom line is that the residual dynamic range (i.e., the difference between the nominal RMS value of the desired signal and that of the largest possible in-band blocker) must be less than the achievable dynamic range of the modulator.

Specifying the maximum allowable IP3 for each modulator is nontrivial because a reasonable estimation is not attainable until these factors have been determined: the linear gain of the RF front-end chain, the amplitudes of in-band blockers after the RF and IF filters, and the IP3 of the mixer. However, a survey of state-of-the-art design examples [5][10][17][21][27] will show that the modulators themselves typically contribute very little to the intermodulation error of the receiver, as long as the feedback DAC's nonlinearity is well taken care of. If a 1-bit DAC is used, then this will not raise much concern. Table 8.3 summarizes the target specifications of the $\Delta\Sigma$ modulators.

Table 8.2 Radio specifications of the multistandard RF receiver.

Standard	GSM	WCDMA	DECT
Modulation scheme	GMSK	QPSK/RRC	GFSK/DBFSK
Access scheme	TDMA/FDMA	CDMA/B-CDMA	TDMA/FDMA
RX bandwidth (MHz)	935–960/1850–1910	1920–1980/2110–2170	1880–1900/1910–1930
Channel bandwidth (MHz)	0.20	3.84	1.40
Channel rate (Mbps)	0.271	3.842	1.152
Channel spacing (MHz)	0.20	5.00	1.728
Sensitivity (dBm) @ BER = 0.001	–100	–110	–83
Nominal signal level (dBm)	–99	–108	–80
Max. In-band blocker level (dBm)	–23	–44	–33
Overall NF (dB)	12	9	18
RF Front-end NF (dB)	8	5	14
Baseband NF (dB)	4	4	4
Max. IP3 (dBV_{rms})	–25	–18	–15

Table 8.3 Target specifications of the $\Delta\Sigma$ modulators.

Standard	GSM	WCDMA	DECT
Dynamic range (dB)	86	54	72
SNDR (dB)	72	52	63
SNR (dB)	76	56	65
Max. IP3 (dBV_{rms})	–25	–18	–15

8.3 System-Level ΔΣ Modulator Design

Power and area are the two biggest concerns in relation to an integrated multistandard RF receiver, so they must be taken into account in the system-level design of the ΔΣ modulators.

As Chapter 5 showed, there are many possible approaches to design ΔΣ modulators, and each has its share of merits and demerits. We can find in Table 5.4 that when it comes to a low-power design, there are three best candidates: the 1-bit/multibit multistage (all stages use 1-bit quantizers except for the *last* one, which uses a multibit quantizer), the 1-bit multistage (each stage uses a 1-bit quantizer), and the high-order 1-bit single-stage.

Based on the target specifications listed in Table 8.3, we can immediately discard the 1-bit/multibit multistage (or *cascaded*) option because using multibit quantizer(s) normally requires the mismatch-shaping circuitry (see Chapter 5), which costs extra power and area, to remove the nonlinearity errors on account of imperfect matching between DAC's unit elements. Hence, it is not suitable to meet the needs of this particular multistandard RF receiver.

Second, the 1-bit multistage structure is appropriate for realizing a loop filter with an order higher than 2 but lower than 7 ($2 < L < 7$), attaining robust stability performance by employing a low-order ($L < 3$) filter in each stage. However, as mentioned in Chapter 5, component mismatch and finite op-amp gain errors cause the quantization noise from the first stage to leak to the output, hence degrading the SNR. For instance, in a fourth-order cascaded (2–2) modulator like the one reported in [10], a relative capacitor mismatch ratio of 0.2% results in a decrease of up to 15 dB in SNR.

By contrast, the high-order 1-bit single-stage structure has the advantages of not requiring the DAC error correction. Also, the high-order structure eases the gain requirement for the op-amps. However, as Chapter 5 showed, a high-order 1-bit single-stage modulator may become unstable. The higher the order, the more difficult it is to design such a modulator with a guaranteed stability. Fortunately, from Equation (5.21), we find that a third-order, rather than fourth- or fifth-order, 1-bit single-stage modulator is adequate to meet the specifications listed in Table 8.3.

IF Frequencies and OSRs

As described in Section 8.2, a high IF greatly relaxes the selectivity requirement for the RF filter and alleviates the imaging problems. An IF of around 100 MHz is often preferred in the digital high-IF receivers [11][17]. For the ΔΣ modulators in this chapter, three different IFs are chosen: 78 MHz for GSM, 138.24 MHz for WCDMA, and 110.59 MHz for DECT.

As mentioned earlier, a third-order 1-bit single-stage modulator structure is adequate to the design requirements in Table 8.3. An OSR of 192 is chosen for GSM to achieve a high SNR more than 100 dB. For DECT, the value of its OSR is set to 64. The OSR is set to only 24 since a channel bandwidth of about 2 MHz is required by the WCDMA standard.

Having determined the OSR values, we can calculate the different sampling rates for the three standards using the *orthogonal hardware modulation* scheme [28], and the results are as follows: 104 Msample/s for GSM, 184.32 Msample/s for WCDMA, and 147.46 Msample/s for DECT. Note that the low-pass modulators are clocked at only half the effective sampling rates, thanks to the two-path IF-to-baseband scheme illustrated in Figure 8.6. For example, the modulators for GSM are actually clocked at 52 Msample/s.

ΔΣ Modulator Design for GSM and DECT

A simplified DFFIR topology (see Chapter 5) is adopted to design the two ΔΣ modulators for GSM, which is shown in Figure 8.8. Note that the internal resonator feedback is removed because simulations show that the SNR requirement can be readily fulfilled without optimizing the NTF's zeros (for GSM only). Also, the simplified DFFIR shown in Figure 8.8 removes the three forward paths, resulting in a smaller and more power-efficient modulator.

Note that the system coefficients (i.e., 1/4, 1/3, and 1/8) are equivalent to capacitance ratios in the circuitry. In practice, it is desirable that these coefficients contain

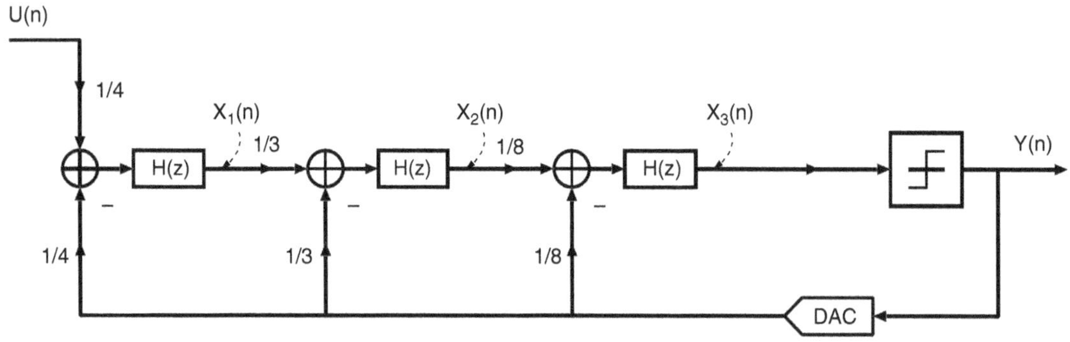

Figure 8.8 ΔΣ modulator topology for GSM.

as few fractional digits as possible. Otherwise, they place stringent accuracy requirements on the implementation of integrated capacitors, which are difficult to meet especially when parasitic capacitances are not negligible.

Additionally, each pair of the integrator input and DAC feedback signals share the same capacitance ratio. This is also favorable in practice because only one input capacitor is needed for each input stage (in the single-ended configuration), resulting in less capacitances and a smaller chip area.

Furthermore, one important requirement of cascaded integrators is that all the integrators have approximately the same average output level. Otherwise, we need to scale the system coefficients to make sure that no integrator's output is beyond a normalized limit (i.e., dynamic range scaling). This is critical for attaining a high dynamic range in a low-voltage design. Figure 8.9 shows the normalized outputs of the three integrators, i.e., X_1, X_2, and X_3. A 50-Hz sinusoid with a peak-to-peak amplitude of −3 dBFS is fed into the modulator. As shown, all three integrators'

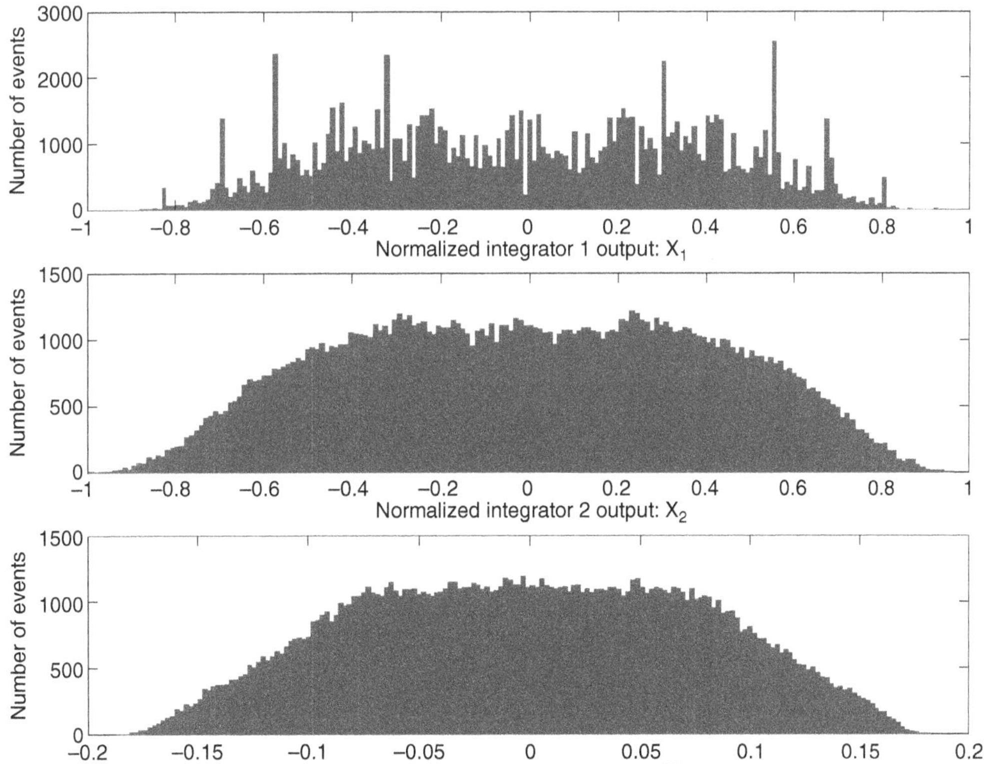

Figure 8.9 Probability distribution of each integrator's output (for GSM only).

output levels are kept within the normalized range between 1 and –1. Thus, no extra dynamic range scaling is needed. In this design, both DECT and GSM standards use the same modulator topology shown in Figure 8.8, but at different OSRs—that is, for DECT the OSR is changed to 64.

ΔΣ Modulator Design for WCDMA

As mentioned earlier, the ΔΣ modulator for WCDMA uses a much lower OSR than 192 (OSR = 24) to achieve a 56 dB SNR, therefore more aggressive noise-shaping is needed. Here, an internal resonator feedback is added to optimize the NTF's zeros, as illustrated in Figure 8.10. Note that the feedback gain or coefficient is set to 1/9. The internal resonator feedback will be activated for WCDMA but deactivated for GSM or DECT.

Sizing the Capacitors

In real circuits, the thermal noise (referred to as $kT/C_{sampling}$ *noise* in SC circuits; see Chapter 1) rather than the quantization noise may limit the ultimate SNR performance of a ΔΣ modulator [22]. The suppression of the thermal noise heavily depends on the sampling capacitors. The smaller the sampling capacitor, the less effectively the thermal noise can be suppressed. For a chain of cascaded integrators, the sampling capacitor(s) of the first integrator ($C_{sampling}$) normally governs the thermal noise suppression. The following formula is often used in practice to determine $C_{sampling}$:

$$20\log\left(\sqrt{\frac{4kT}{OSR\cdot C_{sampling}}}\cdot\frac{1}{V_{ref}}\right)\leq-(SNR+3\,dB) \tag{8.1}$$

For GSM (OSR = 192), if the desired SNR and V_{ref} are equal to 100 dB and 1.25 V, respectively, then $C_{sampling}$ should be no less than 0.6 pF. For WCDMA (OSR = 24),

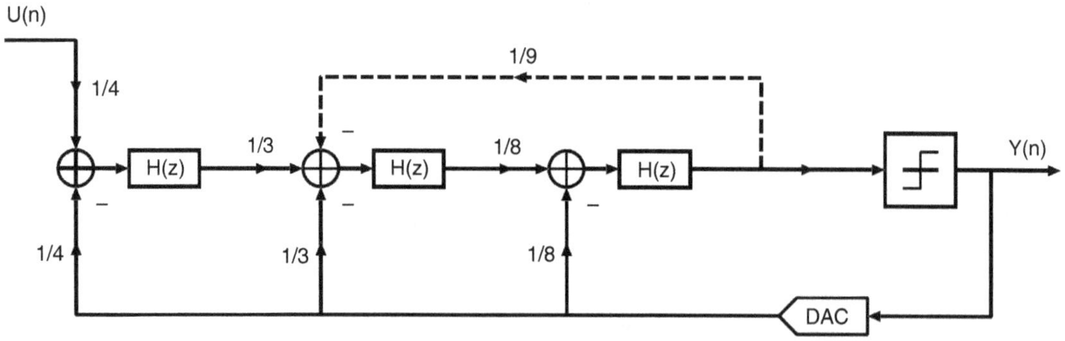

Figure 8.10 ΔΣ modulator topology for WCDMA.

the calculated minimum $C_{sampling}$ is equal to 0.044 pF when the SNR and V_{ref} are set to 70 dB and 1.25 V, respectively. And for DECT, the result is 0.17 pF (OSR = 64) when the SNR and V_{ref} are set to 90 dB and 1.25 V, respectively. Sometimes a smaller reference voltage such as 1.0 V is used to further reduce the thermal noise.

As we can see, $C_{sampling}$ for GSM has the largest capacitance (0.6 pF), hence it should be the bottom line. In this design, $C_{sampling}$ is chosen to be 0.72 pF to provide a design margin. The sizes of the sampling capacitors in the second and third integrators may be scaled down by a factor of two to four for saving chip area.

In this three-in-one configuration, all three RF standards employ the same set of capacitors except that an extra feedback capacitor is needed for WCDMA. The sizing of $C_{integrating}$ (and $C_{feedback}$ for WCDMA only) is determined by the system coefficients (a_i, b_i, c_i, or g_i) derived earlier. Table 8.4 lists all the capacitors used in this design (fully differential). It should be noted that each capacitor is divided into two equal ones in a fully differential design. C_{s1}, C_{s2}, and C_{s3} are the sampling capacitors in the first, second, and third integrators, respectively. C_{i1}, C_{i2}, and C_{i3} are the integrating capacitors in the first, second, and third integrators, respectively. And C_f is the capacitor used in the internal resonator feedback path (for WCDMA only).

Hence, the total capacitance is equal to 6.24 pF (in the transistor-level implementation 6.30 pF is used to achieve a better capacitor matching), and the capacitance spread ratio is equal to 24:1. The unit capacitance is set to 0.06 pF.

Nonidealities in ΔΣ Modulators

Finite Op-Amp Gain

One of the well-defined nonidealities with respect to ΔΣ modulators is the finite op-amp gain problem. It can be found that an SC integrator's transfer function is given by

Table 8.4 Capacitors in each ΔΣ modulator.

C_{s1}	0.36 pF
C_{s2}	0.18 pF
C_{s3}	0.06 pF
C_{i1}	1.44 pF
C_{i2}	0.54 pF
C_{i3}	0.48 pF
C_f	0.06 pF

$$H(z)=\frac{1}{z-\left(1-\frac{1}{A}\right)} \tag{8.2}$$

where A is the dc gain of the op-amp. It is well known that the finite op-amp gain causes not only a gain error but also a phase shift. The phase shift causes the displacement of the NTF zero from its desired position, resulting in an increase of the in-band quantization noise power and hence a lower SNR. Figure 8.11 illustrates the simulated SNR loss as a function of the op-amp dc gain. As the figure shows, if it is assumed the finite op-amp gain is the dominant source of error in the system, then an op-amp gain of at least 1500 V/V, or, equivalently, 63.5 dB is required to keep the SNR loss less than 0.1 dB for all three standards. However, larger op-amp gains are often needed in practice to alleviate other errors such as the charge injection and capacitor mismatch.

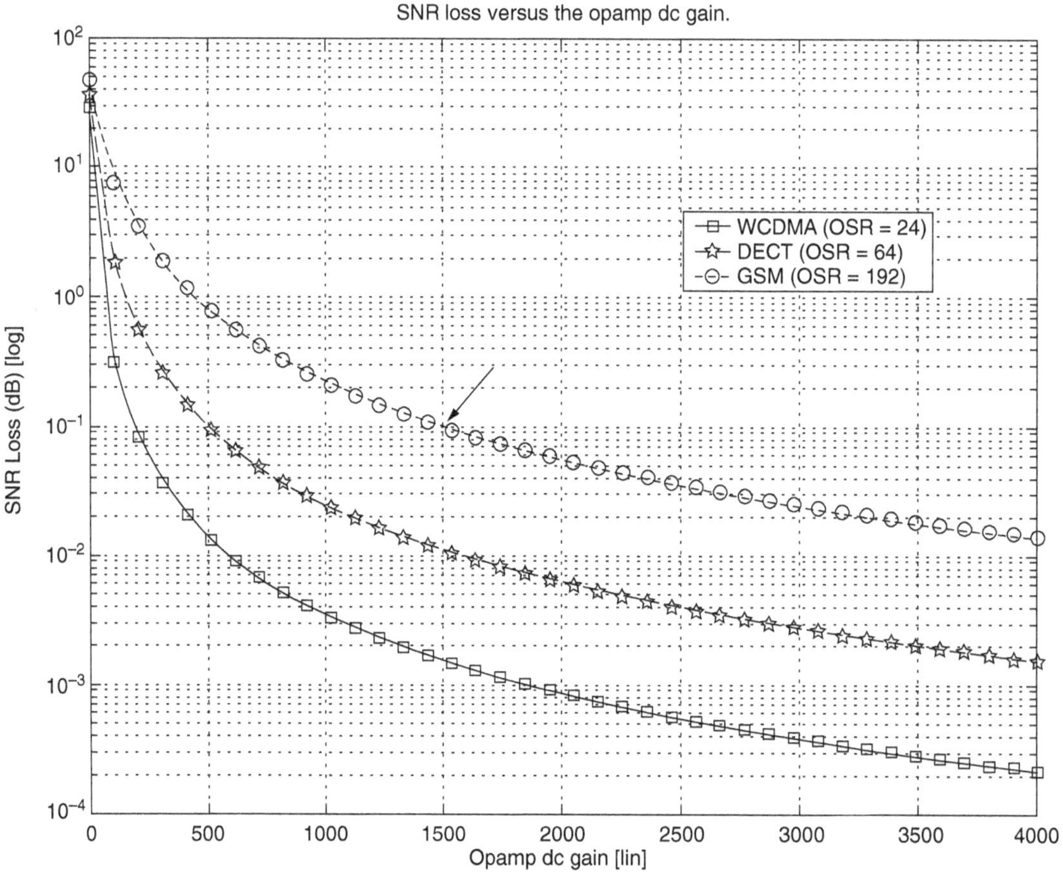

Figure 8.11 SNR loss versus op-amp dc gain.

Finite Op-Amp Bandwidth and Slew Rate

If an SC integrator output voltage cannot settle successfully within half the clock period, then harmonics will appear at its output. Incomplete SC integrator settling is caused by a combination of the finite op-amp bandwidth (linear settling) and the insufficient slew rate (nonlinear settling).

Finite op-amp bandwidth is normally determined by the closed-loop *unity-gain bandwidth* (UGBW), which can be approximated to the result of dividing the transconductance of the op-amp's input devices by the output loading capacitance at the end of the integration phase (i.e., $g_m/C_{eq.load}$). Slew rate is normally determined by how fast $C_{eq.load}$ can be charged (or discharged), hence it is given by $I/C_{eq.load}$ [29].

In the beginning of each settling operation, the slew rate is the limiting factor as $C_{eq.load}$ is being charged by a large input signal that is initiated by the rising edge of the system clock. Once the charging is done (roughly when the op-amp's output reaches the peak voltage and starts falling), the finite op-amp bandwidth takes over.

The finite op-amp bandwidth and the slew rate specifications can be derived by using the behavioral-level simulation [30][31]. Normally, a simple single-pole op-amp model with a constant dc gain is used in the simulation. In this design, the modulator for WCDMA operates at the highest sampling rate with the busiest input signal (i.e., the highest maximum signal bandwidth). Thus, its settling requirement dominates.

In Figure 8.12(a), the contours of peak SNDRs (64 dB, 58 dB, etc.) are illustrated as functions of the op-amp's transconductance (g_m) and the maximum tail current or the slewing current I_{tail}. The optimum design point is often chosen within the flat region (denoted as "Desired design region" in the diagram), where the nonlinear settling characteristics are approximately independent from those of the linear settling [31]. As Figure 8.12(a) shows, a g_m of about 7 ms and an I_{tail} of about 0.5 mA are sufficient for the integrators to settle appropriately.

Figure 8.12(b) shows the contours of constant peak SNDRs as functions of UGBW and slew rate, respectively. As shown, a closed-loop UGBW of about 420 MHz and a slew rate of about 100 V/μs are sufficient to optimize the settling performance. Note that the dc gain of the op-amp model used here is assumed to be 66 dB.

Alternatively, we can estimate the minimum UGBW required to guarantee a 0.1% settling within half the clock period by the utilizing the following formula [30]:

$$f_{UGBW} = \frac{7}{\pi} \times f_{sampling} \tag{8.3}$$

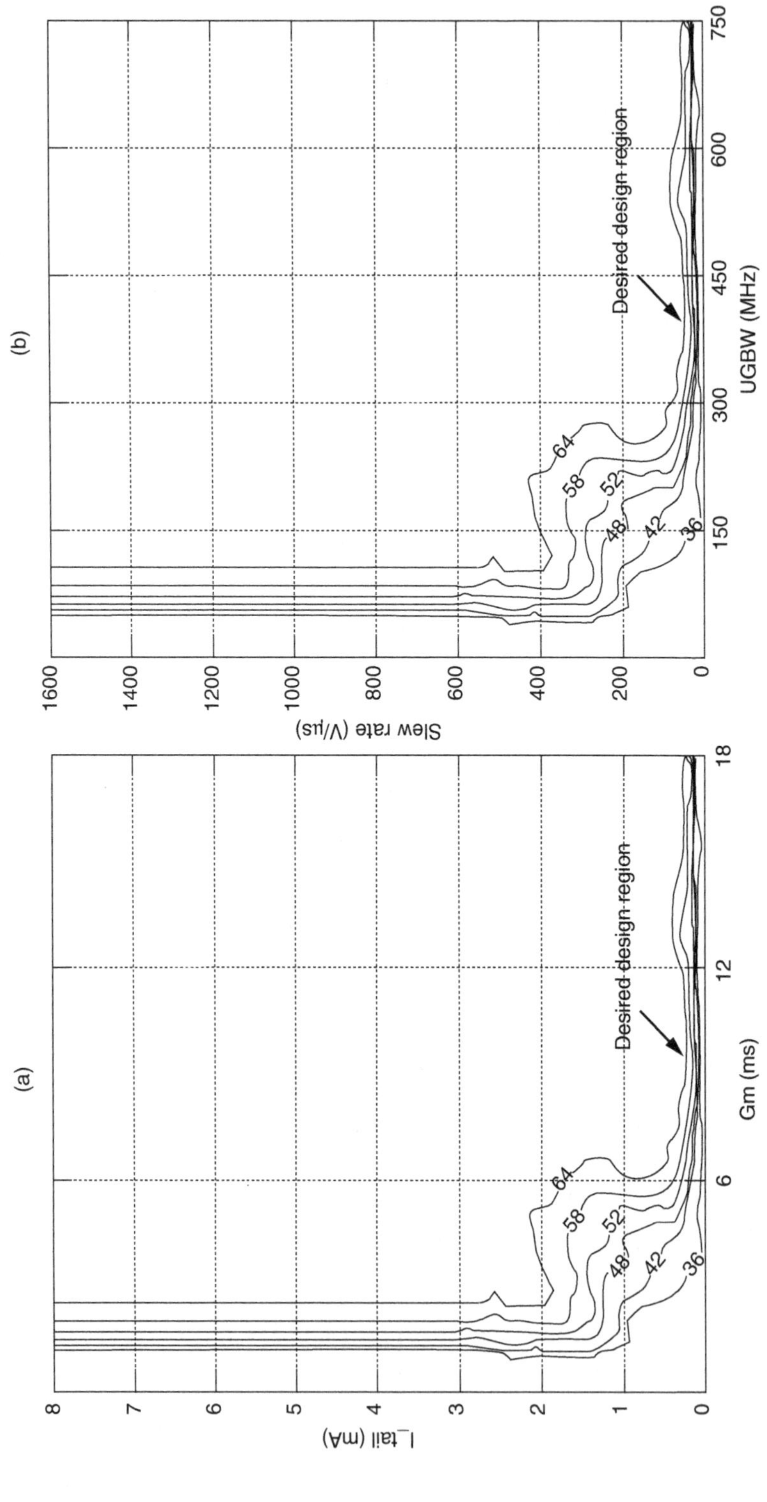

Figure 8.12 Peak SNDRs versus (a) g_m and I_{tail} (b) closed-loop bandwidth and slew rate.

The highest effective $f_{sampling}$ is equal to 184.32 MHz (WCDMA); thus from the preceding formula, a minimum UGBW (f_{UGBW}) as high as 411 MHz is obtained. This result is comparable to that shown in Figure 8.12(b), which is about 420 MHz.

Nonideal Switches

As mentioned in Chapter 1, MOSFET switches (NMOS or PMOS) suffer from nonidealities including the *kT/C* noise, charge injection/clock feedthrough, and nonzero on-resistance. As we saw earlier, *kT/C* noise can be reduced by using larger $C_{sampling}$. Signal-dependent charge injection can be alleviated by using bottom-plate sampling technique [30][32]. Clock feedthrough can be effectively reduced by using a fully differential design.

However, with a 2.5-V supply voltage, the nonzero on-resistance is rather problematic. First, it degrades the settling performance of the SC integrator. If a nonzero on-resistance exists, then the settling time is dominated by the RC constant of the switching network. In effect, the integrator transfer function is altered; hence, the in-band quantization noise power is increased. Second, during the sampling phase, the sampling switch's on-resistance depends on the input signal, resulting in harmonic distortions [30]. The appropriate NMOS size ratio can be roughly estimated to be

$$\left(\frac{W}{L}\right)_{\min} \cong \frac{7}{\mu_N C_{ox}} \cdot \frac{C_{\max} \cdot f_{sampling}}{(V_{gs} - V_{th})} \tag{8.4}$$

The modulator for WCDMA has the worst-case scenario because it must operate at the highest effective $f_{sampling}$, which is equal to 184.32 MHz. According to Table 8.4, the maximum capacitance in the network $C_{\max}$ is equal to 1.44 pF. In addition, the overdrive voltage, $(V_{gs} - V_{th})$, depends on the technology and the supply voltage. In this design, it is approximated to 0.7 V for a 2.5-V power supply. Finally, the product of $\mu_N\ C_{ox}$ is approximated to 1×10^{-4}.

From calculations, the minimum NMOS size ratio (i.e., *W/L*) is about 20/1, which implies that the switch on-resistance cannot be larger than 350 Ω. Since the PMOS transistor is typically two to three times slower than its NMOS counterpart, the minimum PMOS size ratio ranges between 40/1 to 60/1.

Flicker Noise

Thermal noise and flicker noise are the two most common intrinsic noises in MOS devices. As mentioned earlier, thermal noise can be attenuated by using large sampling capacitors in the SC circuit. Unlike the pseudo-white thermal noise, flicker

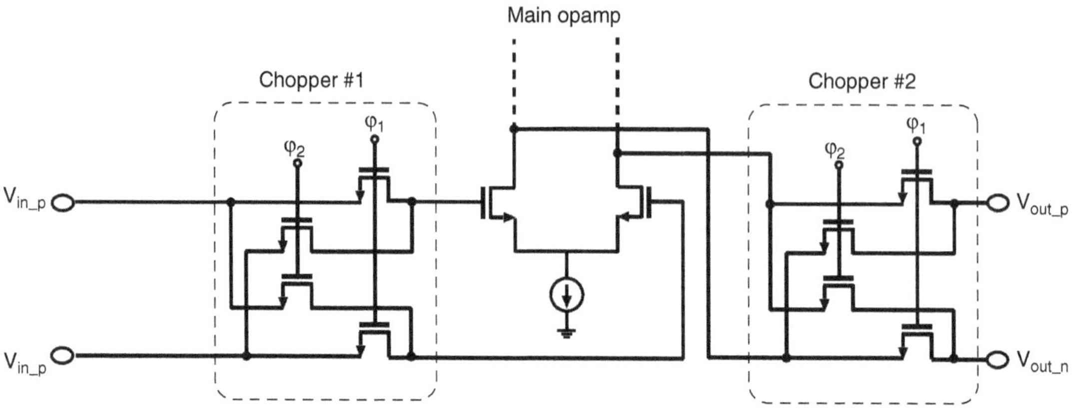

Figure 8.13 Chopping in front of the op-amp.

noise (or $1/f$ noise) has a spectral power density that is inversely proportional to frequency, meaning that most of its power resides within the low-frequency range.

A well-known technique called *chopping* [33] is often used to reduce the in-band flicker noise. In this design, two *choppers* clocked at a quarter of $f_{sampling}$ are placed in front of the op-amp of the first integrator, which effectively pushes the amplifier's $1/f$ noise and dc offsets to higher frequencies (e.g., $f_s/4$ and $3f_s/4$). Figure 8.13 shows the choppers.

Capacitor Mismatch

Capacitor mismatch degrades the accuracy of the modulator coefficients, causing deviations in the NTF and STF responses of the ΔΣ modulator. In standard CMOS technologies, the relative capacitor matching error is kept within 0.2%. To model the effects on the SNR by capacitor mismatch, we first generate a Gaussian-distributed random sequence, E_N, which has a standard deviation of ±0.1%; then we replace each branch capacitor C_i with $C_i\,(1 + E_i)$. Next, the ΔΣ modulator is simulated on the behavior level [1]. The simulation shows that there is no significant degradation of SNR performance because of the capacitor mismatch (typically less than 0.5 dB), which proves the previous statement that the high-order 1-bit single-stage ΔΣ modulator is normally insensitive to the capacitor mismatch errors.

8.4 Circuit Implementation

This section deals with the circuit-level design issues with respect to the third-order ΔΣ modulator illustrated in Figure 8.14. As shown, the key building blocks of the modulator include three SC integrators, a 1-bit quantizer, and a 1-bit DAC. The 1-bit DAC is similar to the one that we saw in Chapter 5.

Seven clock phases are generated to control the operations of the integrators and 1-bit quantization. They include two nonoverlapping phases, ck_1 and ck_2; an early phase, ck_{1e}, which controls the latched comparator in the 1-bit quantizer; two delayed clock phases, ck_{1d} and ck_{2d}; and complements of the two delayed phases, ck_{1db} and ck_{2db}. The clock generation circuitry is shown in Figure 8.15 [34].

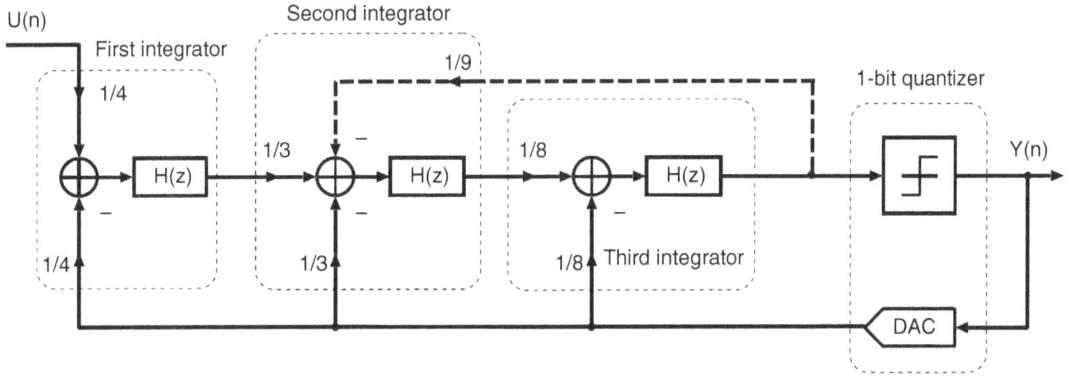

Figure 8.14 Block diagram of the third-order ΔΣ modulator.

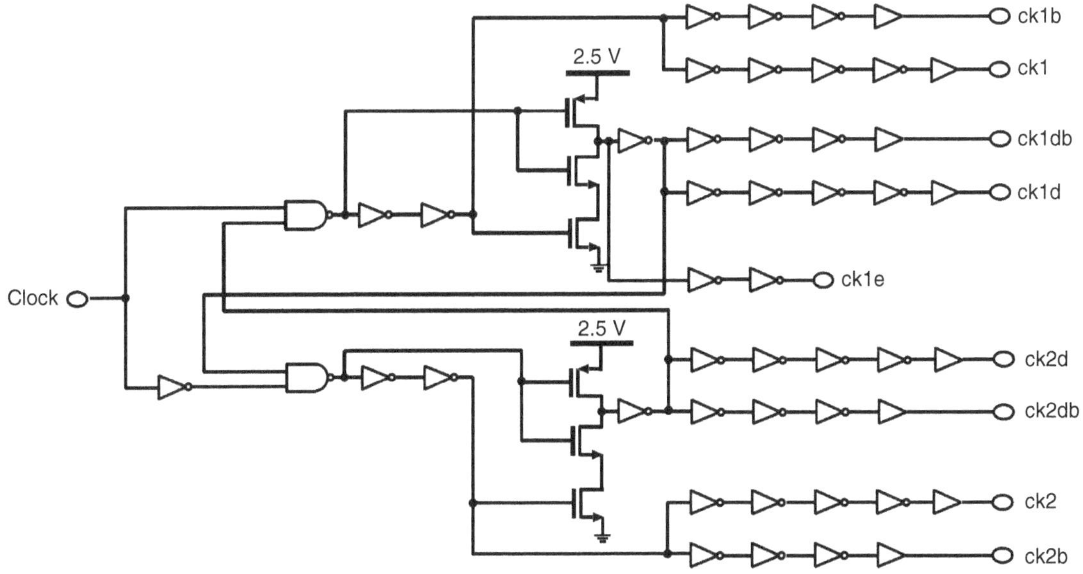

Figure 8.15 Clock generation circuitry.

SC Integrators

The first integrator is shown in Figure 8.16, and it operates as follows. During ck_1, the input voltage V_{in} is sampled onto $C_{sampling}$. During ck_2, a charge proportional to the voltage difference between V_{in} and DAC reference voltage V_{ref} is transferred from $C_{sampling}$ to $C_{integrating}$.

The sampling and integrating capacitances were specified earlier in Table 8.4 and are denoted here in Figure 8.16. The input common-mode voltage, V_{cm_i} is set to 0.9 V for a 2.5-V power supply, thereby allowing switches S_3 and S_4 to be implemented using only NMOS devices. The sizing of the switches is listed in Table 8.5.

The sizing of these switches follows the calculation results in the previous section: The W/L ratio of an NMOS switch should be at least 20/1, and that of a PMOS switch should be 40/1. The term *CMOS* means that the particular switch is a CMOS transmission gate built from an NMOS and PMOS transistors. Here, S_1 and S_2 are built as CMOS transmission gates to reduce the input-dependent errors.

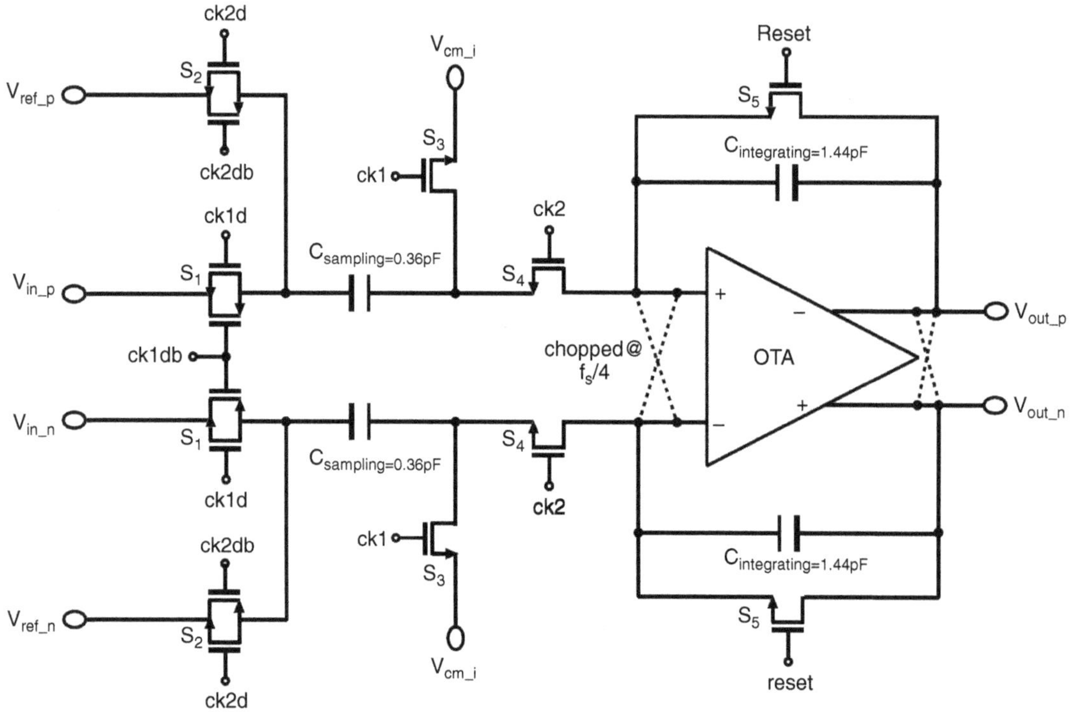

Figure 8.16 The first integrator.

Table 8.5 Switch-sizing for the first integrator.

Switch	Type	Size (μm)
S_1	CMOS	7/0.35 (NMOS); 21/0.35 (PMOS)
S_2	CMOS	7/0.35 (NMOS); 21/0.35 (PMOS)
S_3	NMOS	14/0.35
S_4	NMOS	14/0.35
S_5	NMOS	1/0.35

Additionally, the use of delayed clock phases ck_{1d} and ck_{2d} reduces the charge injection errors [35]. S_5 resets the integrating capacitors during the power-on mode or the switching mode (when the modulator is being transformed to support a new standard). In addition, S_5 serves as a *clipper*, resetting the modulator's output when it is overloaded or unstable [22].

The second integrator is shown in Figure 8.17. The input common-mode voltage, V_{cm_i} is set to 0.9 V, and the midsupply reference voltage, V_{mid} is set to 1.25 V for a 2.5-V supply. The feedback branch that creates the internal resonator for WCDMA is controlled by $en_{\_WCDMA}$ and ck_{1d} through an AND gate and an inverter, and the sizing of the switches is listed in Table 8.6. The third integrator is shown in Figure 8.18. It looks similar to the first integrator. Note that V_x and V_y are marked on purpose, indicating the negative feedback coefficient is obtained by simply reversing the outputs. The sizing of the switches is listed in Table 8.7.

Operational Transconductance Amplifier (OTA)

The operational transconductance amplifier (OTA) (also known as the *operational amplifier* or op-amp) is the most critical element in a $\Delta\Sigma$ modulator. OTAs with high gain-bandwidth products are used to fulfill the speed and accuracy requirements placed by the RF receiver.

Gain-boosting is a proven approach to designing high-gain and high-speed OTAs [36]. The key point here is that the dc gain of the OTA is increased exponentially while the effective UGBW is only decreased linearly. As shown in Figure 8.19, the main amplifier is built based on the telescopic topology. The telescopic topology is chosen over other amplifier types because it is one of the best candidates suitable to meet the needs of high-speed and low-power applications.

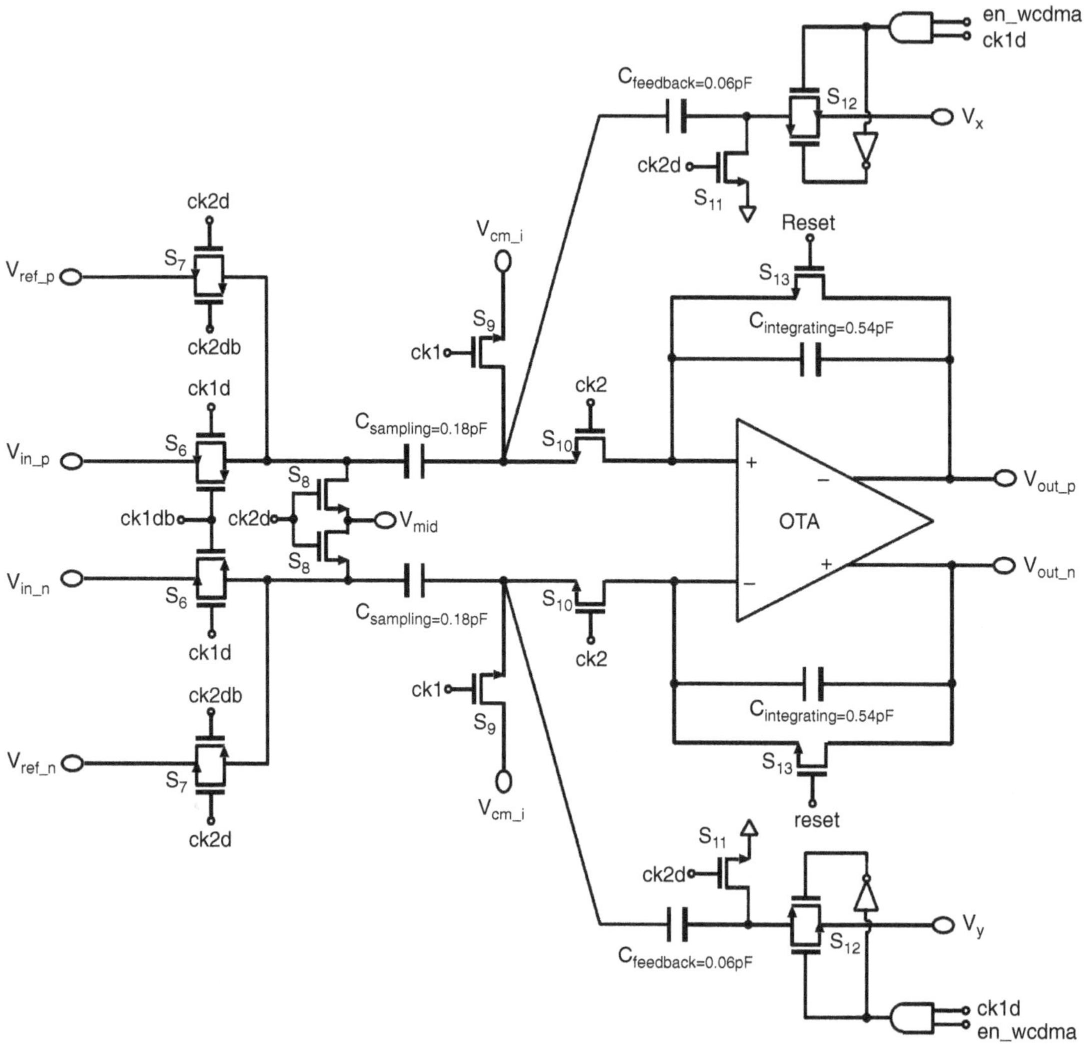

Figure 8.17 The second integrator.

The switched-capacitor common-mode feedback (CMFB) circuit is used to track the output common-mode voltage level. The tail current I_{tail} of the main amplifier in the first integrator is equal to 590 μA. To save power, the value of I_{tail} can be reduced by a ratio of 2 and 4 for its correspondences in the second and third integrators, respectively (the transistors also need to be scaled accordingly). The transistor sizing for the main amplifier in the first integrator is listed in Table 8.8.

The auxiliary amplifiers (i.e., A_1 and A_2 in Figure 8.19) are designed based on the folded-cascode topology to enhance the total dc gain. The top auxiliary amplifier A_1

Table 8.6 Switch-sizing for the second integrator.

Switch	Type	Size (μm)
S_6	CMOS	5/0.35 (NMOS); 15/0.35 (PMOS)
S_7	CMOS	5/0.35 (NMOS); 15/0.35 (PMOS)
S_8	NMOS	10/0.35
S_9	NMOS	10/0.35
S_{10}	NMOS	10/0.35
S_{11}	NMOS	5/0.35
S_{12}	CMOS	5/0.35 (NMOS); 15/0.35 (PMOS)
S_{13}	NMOS	1/0.35

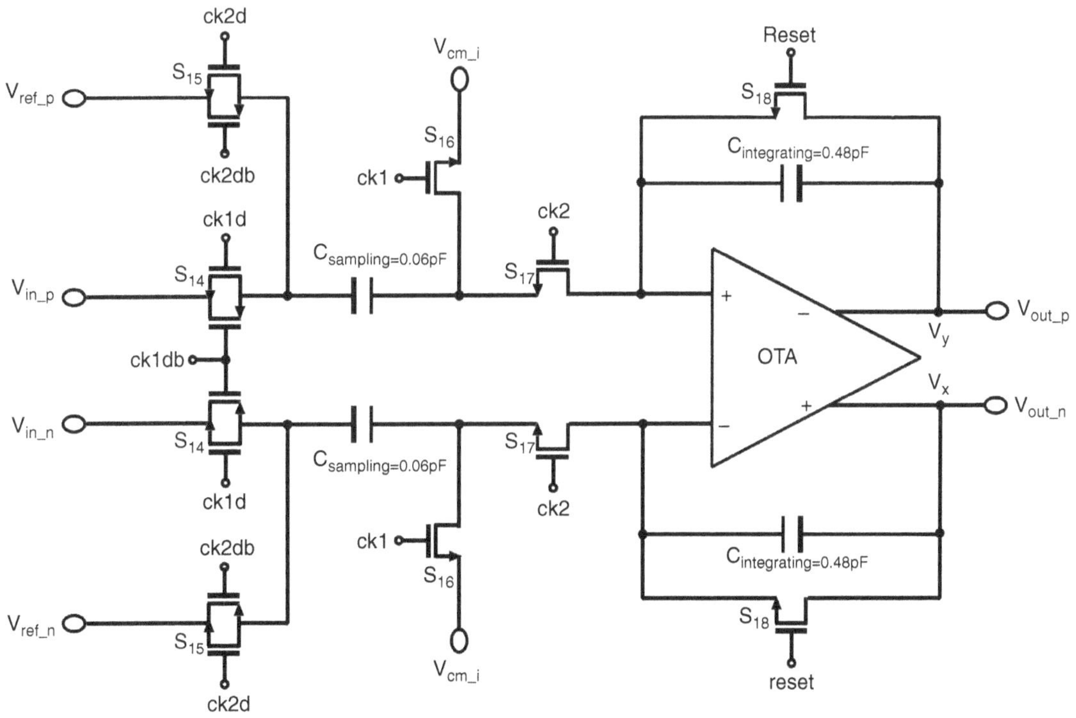

Figure 8.18 The third integrator.

Table 8.7 Switch-sizing for the third integrator.

Switch	Type	Size (μm)
S_{14}	CMOS	5/0.35 (NMOS); 15/0.35 (PMOS)
S_{15}	CMOS	5/0.35 (NMOS); 15/0.35 (PMOS)
S_{16}	NMOS	10/0.35
S_{17}	NMOS	10/0.35
S_{18}	NMOS	1/0.35

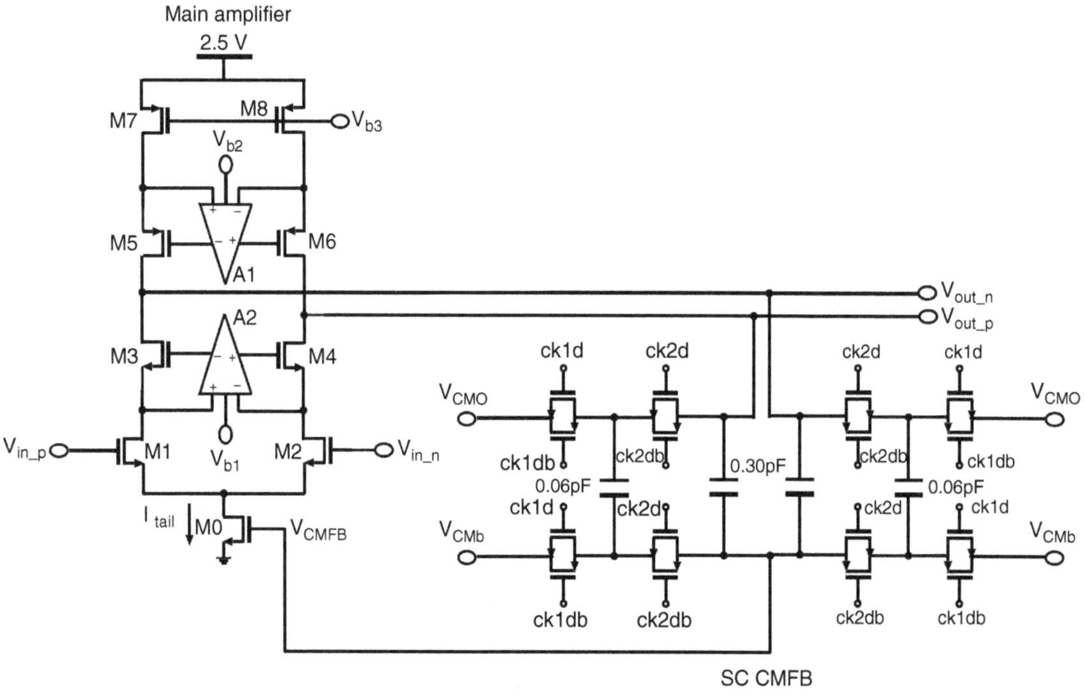

Figure 8.19 Main amplifier.

Table 8.8 Transistor-sizing for the main amplifier.

Devices	Type	Size (μm)
$M_{1,2}$	NMOS	120/0.35
$M_{3,4}$	NMOS	50/0.35
$M_{5,6}$	PMOS	80/0.35
$M_{7,8}$	PMOS	150/0.7

is shown in Figure 8.20. It is on the PMOS side and employs a pair of NMOS input devices to achieve a wide voltage swing [37]. The same strategy is applied to the design of the auxiliary amplifier on the NMOS side A_2 (Figure 8.21), in which a pair of PMOS transistors is adopted as the input devices.

The internal CMFB loop is formed by simply connecting the amplifier's output nodes to the gates of two large transistors, both of which are working in the linear region. In both auxiliary amplifiers, I_{D7}, I_{D9}, and I_{D11} are set to about 40 μA, 20 μA, and 20 μA, respectively. The biasing plan is described as follows. The main amplifier should be driven by a separate biasing network to provide a high *power-supply-rejection ratio* (PSRR). By contrast, the auxiliary amplifiers in each integrator should share the same biasing network to save power. The biasing technique known as Sooch Mirror [38] is adopted in this design to achieve wide output swings. The simulated UGBW and phase margin (PM) of the OTA are equal to 433 MHz and 57 degrees, respectively. The simulated dc gain approximates to 72 dB.

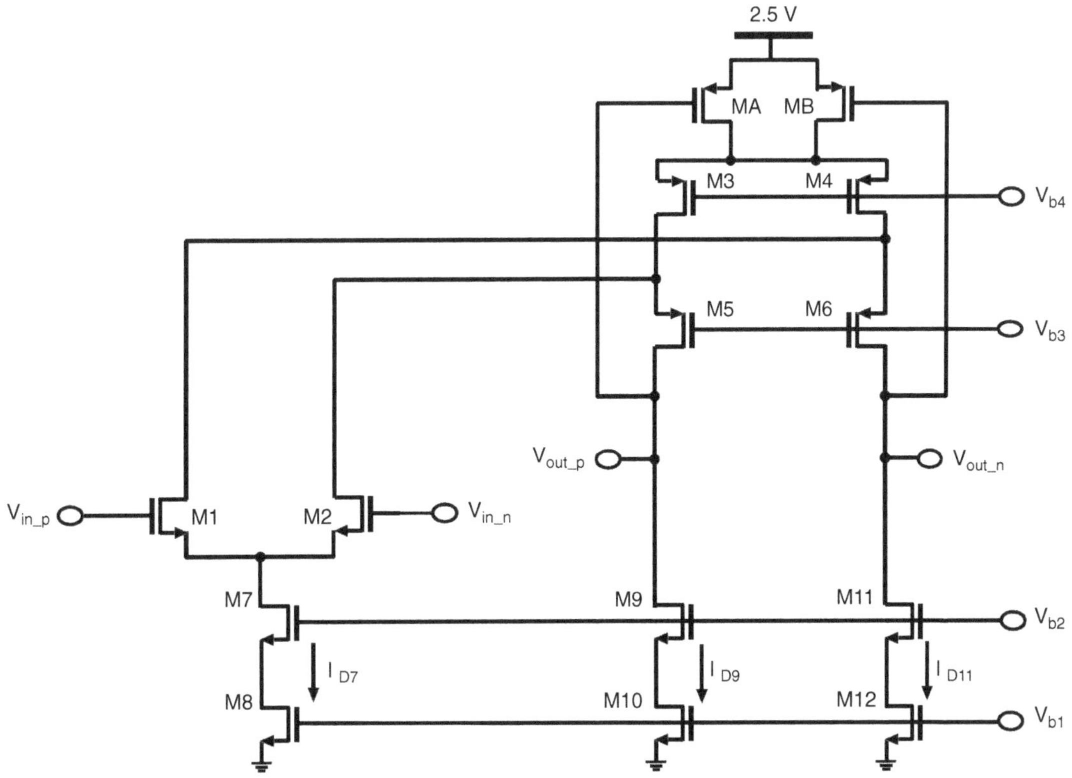

Figure 8.20 Auxiliary amplifier A_1.

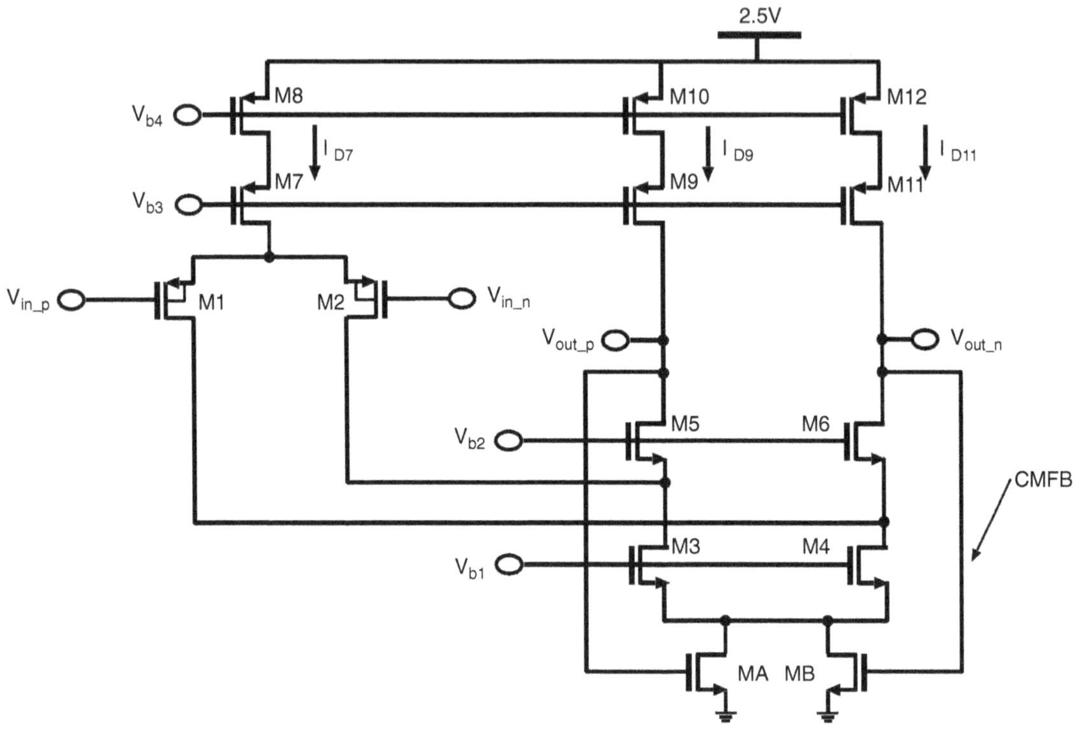

Figure 8.21 Auxiliary amplifier A_2.

Quantizer

The 1-bit quantizer is built from a dynamic regenerative comparator and a static SR latch, as shown in Figure 8.22. The comparator is appropriate for high-speed and low-power applications [39], and it operates as follows. During the *reset* mode (i.e., ck_{1e} is low), the outputs are connected to V_{DD} through M_9 and M_{10}. When ck_{1e} goes high, the comparator enters the *regenerative* mode and transistors M_3–M_8 form a positive feedback loop. As a result, the input difference voltage is amplified to a full-scale rail-to-rail output. Once the comparator makes a decision, the cross-coupled transistors $M_{3,4}$ and $M_{7,8}$ immediately shut down all the connections between V_{DD} and V_{SS}, thereby saving power. This process may be better understood by looking at Figure 8.22: When V_{inp} is high and V_{inn} is low, V_x becomes low and V_y becomes high. As a result, M_3 and M_8 are on (√), whereas M_4 and M_7 are off (×) and hence the comparator is turned off. The transistor sizing for the comparator is listed in Table 8.9.

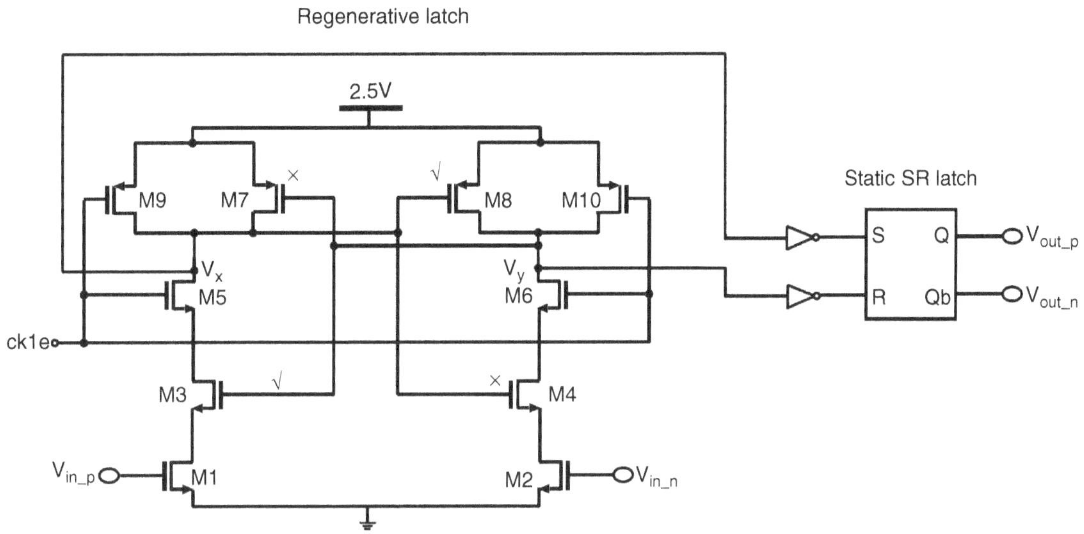

Figure 8.22 Quantizer.

Table 8.9 Transistor sizing for the comparator.

Devices	Type	Size (μm)
$M_{1,2}$	NMOS	16/0.35
$M_{3,4}$	NMOS	10/0.35
$M_{5,6}$	PMOS	10/0.35
$M_{7,8}$	PMOS	25/0.35
$M_{9,10}$	PMOS	10/0.35

8.5 Measurement Results

The $\Delta\Sigma$ modulators described in this chapter have been incorporated into a multistandard RF transceiver chip, which is fabricated in a 0.35-μm, DP5M 2.5-V CMOS process. The chip microphotograph of the $\Delta\Sigma$ modulators and decimation filter is shown in Figure 8.23. The total active silicon area is about 16 mm^2, and the modulator takes up 0.50 mm^2. During the measurement, the acquired output data stream is stored in a file and then loaded into the computer for additional postfiltering and spectral analysis.

Intermodulation measurement is performed for the GSM case only, whose IP3 specification is the most stringent among the three standards. Two sinusoids, one with a frequency 100 kHz lower than the IF and the other with a frequency 200 kHz

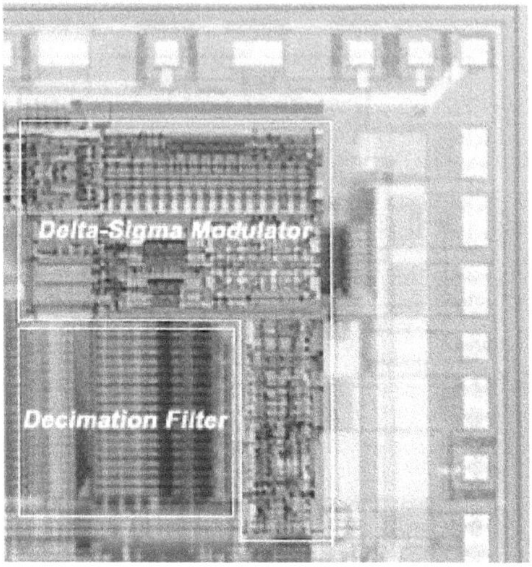

Figure 8.23 Die microphotograph (ΔΣ modulators and decimation filter only).

higher than the IF, are fed into the modulators. The strongest image component is measured to be $-27.5\,\mathrm{dBV}_{rms}$ at the 300-kHz offset from the IF. Hence, the IP3 requirement in Table 8.3 is fulfilled.

We know that the noise-shaping requirements for the WCDMA modulators, such as the low OSR and the high signal bandwidth, are much more stringent than those for the GSM and DECT modulators. Therefore, only the noise-shaping result of the WCDMA modulators is presented here. The fast Fourier transform (FFT) spectrum of the *I*-branch modulator output (before the decimation filter) is shown in Figure 8.24. A sinusoidal input signal with an amplitude of −4 dBFS and oscillating at 100-kHz higher than the IF (in this case, 138.24 MHz) is fed into the modulator.

As shown, the thermal noise floor is roughly flat up to 2 MHz, which is half the signal bandwidth required by the WCDMA standard. Note that only the *I*-branch modulator's output, which is horizontally symmetrical to that of the *Q*-path modulator, is shown.

Figure 8.25 shows the measured SNDRs with respect to different input amplitudes. As shown, the achieved peak SNDRs are 80.1 dB (GSM), 55.3 dB (WCDMA), and 64.9 dB (DECT). The dynamic ranges are 87 dB (GSM), 69 dB (WCDMA), and 74 dB (DECT).

The measurement results of the ΔΣ modulators are listed in Table 8.10. For comparison purposes, the modulator specifications reported in [10] and [17] are also included. It can be seen that the ΔΣ modulators described in this chapter achieve high performance while using simple circuitry that consumes little power.

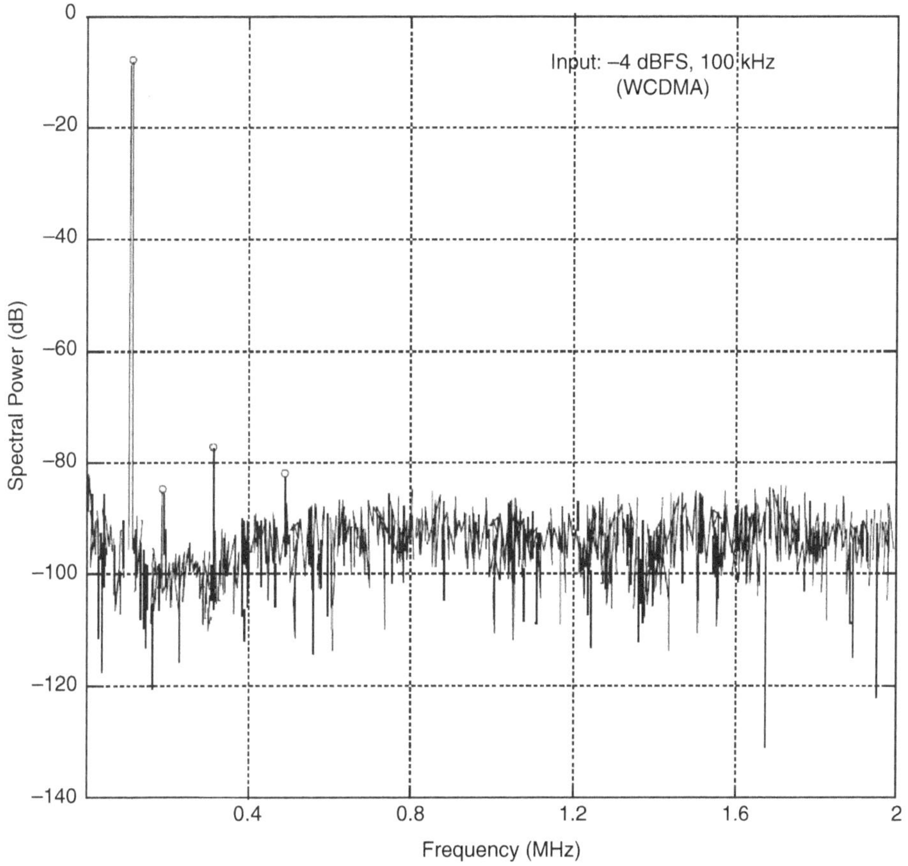

Figure 8.24 Measured output spectrum (WCDMA only).

8.6 Conclusions

In this chapter, we studied the design of delta-sigma ($\Delta\Sigma$) modulators appropriate for multistandard RF receptions. The $\Delta\Sigma$ modulators have been incorporated into a multistandard (GSM/WCDMA/DECT) RF transceiver chip, which is fabricated in a 0.35-μm, DP5M 2.5-V CMOS technology. It has been demonstrated that the high-frequency IF signal digitization can be performed by a pair of low-pass $\Delta\Sigma$ modulators.

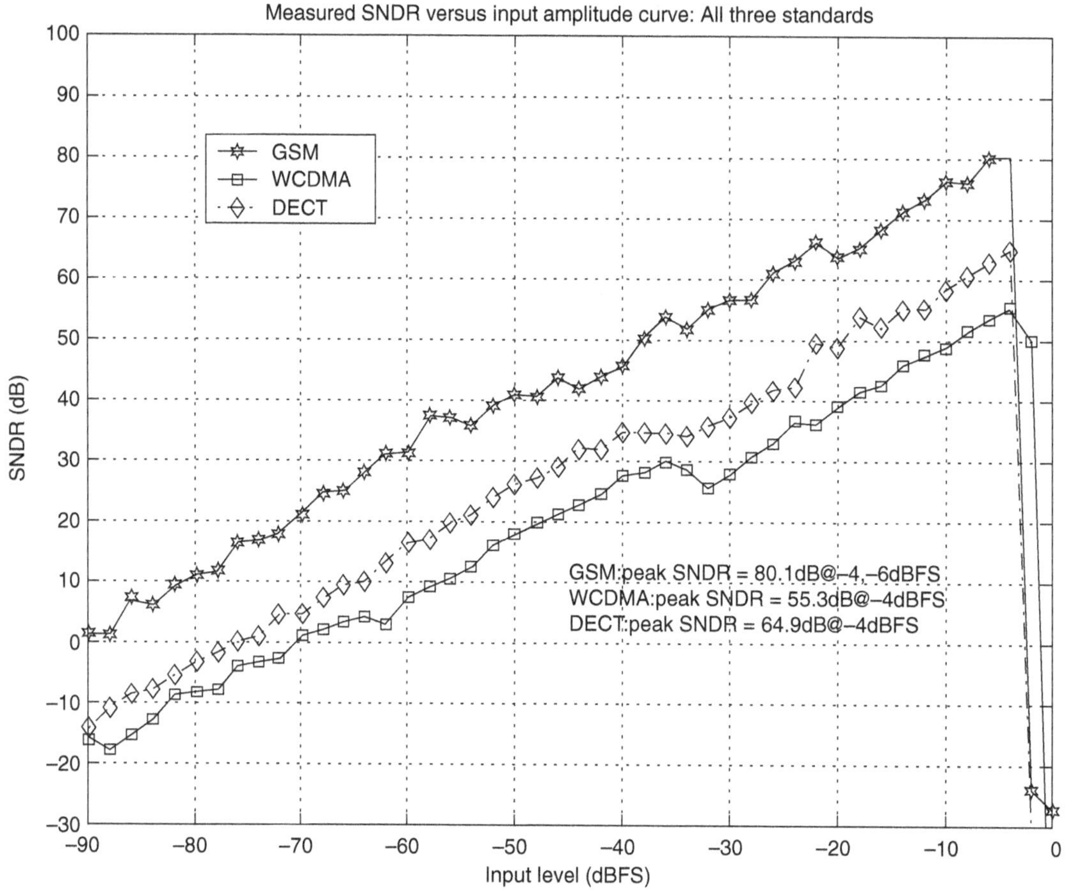

Figure 8.25 Measured SNDR versus input (all three standards).

The existing RF receiver systems were reviewed in Section 8.2. The system-level design of ΔΣ modulators was presented in Section 8.3. The circuit-level implementation of the modulators was explained in Section 8.4. Finally, the performance of the ΔΣ modulators was summarized and compared to the other solutions in Section 8.5.

There are three suggestions for future work. First, the relationship between the IF filter and ΔΣ ADC needs to be investigated to a more exhaustive extent, as the accuracy and dynamic range requirements for the ΔΣ ADC can be relaxed by a selective IF band-pass filter. Second, unity-gain buffers (UGB) [29] may be employed to substitute op-amps to reduce power consumption without resulting in speed and accuracy penalties. Third, it may be feasible to merge the automatic gain controller (AGC) and second mixer with the ΔΣ ADC to achieve higher integration, as their dynamic range requirements are essentially the same.

Table 8.10 Performance of ΔΣ modulators: A comparison.

Ref.	[10]	[17]	This work
Standards	GSM/DECT	GSM/WCDMA	GSM/WCDMA/DECT
Receiver topology	Wideband IF double-conversion	Digital high-IF	Digital high-IF
OSR	128 (GSM), 32 (DECT)	192 (GSM), 24 (WCDMA)	192 (GSM), 24 (WCDMA), 64 (DECT)
Dynamic range (dB)	96 (GSM), 82 (DECT)	86 (GSM), 54 (WCDMA)	87 (GSM), 69 (WCDMA), 74 (DECT)
SNR (dB)	92 (GSM), 76 (DECT)	76 (GSM), 53 (WCDMA)	83 (GSM), 56 (WCDMA), 67 (DECT)
SNDR (dB)	90 (GSM), 75 (DECT)	72 (GSM), 52 (WCDMA)	80 (GSM), 55 (WCDMA), 65 (DECT)
IP3 (dBV_{rms})	–26 (GSM), –12(DECT)	–26 (GSM), –18 (WCDMA)	–27.5 (GSM), –19.4 (WCDMA), –16.8 (DECT)
Modulator topology	4th-order, 1-bit, cascaded 2–2	DQIR with extra forward paths, third-order, 1-bit, single-stage	Simplified DFFIR without the three forward paths, third-order, 1-bit, single-stage
Total capacitance (pF)	>15.0	5.41	6.30
Capacitance spread	—	48:1	24:1
Power consumption (mW)	>18.0 each mode	11.5 (GSM), 13.5 (WCDMA)	12.7 (GSM), 14.1 (WCDMA), 13.5 (DECT)
Power supply (V)	3.3	2.5	2.5
Technology	0.35-μm CMOS double-poly	0.25-μm CMOS double-poly	0.35-μm CMOS double-poly
Active area (mm^2) (modulator only)	—	0.36	0.50

References

[1] M. Liu, *The design of delta-sigma modulators for multi-standard RF receivers*, master's thesis, Oregon State University, Corvallis, OR, June 2003.

[2] P. R. Gray and R. Meyer, "Future directions of silicon ICs for RF personal communications," *Proceedings of the IEEE Custom Integrated Circuits Conference*, pp. 83–90, May 1995.

[3] I. A. Koullias et al., "A 900-MHz transceiver chip set for dual-mode cellular radio mobile terminals," *IEEE Solid-State Circuits Conference, Digest of Technical Papers*, pp. 140–141, February 1993.

[4] A. A. Abidi et al., "The future of CMOS wireless transceivers," *IEEE Solid-State Circuits Conference, Digest of Technical Papers*, pp. 118–119, February 1997.

[5] J. C. Rudell et al., "Recent developments in high integration multi-standard CMOS transceivers for personal communication systems," *IEEE Proceedings of International Symposium for Low-Power Electronics and Devices*, pp. 149–154, August 1998.

[6] R. van Veldhoven et al., "A 3.3-mW $\Sigma\Delta$ modulator for UMTS in 0.18-μm CMOS with 70-dB dynamic range in 2-MHz bandwidth," *IEEE Journal of Solid-State Circuits*, Vol. 37, pp. 1645–1652, December 2002.

[7] X. Li and M. Ismail, *Multi-standard CMOS wireless receivers: Analysis and design*, Kluwer Academic Publishers, Berlin, Germany 2002.

[8] E. Siragusa et al., "A digitally enhanced 1.8 V 15 b 40 Ms/s CMOS pipelined ADC," *IEEE Solid-State Circuits Conference, Digest of Technical Papers*, pp. 452–453, February 2004.

[9] H. C. Liu et al., "A 15-b 40-MS/s CMOS pipelined analog-to-digital converter with digital background calibration," *IEEE Journal of Solid-State Circuits*, Vol. 40, pp. 1047–1055, May 2005.

[10] A. Feldman et al., "A 13-bit, 1.4-MS/s, 3.3-V sigma-delta modulator for RF baseband channel applications," *Proceedings of the IEEE Custom Integrated Circuits Conference*, pp. 229–232, May 1998.

[11] F. Piazza et al., "A 0.25 mm CMOS transceiver front-end for GSM," *Proceedings of the IEEE Custom Integrated Circuits Conference*, pp. 413–416, May 1998.

[12] U. Rohde and T. Bucher, *Communication receivers: Principles and design*, McGraw-Hill, New York, 1988.

[13] B. Razavi, *RF Microelectronics*, Prentice-Hall, Upper Saddle River, NJ, 1998.

[14] A. A. Abidi, "Direct-conversion radio transceiver for digital communications," *IEEE Journal of Solid-State Circuits*, Vol. 30, pp. 1399–1410, December 1995.

[15] L. Longo et al., "A cellular analog front with a 98-dB IF receiver," *IEEE Solid-State Circuits Conference, Digest of Technical Papers*, pp. 36–37, February 1993.

[16] A. Hairapetian, "A 81-MHz IF receiver in CMOS," *IEEE Solid-State Circuits Conference, Digest of Technical Papers*, pp. 56–57, February 1996.

[17] T. Burger and Q. Huang, "A 13.5-mW 185-Msample/s ΔΣ modulator for UMTS/GSM dual-standard IF reception," *IEEE Solid-State Circuits Conference, Digest of Technical Papers*, pp. 44–45, February 2001.

[18] L. J. Breems et al., "A 1.8-mW CMOS sigma-delta modulator with integrated mixer for A/D conversion of IF signals," *IEEE Journal of Solid-State Circuits*, Vol. 35, pp. 468–475, April 2000.

[19] L. J. Breems et al., "A quadrature data-dependent DEM algorithm to improve image rejection of a complex ΔΣ modulator," *IEEE Journal of Solid-State Circuits*, Vol. 36, pp. 1879–1886, December 2001.

[20] A. Tabatabaei et al., "A two-path bandpass sigma-delta modulator with extended noise shaping," *IEEE Journal of Solid-State Circuits*, Vol. 35, pp. 1799–1809, December 2000.

[21] R. Schreier et al., "A 10–300-MHz IF-digitizing IC with 90–105-dB dynamic range and 15–333-kHz bandwidth," *IEEE Journal of Solid-State Circuits*, Vol. 37, pp. 1636–1644, December 2002.

[22] S. R. Norsworthy, R. Schreier, and G. C. Temes, *Delta-sigma data converters: Theory, design and simulation*, IEEE Press, New York, 1997.

[23] S. Jantzi et al., "A fourth-order bandpass sigma-delta modulator," *IEEE Journal of Solid-State Circuits*, Vol. 28, pp. 282–291, March 1993.

[24] "ETS 300 577–579, GSM: Digital cellular telecommunications systems," ETSI, 1997.

[25] "EN 300 176-1 V1.3.2, DECT: Approval test specification; Part 1: Radio," ETSI, 1999.

[26] Third Generation Partnership Project. [Online]. Available at www.3gpp.org.

[27] E. J. van der Zwan et al., "A 10.7-MHz IF-to-baseband sigma-delta A/D conversion system for AM/FM radio receivers," *IEEE Journal of Solid-State Circuits*, Vol. 35, pp. 1810–1819, December 2000.

[28] P. J. Quinn et al., "A 10.7-MHz CMOS SC IF filter using orthogonal hardware modulation," *IEEE Journal of Solid-State Circuits*, Vol. 35, pp. 1865–1876, December 2000.

[29] R. Gregorian and G. C. Temes, *Analog MOS integrated circuits for signal processing*, John Wiley & Sons, New York, 1986.

[30] W. Sansen et al., "Transient analysis of charge transfer in SC filters—gain error and distortion," *IEEE Journal of Solid-State Circuits*, Vol. 22, pp. 268–276, February 1987.

[31] L. A. William, III, *Modeling and design of high-resolution sigma-delta modulators*, Ph.D. dissertation, Stanford University, Palo Alto, CA, 1993.

[32] T. Brooks et al., "A cascaded sigma-delta pipelined A/D converter with 1.25 MHz signal bandwidth and 89 dB SNR," *IEEE Journal of Solid-State Circuits*, Vol. 32, pp. 1896–1906, July 1997.

[33] K. Hsieh et al., "A low-noise chopper stabilized switched-capacitor filtering technique," *IEEE Journal of Solid-State Circuits*, Vol. SC-16, pp. 708–715, December 1981.

[34] Y. Geerts et al., "A 3.3-V, 15-bit, delta-sigma ADC with a signal bandwidth of 1.1 MHz for ADSL applications," *IEEE Journal of Solid-State Circuits*, Vol. 34, pp. 927–936, July 1999.

[35] K. Martin et al., "A differential switched-capacitor amplifier," *IEEE Journal of Solid-State Circuits*, Vol. SC-22, pp. 104–106, February 1987.

[36] K. Bult and G. Geelen, "A fast-settling CMOS op amp for SC circuits with 90-dB DC gain," *IEEE Journal of Solid-State Circuits*, Vol. 25, pp. 1379–1383, December 1990.

[37] B. Razavi, *Design of analog CMOS integrated circuits*, McGraw-Hill, New York, 2001.

[38] N. Sooch and AT&T Bell Lab., "MOS cascode current mirror," U.S. Patent 4550284, 1985.

[39] T. Cho et al., "A 10b, 20Msample/s, 35mW pipelined analog-to-digital converter," *IEEE Journal of Solid-State Circuits*, Vol. 30, pp. 166–172, March 1995.

Index

D

E

F

T

U

V

W

X

Y

Z

www.ingramcontent.com/pod-product-compliance
Lightning Source LLC
Chambersburg PA
CBHW080923220326
41598CB00034B/5658
* 9 7 8 0 7 5 0 6 7 9 0 7 7 *